Photoactive Inorganic Nanoparticles

AF763024

Photoactive Inorganic Nanoparticles

Surface Composition and Nanosystem Functionality

Edited by

Julia Pérez Prieto
Molecular Science Institute, Departamento de Química Orgánica, University of Valencia, Valencia, Spain

María González Béjar
Molecular Science Institute, Departamento de Química Orgánica, University of Valencia, Valencia, Spain

ELSEVIER

Elsevier
Radarweg 29, PO Box 211, 1000 AE Amsterdam, Netherlands
The Boulevard, Langford Lane, Kidlington, Oxford OX5 1GB, United Kingdom
50 Hampshire Street, 5th Floor, Cambridge, MA 02139, United States

Copyright © 2019 Elsevier Inc. All rights reserved.

No part of this publication may be reproduced or transmitted in any form or by any means, electronic or mechanical, including photocopying, recording, or any information storage and retrieval system, without permission in writing from the publisher. Details on how to seek permission, further information about the Publisher's permissions policies and our arrangements with organizations such as the Copyright Clearance Center and the Copyright Licensing Agency, can be found at our website: http://www.elsevier.com/permissions.

This book and the individual contributions contained in it are protected under copyright by the Publisher (other than as may be noted herein).

Notices
Knowledge and best practice in this field are constantly changing. As new research and experience broaden our understanding, changes in research methods, professional practices, or medical treatment may become necessary.

Practitioners and researchers must always rely on their own experience and knowledge in evaluating and using any information, methods, compounds, or experiments described herein. In using such information or methods they should be mindful of their own safety and the safety of others, including parties for whom they have a professional responsibility.

To the fullest extent of the law, neither the Publisher nor the authors, contributors, or editors, assume any liability for any injury and/or damage to persons or property as a matter of products liability, negligence or otherwise, or from any use or operation of any methods, products, instructions, or ideas contained in the material herein.

British Library Cataloguing-in-Publication Data
A catalogue record for this book is available from the British Library

Library of Congress Cataloging-in-Publication Data
A catalog record for this book is available from the Library of Congress

ISBN: 978-0-12-814531-9

For Information on all Elsevier publications
visit our website at https://www.elsevier.com/books-and-journals

ELSEVIER
Book Aid International
Working together to grow libraries in developing countries
www.elsevier.com • www.bookaid.org

Publisher: Matthew Deans
Acquisition Editor: Simon Holt
Editorial Project Manager: Joshua Mearns
Production Project Manager: R. Vijay Bharath
Cover Designer: Greg Harris

Typeset by MPS Limited, Chennai, India

Contents

List of contributors

Manuel Ahumada Bionanomaterials Chemistry and Engineering Laboratory, Division of Cardiac Surgery, University of Ottawa Heart Institute, Ottawa, ON, Canada; Center for Applied Nanotechnology, Faculty of Science, Universidad Mayor, Santiago, Chile

Emilio I. Alarcon Bionanomaterials Chemistry and Engineering Laboratory, Division of Cardiac Surgery, University of Ottawa Heart Institute, Ottawa, ON, Canada; Department of Biochemistry, Microbiology, and Immunology, University of Ottawa, Ottawa, ON, Canada

Kenneth Cerdán-Gómez Department of Organic Chemistry, University of Cordoba, Cordoba, Spain

Jeffrey Comer Institute of Computational Comparative Medicine, Nanotechnology Innovation Center of Kansas State, and Department of Anatomy and Physiology, Kansas State University, Manhattan, KS, United States

Ayda Ali Elhage Department of Chemistry and Biomolecular Sciences and Centre for Advanced Materials Research (CAMaR), University of Ottawa, Ottawa, ON, Canada

Olga García Instituto de Ciencia y Tecnología de Polímeros, Consejo Superior de Investigaciones Científicas ICTP-CSIC, Madrid, Spain

Valeria Grazú Institute of Materials Science of Aragon, University of Zaragoza, CSIC and CIBER-BBN, Zaragoza, Spain

Anabel Estela Lanterna Department of Chemistry and Biomolecular Sciences and Centre for Advanced Materials Research (CAMaR), University of Ottawa, Ottawa, ON, Canada

Caitlin Lazurko Bionanomaterials Chemistry and Engineering Laboratory, Division of Cardiac Surgery, University of Ottawa Heart Institute, Ottawa, ON, Canada; Department of Biochemistry, Microbiology, and Immunology, University of Ottawa, Ottawa, ON, Canada

Xi Li International Joint Research Laboratory of Nano-Micro Architecture Chemistry (NMAC), College of Chemistry, Jilin University, Changchun, P.R. China

Zheng Li International Joint Research Laboratory of Nano-Micro Architecture Chemistry (NMAC), College of Chemistry, Jilin University, Changchun, P.R. China

Marta Liras IMDEA Energy Institute, Móstoles, Spain

Rafael Luque Department of Organic Chemistry, University of Cordoba, Cordoba, Spain

Beatriz Martín Institute of Materials Science of Aragon, University of Zaragoza, CSIC and CIBER-BBN, Zaragoza, Spain

Weiyi Ouyang Department of Organic Chemistry, University of Cordoba, Cordoba, Spain

María Antonietta Parracino Italian Institute of Technology, Genova, Italy

Horacio Poblete Center for Bioinformatics and Molecular Simulation, Universidad de Talca, Talca, Chile; Nucleo Científico Multidisciplinario, DI, Universidad de Talca, Talca, Chile; Millennium Nucleus of Ion Channels-Associated Diseases (MiNICAD), Universidad de Talca, Talca, Chile

Luca Salassa Donostia International Physics Center (DIPC), Donostia, Spain; Ikerbasque, Basque Foundation for Science, Bilbao, Spain

Alain R. Puente Santiago Department of Organic Chemistry, University of Cordoba, Cordoba, Spain

Hélder A. Santos Drug Research Program, Division of Pharmaceutical Chemistry and Technology, Faculty of Pharmacy, University of Helsinki, Helsinki, Finland; Helsinki Institute of Life Science (HiLIFE), University of Helsinki, Helsinki, Finland

Juan C. Scaiano Department of Chemistry and Biomolecular Sciences and Centre for Advanced Materials Research (CAMaR), University of Ottawa, Ottawa, ON, Canada

Alessio Terenzi Donostia International Physics Center (DIPC), Donostia, Spain

Ying-Wei Yang International Joint Research Laboratory of Nano-Micro Architecture Chemistry (NMAC), College of Chemistry, Jilin University, Changchun, P.R. China

Feng Zhang Drug Research Program, Division of Pharmaceutical Chemistry and Technology, Faculty of Pharmacy, University of Helsinki, Helsinki, Finland

Preface

The design of nanoparticles for specific applications and selection of the most convenient capping is not a trivial task. Photoactive nanoparticles, in particular, have a special, almost unique, place among nanomaterials. This book will mainly focus on a selection of the functional capping to coat photoactive nanoparticles while describing the particular features that make them relevant when applied to smart nanomaterials.

First, a chapter focused on understanding the basic aspects of surface chemistry at the nanoscale (Chapter 1) is followed by an additional chapter centering on computational techniques for modeling nanosurfaces (Chapter 2). Surface composition plays a pivotal role in determining the fate and activity of nanostructures in living organisms and catalytic reactions.

Second, the chapters are structured according to the kind of capping and the particular features of each capping are described (e.g., structure, functionality, solubility, (photo)physics, toxicity, and so on). Special emphasis has been placed on how important these specific features are in the preparation of smart nanomaterials. Each chapter covers cutting-edge research carried out with the selected capping and gives examples of their application according to the kind of nanoparticle employed (such as rare-earth doped, semiconducting, and metallic nanoparticles). This selection includes examples of the different applications reported for the prepared functional nanomaterials.

Up to now, the intrinsic properties of the nanomaterial have been the focus of attention. However, selection and activity of the capping agent is crucial for a successful design and its further application. Therefore this book aims to provide information and highlight the importance of the other protagonist—the functional capping. In short, it takes two to tango. Both, the nanoparticle and the capping are relevant to design smart photoactive nanomaterials.

This book is intended for graduate students, academics, and professionals who are newcomers to this field and also for veteran researchers in the area. Each chapter, authored by active researchers, starts with a basic introduction and goes on to in-depth explanations and examples, highlighting the latest results.

Julia Pérez Prieto and María González Béjar

Acknowledgment

The editors would like to express their gratitude to the authors of this book for their esteemed contribution.

In addition, the editors would like to express their gratitude to MINECO [CTQ2017-82711-P, partially co-financed with FEDER funds; Maria de Maeztu: MDM-2015-0538; RTC-2016-5114-5, RyC (MGB)] and Fundación Ramón Areces.

Chapter 1

Fundamental concepts on surface chemistry of nanomaterials

Manuel Ahumada[a,b], Caitlin Lazurko[a,c] and Emilio I. Alarcon[a,c]
[a]Bionanomaterials Chemistry and Engineering Laboratory, Division of Cardiac Surgery, University of Ottawa Heart Institute, Ottawa, ON, Canada, [b]Center for Applied Nanotechnology, Faculty of Science, Universidad Mayor, Santiago, Chile, [c]Department of Biochemistry, Microbiology, and Immunology, University of Ottawa, Ottawa, ON, Canada

Chapter Outline

1.1 Introduction

Nanomaterials have provided a variety of new structural building blocks for developing novel functional materials for catalysis, energy, and biomedical uses. Integrating nanoparticles (NPs) into composite materials must, however, encompass methods to prevent NP surface oxidation in order to avoid toxic side effects in living organisms. Thus, strategies aimed to safely modify the nanosurface architectural landscape by chemical linkage of functional molecules to materials have ramped up interest during the last decade alone. While the concept of immovable surface protection for nanomaterials, when exposed to complex environments, is far from reality, getting a glimpse of the composition of complex mixtures and how surfaces interact with molecules will

Photoactive Inorganic Nanoparticles. DOI: https://doi.org/10.1016/B978-0-12-814531-9.00001-4
© 2019 Elsevier Inc. All rights reserved.

allow elaborating a better idea of the relevance of the dynamic processes in nanosurfaces.

In humans, body water (70%) is distributed intracellularly and as extracellular filler. The intravascular and extravascular locus for the extracellular water accounts for most of the intrinsic active transport of molecules within organs and tissues. Let us consider, for instance, the composition of the most vital biofluid for humans: blood. In terms of cell composition for a healthy individual, human blood is dominated by red blood cells ($\approx$93.6%), followed by platelets ($\approx$6.2%), and well behind white blood cells ($\approx$0.20%). Around 55% of the human blood volume corresponds to a fraction "rich" in protein, called plasma. This portion is mainly composed of water ($\approx$90%), proteins (7%–8%), and the remaining 2%–3% is organic and inorganic matter. Despite the small percentage of protein content in human plasma, less than 10%, some researchers have reported over 3000 different proteins that are contained in it [1–3]. As the techniques used for separation are also affected by protein glycosylation and enzymatic degradation post-harvesting, the actual total protein number should be something in between. In any case, the presence of numerous different proteins in the human plasma schematically depicts the complex environment to which nanomaterials are exposed once incorporated within our bodies. Human plasma is only one of the many other bio-fluids in which proteins are abundant in the hundreds if not thousands, including: interstitial fluid [4], urine [5,6], amniotic fluid [7], synovial fluid [8], and cerebrospinal fluid [9–11]. Although a detailed discussion on the exact biochemical composition of different biofluids is beyond the scope of this chapter. Understanding that there is much more than a single dominant protein in biological systems is essential for understanding the behavior of nanomaterials in living organisms. As a result, the surface composition of a nanomaterial will not remain unaltered, as proteins dynamically changes and adapts in response to physiological changes.

Rational designing of nanomaterials must consider providing effective surface composition for a given nanomaterial. However, most nanomaterials are designed using common protecting agents, including citrate and thiols, which have proven to be suboptimal under physiological conditions [12]. Despite this, surface protection can be partially improved by using polymers, polypeptides, proteins, or glycosides. However, their use is still far from being realistically scalable for commercial NP production. Thus, a better understanding of the parameters and physical features that link the surface composition of nanomaterials including association modes and constants will lead to designing novel functional nanostructures with improved biological function. In this chapter, we will revise fundamental concepts on the association of molecules to nanosurfaces that will be instrumental for better understanding the most relevant underlying mechanisms involved in the binding of molecules to nanomaterials.

1.2 Physical chemistry of the association of molecules to nanoparticles

The association of molecules to metal NP is commonly referred as an adsorption process. Particularly in these systems, the adsorption comes from the "contact" between the solute and the surface of the NP. The degree of adsorption of a given solute on the surface will depend upon several factors such as: solute composition, structure, charge, volume, and concentration [13]. In addition, parameters such as: available surface, NP charge, type of metal, capping agent, porosity, impurities, and roughness play an important role in determining the associativity of molecules to nanostructures [14]. In sum, the adsorption of molecules to NP is intrinsically a complex process to analyze from the physical chemistry point of view, as per se in the following pages we will use some simplifications to tackle the association phenomena.

Overall, the interaction between a given molecule and the surface of a NP can occur due to either chemical (chemisorption) or physical (physisorption) interactions. Table 1.1 shows some of the main characteristics/differences of

TABLE 1.1 Comparative summary for the main characteristics of chemical and physical adsorption.

	Chemisorption	Physisorption
Driven force	Valence force	Van der Waals forces
Specificity[a]	Highly specific	Not specific
Changes in electronic states[b]	Lead to changes in electronic states and generate new bonds	No changes
Solute irreversible modifications	Possible	No
Energy[c]	Activation	Adsorption
Number of maximum layers[d]	1	+ 1
Effect of temperature	Depends on the activation energy of the chemisorption process being considered	Desorption is favored at high temperatures

[a]*Specificity of the solute for the surface of the nanoparticle to be bound.*
[b]*Formation of new chemical bonds between the solute and the surface or just electrostatic interactions.*
[c]*Type of energy considered in the model. In chemisorption the energy requirement is similar to that in a chemical reaction. Physisorption does not have significant differences in the interaction energies between the solute and adsorbent; however, commonly it is larger than the energy of condensation of the solute.*
[d]*The number of layers formed around the nanoparticle will depend on the number of available binding sites at the surface for chemisorption (monolayer), and short-/long-range electrostatic interaction between the surface and the solute for the physisorption (multilayer).*

these two types of adsorptions. Chemisorption is normally considered a type of non-reversible binding process. To didactically illustrate some of the main features of physisorption, we have considered these processes in Table 1.1.

From a thermodynamic angle, one can describe the interaction of a molecule with a nanosurface in terms of the Gibbs free energy (ΔG) [see Eq. (1.1)]:

$$\Delta G = \Delta H - T\Delta S < 0 \tag{1.1}$$

where ΔH corresponds to the enthalpy, T corresponds to temperature, and ΔS corresponds to entropy. Thus in a qualitative analysis it is expected that the difference $\Delta H - T\Delta S$ is negative to allow spontaneous association of the molecule to the nanosurface. While the formation of a chemical bond between the molecule and the NP surface results in a net gain in enthalpy, close proximity to the surface will restrict its mobility, this being another example of the archetypical enthalpy–entropy compensation effect now being applied at the nanoscale [15].

From a kinetic point of view, considering the reaction presented in Eq. (1.2), the association constant (K_a) can be expressed as:

$$xNP + yL \underset{k_d}{\overset{k_b}{\leftrightharpoons}} NP_xL_y \tag{1.2}$$

$$\text{And once equilibrium has been reached } K_a = \frac{k_b}{k_d} = \frac{[NP_xL_y]}{[NP]^x[L]^y} \tag{1.3}$$

where L corresponds to the molecule or ligand, x is equal to 1 for a single NP, y is the number of molecules bound per NP, k_b and k_d correspond to the rate constants for the binding and dissociation processes, respectively. Thus rewriting Eq. (1.1) by incorporating association parameters becomes:

$$\Delta G = -RT\ln K_a \Leftrightarrow \Delta G = -RT\ln\frac{k_b}{k_d} \tag{1.4}$$

From Eq. (1.4), to have spontaneous binding (i.e., ΔG negative) k_b must be larger than k_d. The association constant (K_a) is a key parameter for predicting a variety of catalytic and biological functions of nanomaterials. Thus the "degree" of association is normally determined by using adsorption isotherms. The shape of the adsorption isotherm will depend on both the molecule and the NP. Fig. 1.1 depicts some representative examples for adsorption isotherm curves of molecules binding to nanosurfaces.

Furthermore, the association of a molecule to the surface of the NP can also be described in terms of the total surface available, which in principle can be seen as a binding site model as follows:

$$Free\ site_{NP} + L \underset{k_d}{\overset{k_b}{\leftrightharpoons}} L_b \tag{1.5}$$

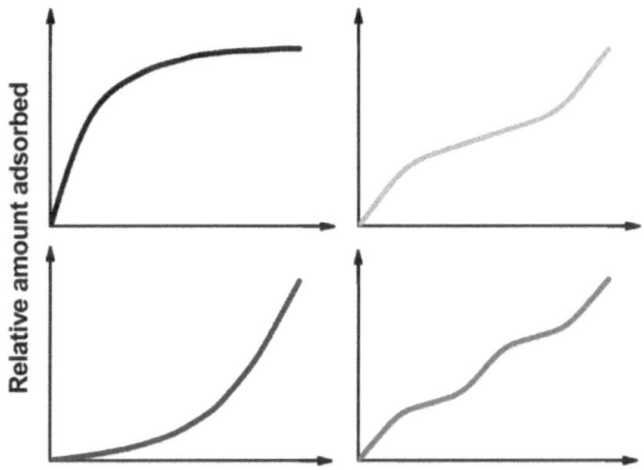

FIGURE 1.1 Representative adsorption isotherm shapes for association of molecules to nanoparticles surfaces.

where L is the solute/ligand free in solution, *Free site*$_{NP}$ corresponds to the available surface on the NP to bind, and L_b is the solute bound to the NP. Therefore K_a is given by:

$$K_a = \frac{k_b}{k_d} = \frac{[L_b]}{[Freesite_{NP}][L^0]} \tag{1.6}$$

with L^0 being the total analytical concentration of the solute. Experimentally to determine the association constant, Eq. (1.6) can be expressed in terms of the fraction of solute bound to the NP surface (with $[L^0] = \alpha$) according to:

$$K_a = \frac{\alpha}{(n[NP] - \alpha[L^0])(1 - \alpha)} \tag{1.7}$$

where n corresponds to the number of NP free sites. Then, Eq. (1.7) can be reorganized to Eq. (1.8), where K_a and n can be determined from the intercept and the slope, respectively.

$$\frac{1}{(1 - \alpha)} = -K_a\left[L^0\right] + K_a n\frac{[NP]}{\alpha} \tag{1.8}$$

also, K_a can be expressed in terms of the average of the fraction of occupied NP bound sites (θ), where $[L^0] = \theta\ n\ [NP]$; and $[L] = [L^0] - \theta\ n\ [NP]$, which upon reorganizing Eq. (1.6) gives:

$$K_a = \frac{\theta}{([L^0) - \theta n[NP])(1 - \theta)} \tag{1.9}$$

Furthermore, mathematical operations on Eq. (1.9) produce a friendlier method for determining the association constant from the isotherm,

$$(1-\theta)[NP] = \left(\frac{1}{K_a n}\right) + \left(\frac{1}{n}\right)\left(\frac{1-\theta}{\theta}\right)[L^0] \tag{1.10}$$

Both Eqs. (1.8) and (1.10) correspond to classical equations used to establish an adsorption isotherm. Amongst the common tools used to measure adsorption isotherms with biomolecules, fluorescence stands out. However, as nanomaterials are normally lacking fluorescence emission, monitoring changes in properties of the molecule upon binding to the nanosurface [Eq. (1.8)], prevails over monitoring the NP [Eq. (1.10)] for determining association constants (*vide infra*). Thus from the equations above, adsorption isotherm expressions such as Langmuir, Freundlich, and Brunauer–Emmett–Teller (BET) can be easily obtained. While discussing each of those isotherms is beyond the scope of this chapter; in the following sections, we will briefly revise some relevant methodologies for evaluating association. However, readers willing to learn more about association models can consult a recent review of our group [16], which covers fundamental aspects for the association of molecules to nanosurfaces.

As previously mentioned, the association of molecules to NPs depends on several factors related to the NP, solute/molecules, and solvent. Surface topology, namely, roughness of the NP and the interactions between the molecules once bound are also key factors for the association. Roughness can cause differences in the available binding sites on the same NP, which is particularly important for larger NPs. A homogeneous surface will allow for equal binding sites, meaning that both the ΔH and k_b/k_d are independent of θ. However, a more heterogeneous surface implies that not all the regions on the NP surface are equivalent, meaning that θ value will have an impact on the binding mechanism, that is, non-Langmuir isotherms. Note that in the latest scenario, the activity of the free sites decreases with θ, because the most active regions on the surface are occupied first. Further, interactions between the molecules on the NP surface will also play a role in the resulting association process. Thus, for example, if the affinity of a molecule increases upon having another previously bound to the surface, the resulting isotherm will show cooperative features. Assuming that interactions between molecules adsorbed onto the surface are the result of attractive forces such as hydrophobic interactions, increasing the number of preadsorbed molecules per NP will ultimately lead to an increase in the binding. In the case that binding of molecules to the nanosurface leads to repulsive interactions, a net decrease in the affinity of molecules for the NP surface will take place in what is called "anticooperativity."

Surface composition of nanomaterials has been shown to be a crucial factor in providing effective and nontoxic nanomaterials. While many uncoated NPs are unstable due to their high surface energy and tendency to oxidize, capping agents are used to stabilize the NP, promote NP dispersion, and control the NP's interaction with its environment [17–19]. Protecting agents can be added (1) during the formation of the NP or (2) by capping agent exchange, where the desired capping agent replaces a previously associated molecule on the surface. Capping agent exchange allows custom surface engineering of nanostructures, which is essential for biological applications. Independent to the route used to engineer the NP surface, the association of a ligand will be determined mainly by its affinity to the surface (i.e., k_b and k_d). Therefore an ideal capping agent should rapidly bind to the surface (k_b) and remain bound (k_d). As expected, to exchange a capping agent, the new surface molecule must have a much larger k_b than the molecule occupying the NP surface. In thermodynamics terms, it is anticipated that the most stable composite (molecule-capped NP) has the lowest energy (negative ΔG) and is therefore the most favorable.

1.3 Corona formation and its role in surface composition

In a biological setting, NPs are modified by the adsorption of biomolecules, such as proteins, that compete with previously adsorbed molecules for NP surface interaction and can act as capping agents on the NP. Under these conditions, NPs adsorb a mixture of different biomolecules that are aggregated onto their surface to form the protein corona [19–26].

The formation of the protein corona is a dynamic process of adsorption and desorption of biomolecules that is governed by the affinity of these molecules for the surface [25]. The biological identity of the NP is not necessarily controlled by the most abundant biomolecule adsorbed on the surface of the particle, but by the molecules with the highest affinity and specificity for the particle [23,25,26]. As described by the "Vroman effect," the most abundant proteins are the first to adsorb onto the NP, but are slowly replaced with proteins with slower exchange rates, lower concentrations, and higher affinities for the particle [22–24,26–28]. The binding of low concentration proteins to the NP can concentrate these proteins, which causes the protein concentration around the NP to differ from their concentration in the bulk of the solution [26].

The aforementioned processes can be explained through kinetic theory using the frequency of collisions. While multiple collisions between proteins and NPs occur in a short period of time (in blood it is $10^6\ s^{-1}$) [14], only those with high k_b and low k_d can eventually adsorb onto the surface. Then the relationship between the adsorption/desorption rate with the energy will be given by [29],

$$K_a = \frac{k_b}{k_d} = \exp\left(\frac{\Delta G_{ads}}{RT}\right) \tag{1.11}$$

Thus, the association process of the protein to the surface of NP can be described using an energy diagram. Fig. 1.2 depicts a typical free-energy diagram for a capped NP interacting with a second capping agent separated by a surface-to-surface distance, *d*.

Fig. 1.2 represents the energy change throughout a typical capping agent replacement. In a biological application, the starting point will be a

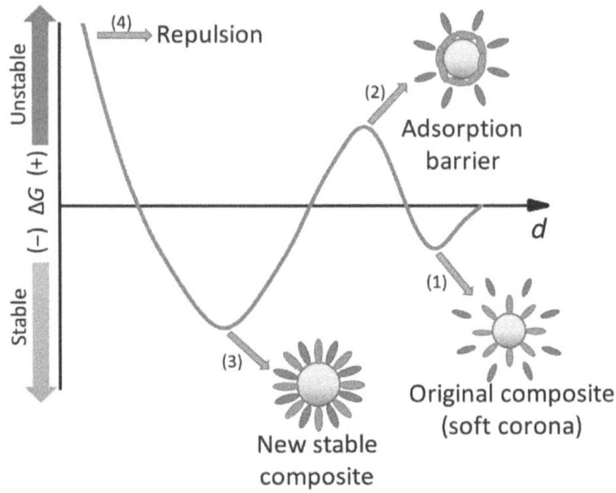

FIGURE 1.2 Free energy diagram for a capped nanoparticle interacting with a second capping agent separated by a surface-to-surface distance. Representation: Nanoparticle (yellow in web version), first capping agent (blue in web version), and second capping agent (redin web version).

previously synthetized NP stabilized with a capping agent (inorganic/organic molecule), and then the addition of a new capping agent (biomolecule) (see Step 1 in Fig. 1.2). At this distance, the attractive energies are more favorable than the repulsion energies. Thus, a metastable state with a negative ΔG is reached; particularly in biological systems this state is known as soft corona, which will be further described below. Once the second agent is situated closer to the NP, the repulsion energies of the original capping agent promote the formation of a high-energy barrier restricting the binding of the new molecule (Step 2). In Step 3, the closer distance allows the association of the new molecule, mainly due to its high affinity of the molecule for the NP surface. Note that in order to fulfill the condition of minimum energy in this step, the affinity for the NP must be greater than repulsion energies. At minimum distances the repulsion energies will dominate the system (Step 4). Eventually, if the affinity (k_b) of the second capping agent is greater than the parent, the replacement is likely to occur in Step 3, which for proteins is called hard corona [30].

The protein corona is composed of both hard and soft layers (Fig. 1.3) [19,20,23]. The hard corona is the protein layer nearest to the NP surface. It is formed when biomolecules with high exchange rates bind tightly to the surface of the NP, rendering a long-lived hard corona, which ultimately dictates the biological identity of the NP [19,20]. The molecules in the hard corona are tightly bound to the NP, but they do not bind irreversibly, so over time the biomolecules with the highest affinity will replace the biomolecules in the hard corona that have a lower affinity but are more abundant

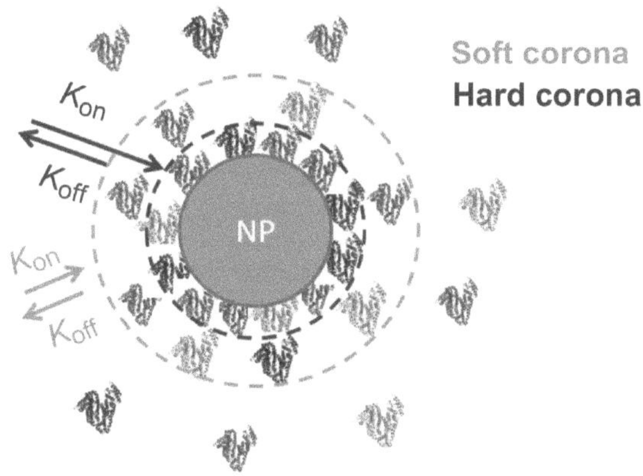

FIGURE 1.3 The protein corona effect. The protein corona is composed of the hard (red in web version) and soft (orange in web version) corona. The hard corona has long exchange times and largely defines the biological identity of the nanoparticle (NP). The soft corona has fast exchange times and is more loosely associated than the hard corona.

[19,20,24]. The soft corona forms on top of the hard corona, and as these biomolecules are more loosely associated than the molecules in the hard corona they exchange more rapidly [19,20].

As molecules exchange on the surface of the NP, biomolecules in the protein corona constantly exchange, which dynamically affects the biological identity of the NP [24]. This dynamic nature of the corona results from the range of binding affinities and equilibrium constants of the different biomolecules that affects the exchange rate, and form the variation in the binding and dissociation rates and protein profile of the biological environment [26]. While it might be expected that an increase in the concentration of the biomolecule will lead to an increase in the exchange rate between the molecules bound to the NP surface with those in Layer 2, this will hold true only for nonspontaneous and diffusion controlled processes (see Fig. 1.4). This can be explained, if one considers that at equilibrium, there are going to be higher concentrations of the capping agent in Layer 3. Therefore the exchange rate in regions far from the surface will be higher, but it will not affect the association at short distances. Under complex multiequilibrium scenarios, it is difficult to determine an accurate association constant (K_a), mainly because of a complex stoichiometry. Thus, only an apparent K_a value ($K_{NP-S} = \prod K_i$) can be estimated [31].

The adsorption of different proteins onto the NP surface influences the biological fate of the particle. Thus, for example, once certain proteins have been adsorbed onto the NP surface, the NP can be recognized by cell receptors, facilitating cellular uptake and biodistribution, for instance [20,26]. Proteins may also trigger signaling pathways, which can affect NP biodistribution. These physiological interactions, as a sum, are the key for biomedical

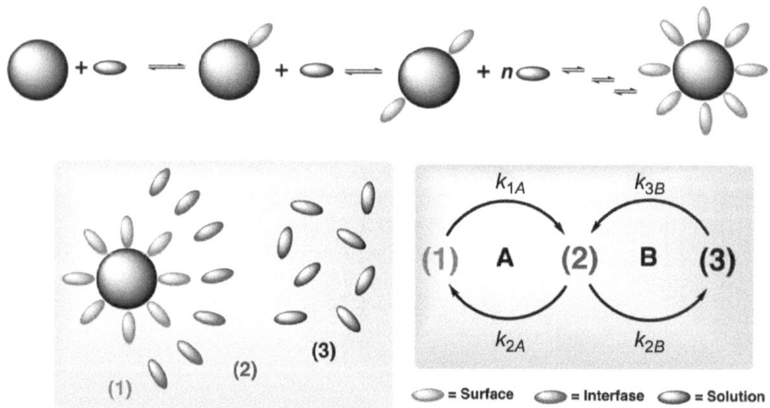

FIGURE 1.4 Top: Cartoon representation of a multi-step capping agent binding to a nanoparticle surface. The bound capping agent is shown in red color and the free, non-bound one in grey. Bottom left: Representative diagram showing the three different populations of capping agents once capping of the nanoparticle surface has been entirely achieved as follows: (1) in surface contact, (2) in the near vicinity of the surface (more organized), and (3) molecules found far from the nanoparticles (disorganized). Bottom right: Simplified dynamic model showing capping agent exchange within the molecules located in the different layers shown in the left. *Reproduced with permission of The Royal Society of Chemistry from Ahumada et al. [16].*

uses of NPs and also highlight the relevance of understanding the impact of surface composition [26]. Thus, consequently, discrepancies between in vitro and in vivo experiments are expected for the same NP, mainly due to biomolecule availability and concentration in living organisms [20]. For example, plasma proteins will compete for the NP surface and most likely will displace the existing capping agents [23]. One can go a step further and think the functional effects observed in different animal models, for the same type of NPs might also be linked to intrinsic differences in the plasma/biofluid compositions in different species.

1.4 General considerations for evaluating association of molecules to nanostructures

The archetypical picture of association in biological systems takes place in determined regions of the protein or "host" (i.e., pockets or binding sites), where specific interactions between protein structural components and the "guest" molecule occurs. Most of the time, steric factors allow only one molecule per host, leading to a 1:1 stoichiometry. Association constants (K_a) under those conditions for molecule–protein binding site complexes, range from $10^{3.5}$ to $10^{16}\,M^{-1}$ [32–34]. One could now wonder what sort of binding model should be used for NPs, where association is not confined within pockets or regions protected from the solvent. Overall, the binding of low

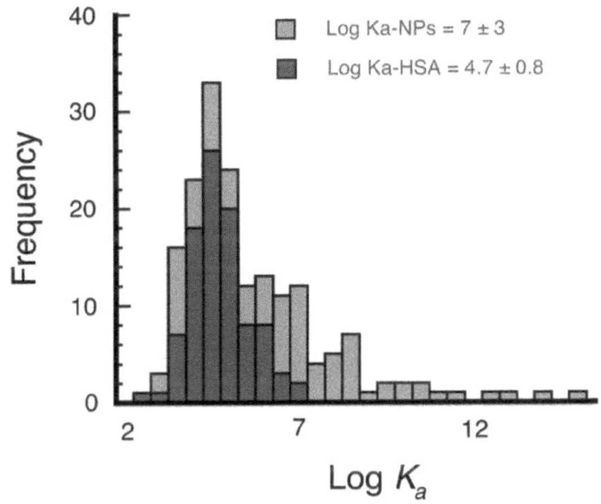

FIGURE 1.5 Representative association constant distributions for molecules to nanomaterials (purple bars in web version) and human serum albumin (HSA, red bars in web version). *Reproduced with permission of The Royal Society of Chemistry from Ahumada et al. [16].*

molecular weight molecules to nanomaterials is a multistep process, where more than one layer of molecules supramolecularly arranges around the NP core (*vide supra*). This intrinsic complexity in associativity, added to the lack of precise ways of determining NP concentration in solution (*vide infra*), has led to differences of several orders of magnitude in association constants for NPs [16]. We recently reported from comparing the association constants (K_a) of low molecular weight molecules that nanomaterials have considerably larger association constants than one of the most important carrier proteins (human serum albumin, HSA) (see Fig. 1.5), with log K_a of 7 ± 3 and 4.7 ± 0.8 ($P < .001$, t-test), for NPs and proteins, respectively. Note that in the plotted data shown in Fig. 1.5 multioccupation phenomena were largely ignored in the calculations and NP total concentration was in most cases considered as "precursor" concentration or calculated assuming monodisperse populations. In the following section, we will briefly revise some important concepts that must be considered when studying association phenomena for colloidal solutions of NPs.

1.5 Role of colloidal stability in nanoparticle-binding capabilities

Surface composition of NPs ultimately dictates their stability in solution. Thus, changes in colloidal charge upon association of a new molecule to the NP surface can effectively occur. In Fig. 1.6, a representative example of what one could expect on changes of the surface charge would look like is presented. In the case shown in Fig. 1.6, the surface-capping agent

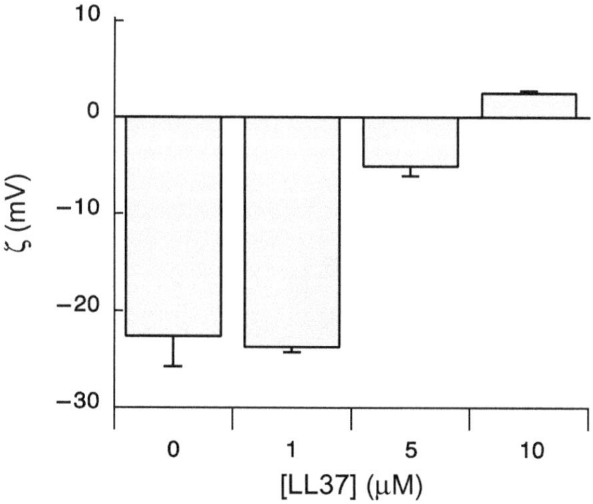

FIGURE 1.6 Changes in zeta potential ζ(mV) for citrate-capped nanosilver, observed upon addition of increasing concentrations of LL37-SH peptide. *Figure adapted with permission of The Royal Society of Chemistry from McLaughlin et al. [35].*

citrate from spherical silver NPs was replaced upon addition of the CSG-LLGDFFRKSKEKIGKEFKRIVQRIKDFLRNLVPRTES-NH_2 (LL37-SH) peptide, resulting in a decrease in the colloidal stability, as demonstrated by the change in electrophoretic mobility or zeta potential [35]. In this particular work, we manipulated the decrease in colloidal stability of nanosilver to produce a sprayable formulation of nanosilver, which forms a protective nontoxic, yet antimicrobial barrier [35].

However, it is not only large peptides that can promote changes in the NP surface charge. Thus, for example, the micromolar concentration of the pentapeptide CLKRS was able to reduce and modify nanosilver colloidal stability [36]. More recently, collagen mimetic peptides containing a −CLK terminal-anchoring system have proven to be effective as structural backbone for novel biomimetic matrices [37]. In summary, if surface modification leads to changes in the colloidal stability of the NPs, one should be fully aware of the possibility of having aggregates in solution that will increase light scattering, used for calculating NP diameter, and ultimately modify the total surface area available for binding.

1.6 General considerations on measuring association of molecules to nanoparticles

In the previous section, we elaborated on the thermodynamics, general considerations for association, as well as the impact of NP stability in the evaluation of binding. Before moving to explore the effect of NP polydispersity, we will explore some selected techniques commonly used for evaluating the association of molecules to NPs. The reader can find a more extensive discussion on this topic in a recent review published by our team [16].

Absorption spectroscopy has demonstrated to be a solid technique for evaluating association of molecules to nanostructures. However, particularly for large nanostructures ($>$ 100 nm), it does require that either the NP or the ligand absorbs light in the visible region. So, NPs that have plasmonic absorbance and ligands with highly absorbing chromophores in their structure are well suited for this. Implicitly there must be a measurable difference, within the experimental error, for the free and bound state of the components. Some representative examples for using changes in the absorption properties of the components in the system are found in references [38–40]. While several different approaches can be used for determining the association constant for a given system, normally Benessi–Hilderbrand and Scatchard models are used, despite their pitfalls and associated errors [31]. For example, while the Benessi–Hilderbrand model is only valid for 1:1 complexes and it is not sensitive to deviations from multiple equilibriums, it has been extensively used for determining association parameters for nanomaterials [16].

Another alternative for quantifying association of molecules to nanostructures is by measuring changes in the fluorescence emission. In this case, as opposed to the absorption measurements, the binding molecule is normally the emitter. Fluorescence measurements will be by monitoring either quenching or enhancement of fluorescence signals [39,41–55]. Static quenching of the emission fluorescence is equivalent to complex formation in the ground state and can be analyzed using a Stern–Volmer equation, some mathematical operations to the Stern–Volmer equation allow for producing an expression that renders K_a. Such derivations assume a quenching efficiency of 1.0 (total). Thus, for example, this approach has been used to determine the association of proteins such as bovine serum albumin to nanosilver [42]. As per using the fluorescence enhancement emission, most of the literature available is centered around the use of gold nanoclusters in the presence of proteins [44,45].

When using fluorescence emission as a tool for determining association, one should consider corrections in the measured emission due to inner filter effects [56]. Such interferences will produce attenuation of the fluorophore emission signal, with the concomitant overestimation of the association extent. In addition, for samples in solution, absorbance at the excitation wavelength should not exceed 0.05 to avoid deviations from linearity.

1.7 Nanoparticle concentration and polydispersity

Up to now, precise evaluation of NP concentration in solution remains elusive, mostly because of the lack of consensus on how NP concentration should be actually calculated. If we consider that normally NP concentrations are relatively small ($\approx 10^{-9}-10^{-12}$ M), small variations between batches of NPs will have a tremendous impact on the total number of available NPs per unit volume. Such discrepancies in the NP concentration are directly linked to differences of up to several orders of magnitude for the same NP–molecule system [16]. The most substantial difference between NPs and solutions of molecules is that not all the NPs are created equally, even for those that are normally described as monodisperse. Basically as NPs are the assembly of individual atoms, larger NPs will use more atoms than smaller ones, which will affect the total number of formed NPs per liter, as the number of total atoms is fixed to start with.

Accounting for polydispersity effects, the estimation of the NP concentration is possible if the NP size distribution of a given sample is considered. Very recently, our team has reported a new algorithm (Nanoparticle Polydispersity Corrector, NANoPoLC), which allows incorporating adjustments to the concentration based on the sample size polydispersity. This algorithm was developed to work with data output from dynamic light scattering as well as measurements from large sample numbers of NP populations obtained by transmission microscopic measurements. In the following section, we will briefly elaborate on how NANoPoLC was derived and some representative examples on the differences in concentrations calculated are also included. For a full discussion on the details of our algorithm, see Reference [57].

Let us consider that the probability of finding a NP of diameter d can be defined as:

$$\int_{d^-}^{d^+} g(s)ds \approx g(d)h \tag{1.12}$$

where the value $g(d)h$ is the percentage of NPs with diameter, d, from the total number of NPs in a given sample. g corresponds to the function, in our derivation any distribution applies, which describes the population. After a series of mathematical operations, and upon using the Riemann sum and the "Trapezoidal rule," the NP concentration [NP] can be calculated as follows:

$$[NP] = \frac{2[S]}{\sum[F(d_i)g(d_i) + F(d_{i+1})g(d_{i+1})]h_i} \tag{1.13}$$

where $[S]$ denotes the initial concentration of the NP precursor and $F(d_i)$ is the total number of atoms for a NP precursor of diameter d. Note that the function $F(d_i)$ is a unique feature of this equation that links the volume of the NP. To illustrate the impact of using this correction in the calculation of the NP concentration, let us compare the calculated concentration of two samples of spherical nanosilver prepared with the same total silver concentration but with different mean size diameter (see Table 1.2).

The values shown in Table 1.2 show striking differences with the concentration calculations carried out assuming perfectly monodisperse nanosilver concentrations. Particularly, even for the values calculated using transmission

TABLE 1.2 Comparison of nanoparticle concentrations for citrate-protected nanosilver colloidal solutions.

Method	10 nm diameter (concentration in nM)	4.3 nm diameter (concentration in nM)
Ionic silver concentration	$2x10^5$	$2x10^5$
Monodispersity	3.5 ± 1.1	44 ± 12
NANoPoLC[a]	0.017 ± 0.003	0.83 ± 0.30
NANoPoLC[b]	0.007	0.15

[a]*Values were calculated from averaging nine independent batches of nanoparticles measured in dynamic light scattering using NANoPoLC algorithm.*
[b]*Values calculated from transmission electron microscopy (1100 individual particles counted, see [57]).*
Source: Adapted with permission from C. Lazurko et al., NANoPoLC algorithm for correcting nanoparticle concentration by sample polydispersity, Nanoscale 10 (2018) 3166–3170 [57].

TABLE 1.3 Association constants of human serum albumin to nanosilver.

	Diameter (nm)	$10^9\ M^{-1}$
NANoPoLC	4.3 ± 1.2	$K_{sv} = 4.3 \pm 0.21$
Monodisperse		$K_{sv} = 0.083 \pm 0.004$
NANoPoLC	10 ± 1.2	$K_{sv} = 176 \pm 11$
Monodisperse		$K_{sv} = 0.86 \pm 0.05$

K_{sv}, Stern–Volmer constant; K_a, association constant.
Source: Adapted with permission from C. Lazurko et al., NANoPoLC algorithm for correcting nanoparticle concentration by sample polydispersity, Nanoscale 10 (2018) 3166–3170.

electron microscopy data, there was an underestimation of the NP concentration, which accounts for the impact of sample polydispersity and size of the population used for determining the average size of the samples. This is an interesting example on how for samples whose surfaces are modified with relatively small protecting agents (i.e., citrate), dynamic light scattering presents a valuable venue for overcoming such limitations, as in this technique thousands of particles are measured and averaged. Further, the extent of the impact of using a corrected NP concentration was illustrated by exploring the binding of human serum albumin to nanosilver. Changes in fluorescence intensity upon association to the nanostructure were measured for the two NP sizes (see Table 1.3)

As the bimolecular rate constant measured was larger than the diffusion limit for all cases (see [57]), the approximation that $K_{sv} \approx K_a$ was considered accurate. Table 1.3 depicts how including the corrections for polydispersity produces a ≈ 52 times difference for the 4.3 nm NPs; this difference was ≈ 200 times for the 10 nm sample. Such disparities might be responsible for differences in biological and catalytic activities of different batches of NPs.

1.8 Conclusions and outlook

The development of novel functional nanomaterials must encompass the rational design of the NP surface. Producing such design requires considering fundamental concepts of surface chemistry during the synthesis of the composite nanomaterial. One should consider the ultimate use or niche of the nanomaterial and where it will be used or resided in order to design a material with a suitable surface composition. Thus, for example, for nanomaterials that will be in close contact with bio-fluids, considerations including the formation of protein corona and dynamic exchange of molecules onto the NP surface are required. In addition, the chosen technique used for evaluating the NP–molecule interactions is as important as the actual binding process. Finally, corrections aimed to incorporate polydispersity are essential if one wants to fully understand processes such as corona formation or capping agent ligand exchange. It is not clear what is next for nanotechnology, but developing high throughput screening tools for nanomaterials as well as algorithms for identifying target molecules in complex biological mixtures with high affinities for nanosurfaces should be certainly in the list. Such advances will take the field of nanomaterials to the next level, where surface engineering of composite materials will allow the safe and effective use of precision medicine and site-specific treatment of diseases.

Acknowledgments

This work was supported by the University of Ottawa Heart Institute (UOHI), the Natural Sciences and Engineering Research Council (NSERC-Discovery), and the Canadian Institutes of Health Research (CIHR-Grant Scheme) to Dr. Alarcon. Dr. Ahumada extends his thanks to CONICYT for Becas Chile postdoctoral fellowship (#74180020).

References

[1] Y. Shen, et al., Characterization of the human blood plasma proteome, Proteomics 5 (15) (2005) 4034–4045.

[2] T. Liu, et al., High dynamic range characterization of the trauma patient plasma proteome, Mol. Cell. Proteomics. 5 (10) (2006) 1899–1913.

[3] N.C. Garbett, et al., Calorimetric analysis of the plasma proteome, Sem. Nephrol. 27 (6) (2007) 621–626.

[4] M.D. Stone, et al., Novel in situ collection of tumor interstitial fluid from a head and neck squamous carcinoma reveals a unique proteome with diagnostic potential, Clin. Proteomics 6 (3) (2010) 75–82.

[5] J. Adachi, et al., The human urinary proteome contains more than 1500 proteins, including a large proportion of membrane proteins, Genome. Biol. 7 (9) (2006) R80.
[6] D.W. Wilkey, M.L. Merchant, Proteomic methods for biomarker discovery in urine, Sem. Nephrol. 27 (6) (2007) 584–596.
[7] C.K. Cho, et al., Proteomics analysis of human amniotic fluid, Mol. Cell. Proteomics. 6 (8) (2007) 1406–1415.
[8] R. Gobezie, et al., High abundance synovial fluid proteome: distinct profiles in health and osteoarthritis, Arthritis. Res. Ther. 9 (2) (2007) R36.
[9] B.R. Wenner, M.A. Lovell, B.C. Lynn, Proteomic analysis of human ventricular cerebrospinal fluid from neurologically normal, elderly subjects using two-dimensional LC-MS/MS, J. Proteome. Res. 3 (1) (2004) 97–103.
[10] J. Zhang, et al., Quantitative proteomic analysis of age-related changes in human cerebrospinal fluid, Neurobiol. Aging. 26 (2) (2005) 207–227.
[11] S.E. Schutzer, et al., Establishing the proteome of normal human cerebrospinal fluid, PLoS. ONE. 5 (6) (2010) e10980.
[12] E.I. Alarcon, et al., The biocompatibility and antibacterial properties of collagen-stabilized, photochemically prepared silver nanoparticles, Biomaterials. 33 (19) (2012) 4947–4956.
[13] X.R. Xia, et al., Mapping the Surface Adsorption Forces of Nanomaterials in Biological Systems, ACS Nano 5 (11) (2011) 9074–9081.
[14] E. Casals, et al., Time evolution of the nanoparticle protein corona, ACS Nano 4 (7) (2010) 3623–3632.
[15] K.F. Freed, Entropy – Enthalpy Compensation in Chemical Reactions and Adsorption: An Exactly Solvable Model, J. Phys. Chem. B 115 (7) (2011) 1689–1692.
[16] M. Ahumada, et al., Association models for binding of molecules to nanostructures, Analyst 142 (12) (2017) 2067–2089.
[17] S. Tenzer, et al., Rapid formation of plasma protein corona critically affects nanoparticle pathophysiology, Nat. Nanotechnol. 8 (10) (2013) 772–U1000.
[18] H.S. Toh, K. Jurkschat, R.G. Compton, The influence of the capping agent on the oxidation of silver nanoparticles: nano-impacts versus stripping voltammetry, Chem-Eur J. 21 (7) (2015) 2998–3004.
[19] M.P. Monopoli, et al., Biomolecular coronas provide the biological identity of nanosized materials, Nat. Nanotechnol. 7 (12) (2012) 779–786.
[20] M.P. Monopoli, et al., Physical-chemical aspects of protein corona: relevance to in vitro and in vivo biological impacts of nanoparticles, J. Am. Chem. Soc. 133 (8) (2011) 2525–2534.
[21] Pd Pino, et al., Protein corona formation around nanoparticles - from the past to the future, Mater. Horiz. 1 (3) (2014) 301–313.
[22] I. Lynch, K.A. Dawson, Protein-nanoparticle interactions, Nano Today 3 (1–2) (2008) 40–47.
[23] T. Cedervall, et al., Understanding the nanoparticle–protein corona using methods to quantify exchange rates and affinities of proteins for nanoparticles, P. Natl. Acad. Sci. USA. 104 (7) (2007) 2050–2055.
[24] A.E. Nel, et al., Understanding biophysicochemical interactions at the nano-bio interface, Nat. Mater. 8 (7) (2009) 543–557.
[25] C. Corbo, et al., The impact of nanoparticle protein corona on cytotoxicity, immunotoxicity and target drug delivery, Nanomedicine. 11 (1) (2016) 81–100.
[26] C. Gunawan, et al., Nanoparticle-protein corona complexes govern the biological fates and functions of nanoparticles, J. Mater. Chem. B 2 (15) (2014) 2060–2083.

[27] L. Vroman, Effect of adsorbed proteins on the wettability of hydrophilic and hydrophobic solids, Nature 196 (4853) (1962) 476–477.

[28] L. Vroman, A.L. Adams, Identification of rapid changes at plasma–solid interfaces, J. Biomed. Mater. Res. 3 (1) (1969) 43–67.

[29] C. Rocker, et al., A quantitative fluorescence study of protein monolayer formation on colloidal nanoparticles, Nat. Nanotechnol. 4 (9) (2009) 577–580.

[30] L.A. Lane, et al., Physical chemistry of nanomedicine: understanding the complex behaviors of nanoparticles in vivo, Annu. Rev. Phys. Chem. 66 (2015) 521–547.

[31] P. Thordarson, Determining association constants from titration experiments in supramolecular chemistry, Chem. Soc. Rev. 40 (3) (2011) 1305–1323.

[32] G.M. Blackburn, A. Garçon, Catalytic antibodies, Biotech. Set., Wiley-VCH Verlag GmbH, 2008, pp. 401–490.

[33] K.N. Houk, et al., Binding affinities of host–guest, protein–ligand, and protein–transition-state complexes, Angew. Chem. Int. Ed. 42 (40) (2003) 4872–4897.

[34] A.J.T. Smith, et al., Beyond picomolar affinities: quantitative aspects of noncovalent and covalent binding of drugs to proteins, J. Med. Chem. 52 (2) (2008) 225–233.

[35] S. McLaughlin, et al., Sprayable peptide-modified silver nanoparticles as a barrier against bacterial colonization, Nanoscale 8 (46) (2016) 19200–19203.

[36] H. Poblete, et al., New insights into peptide-silver nanoparticle interaction: deciphering the role of cysteine and lysine in the peptide sequence, Langmuir 32 (1) (2016) 265–273.

[37] M. Ahumada, et al., Novel specific peptides as superior surface stabilizers for silver nano structures: role of peptide chain length, J. Mat. Chem. B. 5 (2017) 8925–8928.

[38] K.G. Thomas, J. Zajicek, P.V. Kamat, Surface binding properties of tetraoctylammonium bromide-capped gold nanoparticles, Langmuir 18 (9) (2002) 3722–3727.

[39] S. Roy, T.K. Das, Study of interaction between tryptophan, tyrosine, and phenylalanine separately with silver nanoparticles by fluorescence quenching method, J. Applied Spec. 82 (4) (2015) 598–606.

[40] I. Blakey, et al., Interactions of phenyldithioesters with gold nanoparticles (AuNPs): implications for AuNP functionalization and molecular barcoding of AuNP assemblies, Langmuir 26 (2) (2010) 692–701.

[41] E.I. Alarcon, et al., Human serum albumin as protecting agent of silver nanoparticles: role of the protein conformation and amine groups in the nanoparticle stabilization, J. Nano. Res. 15 (1) (2013) 1374.

[42] S. Jiang, et al., Quasi-spherical silver nanoparticles with high dispersity and uniform sizes: preparation, characterization and bioactivity in their interaction with bovine serum albumin, Luminescence 31 (5) (2016) 1146–1151.

[43] S. Zhang, et al., Binding analysis of carbon nanoparticles to human immunoglobulin G: Elucidation of the cytotoxicity of CNPs and perturbation of immunoglobulin conformations, Spectrochim. Acta Part A 154 (2016) 33–41.

[44] L. Shang, et al., Effect of protein adsorption on the fluorescence of ultrasmall gold nanoclusters, Small. 8 (5) (2012) 661–665.

[45] L. Shang, et al., Effect of protein adsorption on the fluorescence of ultrasmall gold nanoclusters, Small. 10 (9) (2014). 1667-1667.

[46] N. Nagarajan, et al., *Photosensitization of Colloidal SnO_2 Semiconductor Nanoparticles with Xanthene Dyes*, J. Chem., (2013) 7.

[47] B.-F. Pan, F. Gao, L.-M. Ao, Investigation of interactions between dendrimer-coated magnetite nanoparticles and bovine serum albumin, J. Mag. Magne. Mat. 293 (1) (2005) 252–258.

[48] P.P. Macaroff, et al., The effect of bovine serum albumin on the binding constant and stoichiometry of biocompatible magnetic fluids, IEEE Trans. Mag. 40 (4) (2004) 3027–3029.

[49] A.C. Tedesco, et al., *Determination of binding constant K_b of biocompatible, ferrite-based magnetic fluids to serum albumin*, J. Applied Phys. 93 (10) (2003) 6704–6706.

[50] A.C. Tedesco, et al., Investigation of the binding constant and stoichiometry of biocompatible cobalt ferrite-based magnetic fluids to serum albumin, J. Mag. Magne. Mat. 3 (2004) 2404–2405. **272–276, Part**.

[51] A. Bhogale, et al., Comprehensive studies on the interaction of copper nanoparticles with bovine serum albumin using various spectroscopies, Coll. Surf. B. 113 (2014) 276–284.

[52] R. Revathi, A. Rameshkumar, T. Sivasudha, Spectroscopic investigations on the interactions of AgTiO2 nanoparticles with lysozyme and its influence on the binding of lysozyme with drug molecule, Spectrochim. Acta Part A 152 (2016) 192–198.

[53] J. Hou, D.M. Szaflarski, J.D. Simon, Quantifying the association constant and stoichiometry of the complexation between colloidal polyacrylate-coated gold nanoparticles and chymotrypsin, J. Phys. Chem. B 117 (16) (2013) 4587–4593.

[54] J.A. Yang, et al., Study of wild-type alpha-synuclein binding and orientation on gold nanoparticles, Langmuir 29 (14) (2013) 4603–4615.

[55] G. Wang, et al., Probing the binding of trypsin to glutathione-stabilized gold nanoparticles in aqueous solution, Coll. Surf. B 135 (2015) 261–266.

[56] B. Valeur, M.N. Berberan-Santos, Steady-state spectrofluorometry, Molecular Fluorescence., Wiley-VCH Verlag GmbH & Co. KGaA, 2012, pp. 263–283.

[57] C. Lazurko, et al., NANoPoLC algorithm for correcting nanoparticle concentration by sample polydispersity, Nanoscale 10 (2018) 3166–3170.

Chapter 2

Computational modeling of the adsorption of capping agent biomolecules to inorganic nanoparticles

Horacio Poblete[a,b,c] and Jeffrey Comer[d]

[a]*Center for Bioinformatics and Molecular Simulation, Universidad de Talca, Talca, Chile,* [b]*Nucleo Científico Multidisciplinario, DI, Universidad de Talca, Talca, Chile.,* [c]*Millennium Nucleus of Ion Channels-Associated Diseases (MiNICAD), Universidad de Talca, Talca, Chile,* [d]*Institute of Computational Comparative Medicine, Nanotechnology Innovation Center of Kansas State, and Department of Anatomy and Physiology, Kansas State University, Manhattan, KS, United States*

Chapter Outline

2.1 Modeling in biomolecules and inorganic nanoparticles adsorption: past and present

Due to their considerable surface-area-to-volume ratio, the physicochemical properties of nanomaterials are controlled by their surface interactions. These interactions can be with small molecules or macromolecules, such as proteins. Moreover, the applications of the study of adsorption of organic molecules at the water–nanomaterial interface can be considered in two broad categories: applications considering nanomaterials in a physiological environment, such as drug delivery in vivo, and more technological applications where the environment is more controlled. In a physiological

Photoactive Inorganic Nanoparticles. DOI: https://doi.org/10.1016/B978-0-12-814531-9.00002-6
© 2019 Elsevier Inc. All rights reserved.

environment, nanomaterials usually possess a "corona" of adsorbed biomolecules such as proteins on their surfaces [1–3]. The biological activity of these nanoparticles is strongly affected by the structure and composition of this corona, influencing cellular uptake [4,5], physiological distribution, and excretion [4,6,7], and their interaction with the immune [8,9] and coagulation systems [10,11]. The second set of applications involve nonliving systems, which mainly focus on developing nanomaterials capable of purifying [12–14] biological molecules and removing contaminants [15–17] from complex mixtures. In these cases, selective adsorption may be desirable to prevent fouling of the material surfaces.

Since its discovery, graphene has been a central material in the field of nanotechnology. In particular, graphene is a promising nanoplatform for numerous biomedical applications including drug/gene delivery, biosensors, cell and tumor imaging, adsorption of enzymes, and cancer photothermal therapy [18]. These and other applications have attracted interest in the manufacturing of graphene and its derivatives (e.g., graphene nanosheets, graphene ribbons, graphene oxide, and reduced graphene oxide). Although oxidation of graphite was first performed during the 19th century and improved more than 50 years ago, the discovery of the unique properties of graphene has brought a new focus on this material [18]. When in contact with biological fluids, graphene oxide, like many other engineered nanomaterials, exhibits a "corona" of adsorbed biomolecules on its surface (see Fig. 2.1), which

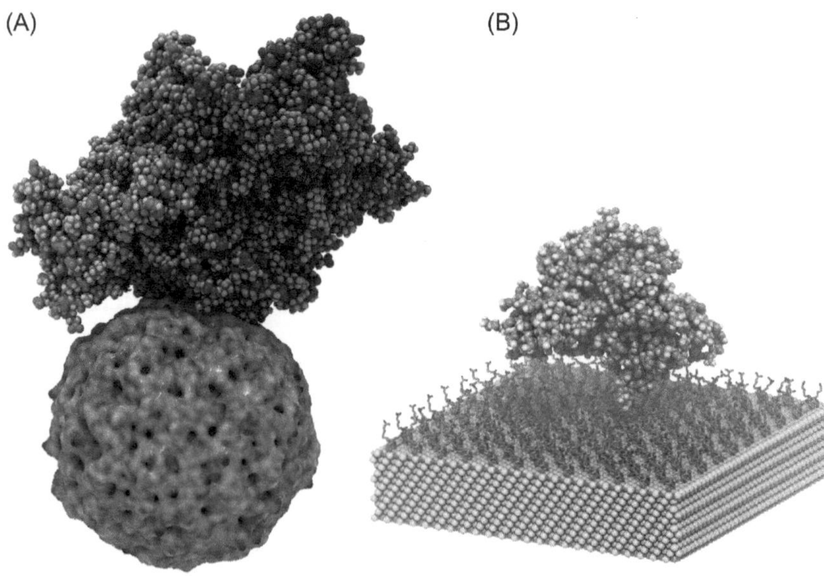

FIGURE 2.1 Molecular modeling of the protein corona on nanomaterials. (A) Two molecules of human serum albumin (HSA) (shown as atomic spheres) adsorbed on the surface of a SiO_2 nanosphere (gray). (B) Adsorption of HSA to a lipoic acid-functionalized gold nanosurface.

determines the rate and pathway of cellular uptake of the material, as well as its physiological distribution, route of excretion, and toxicology. Surface interactions with graphene oxide may stabilize or destabilize various protein structural elements [19], altering its biological activity. For example, a corona of bovine serum albumin has been shown to mitigate the cytotoxicity of graphene oxide by reducing its cell membrane penetration [18]. Interestingly this corona has been demonstrated to play a pivotal role in the nanomaterial's biocompatibility, systemic circulation, biodistribution, and bioavailability [20,21]. It should be noted that the precise mode in which this different corona components (biomolecules other than proteins, lipids, and sugars) interact with the surface and how they influence the biological activity has not been well studied thus far. In nanomedicine, the need has emerged to answer: (1) how is the interaction of engineered nanomaterial toward biomolecules such as proteins? (2) Which properties can give the corona to the nanoparticle? (3) What is the effect of the nanomaterial exposure on humans and other organisms? In the last few years, all these questions have been addressed by the constant development of different areas of study focusing on nanotechnology.

Another example of nanomaterials used in biomedical applications is nano silver, whose antimicrobial properties have been demonstrated [4,22–24]. Proteins [25], peptides [26], and free amino acids [27] can be used to control the structure of silver nanoparticles during synthesis, improving their stability under a variety of conditions. Similar synthesis techniques have been applied to other metals, such as gold; however, the mechanism(s) that underlie such stabilization and the exact role of various amino acid moieties interacting with silver nanostructures remain elusive. The binding affinity of a given peptide for particular crystal facets of this metal appears to mirror the peptide's ability to control nanoparticle synthesis and enhance stability [28–30]. Thus understanding peptide binding to silver surfaces may yield insight into peptide-mediated growth and stabilization of silver nanoparticles [28]. Both experiment and molecular simulation have revealed atomic-level details of peptides–nanosilver complexes. In particular, Fig. 2.2 illustrates a group of peptides that have been tested and proposed as nanosilver-surface stabilizers; from this set of peptides, molecular simulations have predicted high affinities of Arg, Cys, and Met residues for the Ag{111} facet [28]. However, cooperative effects over larger portions of the peptide sequence have been demonstrated to determine affinity for metal surfaces. Indeed, peptides selective for Ag surfaces demonstrate that the affinity is not determined solely by individual amino acids but also by the conformational statistics of the peptide as determined by its sequence [31].

For the potential application of metal nanomaterials in medicine, it is necessary to obtain a comprehensive understanding of the effect of metal nanoparticles on biomolecules. Metal nanomaterials have achieved selective modification of biomolecules, leading to applications in fields such as anchoring biomolecules as biosensors and biomarkers in the synthesis of

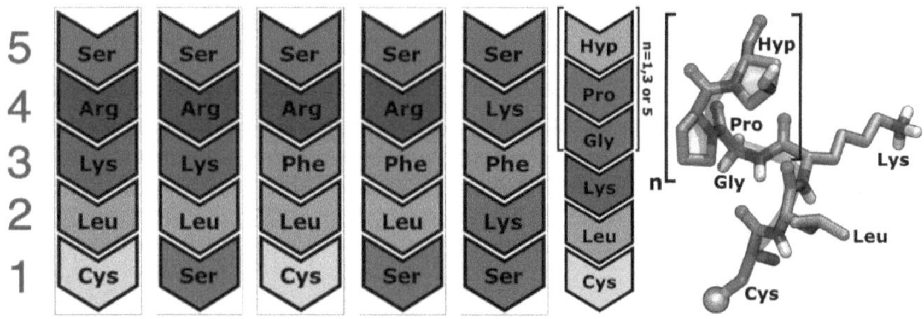

FIGURE 2.2 Structural and sequence representations for the peptide sequences considered for nanosilver-surface stabilization. (Left) Peptide sequences considered. (Right) Molecular model of a peptide shown to efficiently stabilize silver nanoparticles [31]. *Adapted from M. Ahumada, E. Jacques, C. Andronic, J. Comer, H. Poblete, E. Alarcon, Novel specific peptides as superior surface stabilizers for silver nano structures: role of peptide chain length, J. Mater. Chem. B (2017).*

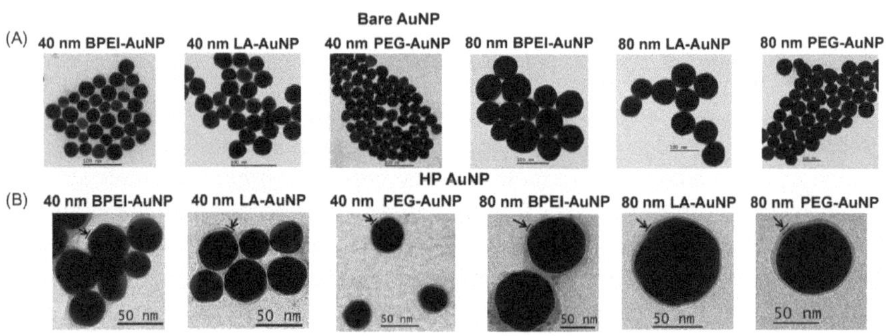

FIGURE 2.3 Generation of a protein corona. (A) Characterization of 40 and 80 nm bare polyethylenimine (BPEI), lipoic acid (LA), and polyethylene glycol (PEG) AuNPs. (B) Characterization of human plasma protein (HP) on 40 and 80 nm bare BPEI, LA, and PEG AuNPs. *Adapted from K.Choi, J.E. Riviere, N.A. Monteiro-Riviere, Protein corona modulation of hepatocyte uptake and molecular mechanisms of gold nanoparticle toxicity, Nanotoxicology 11 (1) (2017) 64–75 [35].*

new antibodies and/or protein engineering for study of neurodegenerative and cardiovascular diseases [32–34]. Interestingly, the formation of a protein layer around metal nanoparticles has been observed, affecting their uptake by cells [4,20] (see Fig. 2.3). With this has come the idea that the intentional formation of a suitable corona might increase the biocompatibility of nanomaterials and allow for controlled cellular uptake [19].

Computational modeling has become an invaluable approach to understand the interaction between biomolecules and nanomaterials due to its high spatial and temporal resolution. This field has shown strong growth in the last few decades, and has increasingly been a source of guidance in nanomaterial design. Furthermore, computational modeling can complement experimental assays and assist in the interpretation of experimental data. Fig. 2.4 shows one

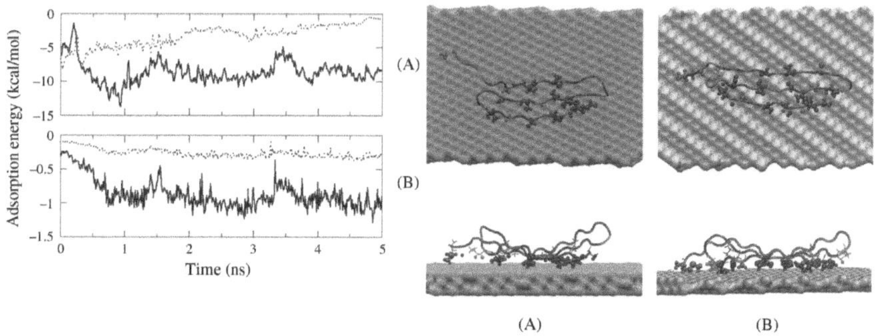

FIGURE 2.4 Adsorption of gold-binding peptides on gold surface modeled by molecular dynamics simulations [36]. (Left) Adsorption energy as a function of time during the simulation. (A) Adsorbed peptide on Au(111). (B) Adsorbed peptide on Au(211). *Adapted from R.Braun, M. Sarikaya, K.Schulten, Genetically engineered gold-binding polypeptides: structure prediction and molecular dynamics, J. Biomater. Sci. Polym. Ed. 13 (7) (2002) 747–757.*

of the first studies using molecular simulation to explore the interface between nanotechnology and biomolecules, pioneered by Klaus Schulten and coworkers [36] and focusing on the interaction between gold and gold-binding peptides. Here the authors showed the basis of the use of atomistic molecular dynamics to describe the adsorption processes of three experimentally identified 42-residue peptides (GBP1, GBP2, and GPB3). These peptides were modeled over two crystallographic facets of gold, Au(111) and Au(211), in an explicit-solvent environment. These simulations were remarkable for the year 2002 due to the system size, level of detail, and the simulation time of 5 ns, which at that time required months of simulation. Despite this, it is known that the degree of conformational sampling reached by the authors is not enough to sample all the different conformations that the peptides can adopt in the adsorption process. In general, the problem of obtaining sufficient sampling is a major consideration for simulations of large, flexible molecules, including most biomolecules, with molecular dynamics.

Historically, the computational study of metal surfaces has mostly involved quantum mechanical models. Unfortunately, due to scaling of quantum mechanics methods, the computational costs of simulating these surfaces remain quite high, even when many approximations are used and no explicit water molecules are included. Moreover, conformational sampling of molecules such as peptides and proteins is prohibitively costly for quantum mechanical methods. Classical atomistic models, which we have already discussed, provide some vast improvement in performance, at the cost of a more approximate representation of interatomic forces and the need to develop reliable empirical force fields, fitted to experimental data and the results of quantum mechanical calculations. However, currently, simulations of proteins using fully atomistic models typically only reach timescales of a few microseconds. For some applications, coarser models are required to obtain longer

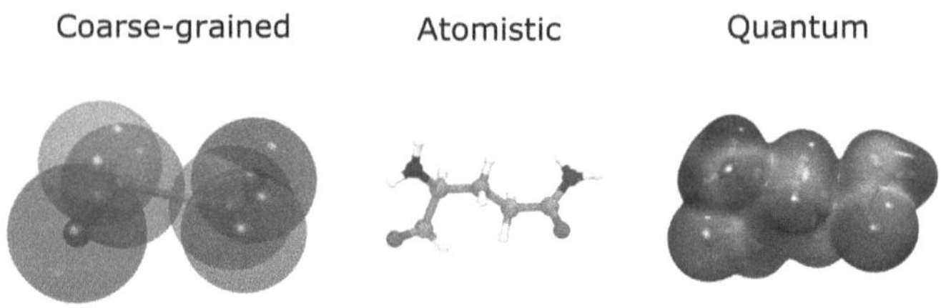

FIGURE 2.5 Representation of the amino acid glutamine (Gln) at different levels of description. (Left) Coarse-grained structure based on the SIRAH force field [40]. (Center) Typical atomistic representation for classical molecular dynamics. (Right) Electrostatic potential calculated from the electronic density of Gln calculated by quantum chemistry methods.

timescales and observe greater configurational changes. In these so-called coarse-grained (CG) methods several atoms are represented as single spheres that contain the sum of the atomic mass and the total charge of the group of atoms. However, it can be difficult to represent protein and peptide structuring at this scale accurately. Several studies focusing on the biomolecules adsorption to inorganic surfaces have been reported [37–39]. The greater simulation length scales and timescales that can be reached with CG have been the reason for all the efforts that many groups have made to develop reliable CG force fields. Fig. 2.5 shows a comparison of the amino acid glutamine in CG, atomistic, and quantum representations, respectively. Which type of modeling is the best to describe the adsorption of biomolecules to inorganic nanoparticles? The answer depends on the scientific problem at hand. Often, a multiscale approach, using methods that work best on different scales, can give the best understanding of the system being studied and reveal when one model might be producing artifactual results.

As a general rule, computational cost increases as the accuracy of the method increases; however, it should be noted that well-parameterized empirical methods could sometimes outperform lower-level methods in particular contexts.

2.2 Using molecular dynamics to characterize nanoparticle surface adsorption of capping agents

How does one understand adsorption processes? Which are the atomic forces that control the molecular adsorption on the surface? What is the effect of the surface on the structure/function of a protein? As it was mentioned above, a mixture of experimental approaches and computational techniques can address these and other questions. Elucidating the relationship between structure and properties has been the principal goal for decades for the computational simulation methods.

Atomistic molecular simulation offers the unique ability to explore interactions between biomolecules and engineered nanoparticles in a realistic aqueous environment with atomic detail and femtosecond temporal resolution. Moreover, with sufficiently accurate force fields for the materials of interest, the method is capable of yielding precise agreement with the experiment for free energies of binding [41,42] and many other measurements determined by nuclear magnetic resonance [43]. While docking, another tool of computational pharmaceutical science with atomic resolution, features a lower computational cost, it is not capable of quantitative precise estimation of free energy, provides no information on kinetic parameters, and cannot model the extensive conformational flexibility of many nanomaterials, such as those considered here having long-chain functional groups.

.A major limitation of brute-force atomistic molecular dynamics is the timescale that can be accessed, usually limited to many microseconds. Fig. 2.6 shows a general outline of the molecular dynamics/free-energy calculations processes; therefore nearly all calculations of free energy involve specialized techniques that permit probing transitions that take place on much longer timescales. Under the assumption that a transition, such as the formation of a bound complex, can be captured by a few slow degrees of freedom, important sampling strategies permit calculation of the free-energy landscape (see Fig. 2.7) along the transition pathway [45]. However, these strategies have traditionally provided only the static picture of such transitions. Recent work done by the Comer group [28,46–50] and others [51–53] has succeeded in

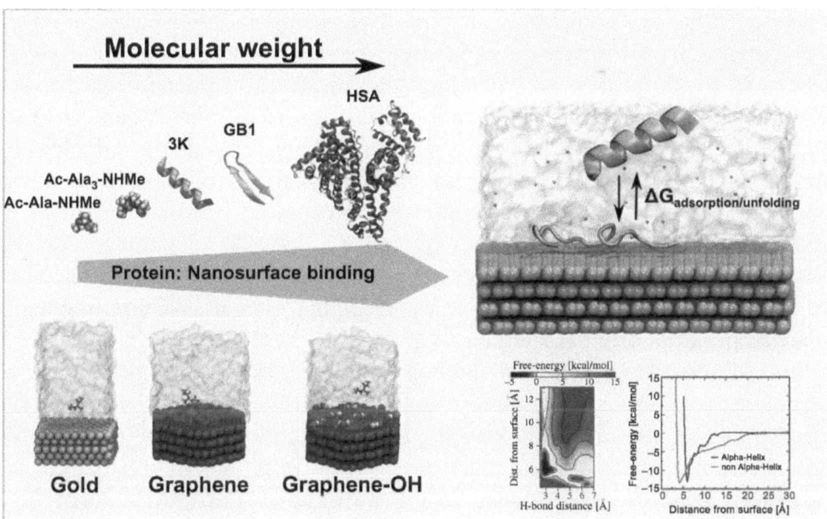

FIGURE 2.6 (Left top) Examples of peptides and proteins with well-defined structure in aqueous environment. (Left bottom) Models of surfaces used in simulations of nanoparticle–biomolecule interactions. (Right top) Diagram outlining calculation of the adsorption free energy of a structured peptide. (Right bottom) Results of free-energy calculations for adsorption and unfolding of a peptide on a graphene surface.

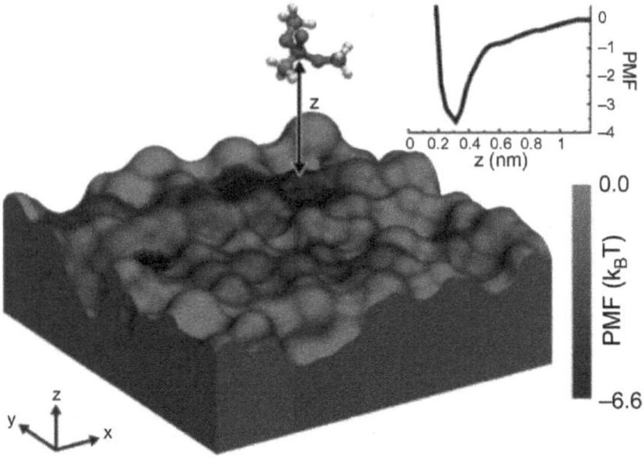

FIGURE 2.7 Mapping the adsorbate-surface free-energy landscape in three dimensions. For clarity, a two-dimensional representation is shown in which the color of the surface indicates the minimum free energy along a line parallel to the z-axis at that (x,y) position. The arrow appears above a dark region—a small depression that is the global free-energy minimum and the most favorable binding location. The inset in the upper-left corner shows the integration over the three-dimensional data to obtain a mean one-dimensional free-energy profile along the z-axis. *Adapted from R. Carr, J. Comer, M.D. Ginsberg, A. Aksimentiev, Microscopic perspective on the adsorption isotherm of a heterogeneous surface, J. Phys. Chem. Lett. 2 (14) (2011) 1804–1807 [44].*

extracting kinetics from simulations where importance of sampling is invoked. With no additional computational cost, free energies can be supplemented with kinetics to yield complete diffusive models of transitions along slow degrees of freedom. Beyond extending the timescales of phenomena that can be studied through atomistic simulation, these diffusive models can also help in developing intuitive pictures of complex systems, effectively reducing the number of degrees of freedom from the thousands or millions in atomistic simulations to just a few that capture the process of interest. The reduced models so created can then be explored at little computational cost by Brownian dynamics simulation on timescales much longer than those accessible by brute-force molecular dynamics, revealing phenomena that might not be apparent in the original simulations [50].

In addition to the different methods used to describe the adsorption of biomolecules to inorganic nanosurfaces (of different types surface, see Fig. 2.8), it is important to consider that the surfaces themselves and the force fields representing their must be modeled in an accurate way. To this end, Heinz and coworkers added to the well-known CHARMM force field [54] a generalized section in which they included metals as Au, Pd, Ag, and Pt called CHARMM-METAL [55]. In more detail, this metals force field was able to describe the interaction of peptides with an aqueous–metal interface, and it was quickly incorporated into the INTERFACE force field [56]. Based only on Lennard–Jones interfacial interactions, this force field was

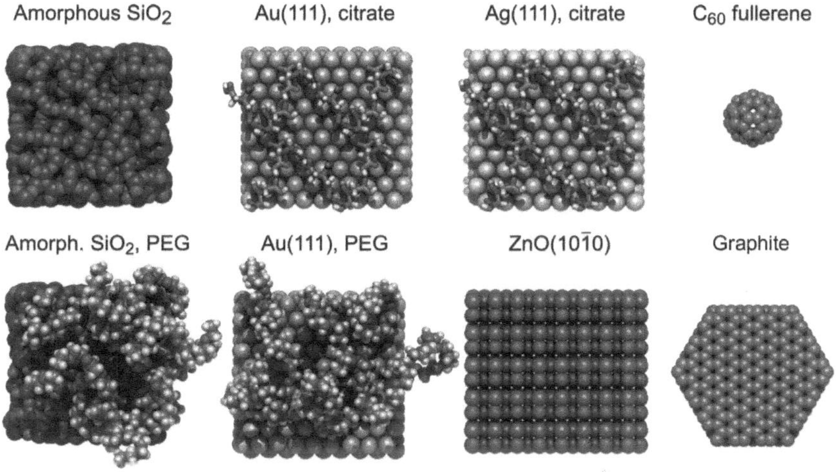

FIGURE 2.8 Example of parameterized surfaces. Numbers in parentheses indicate the crystal plane. The configuration of the citrate molecules [58] and the density of PEG chains [59] were estimated from the literature. Atom color key for electronic version: Si, dark gray; O, red; O (deprotonated silanol), purple; C, green; H, white; Au, light yellow; S, orange; Ag, light gray; Zn, blue-gray. The force fields for Au [60] and Ag [61] include interstitial particles to better replicate the calculated electron density, which we show here as translucent blue spheres.

extensively used to simulate noble-metal surfaces, and CHARMM-METAL force field was also able to describe dynamic or disordered metal interfaces. Remarkably, also in 2008, Corni and coworkers developed the GolP force field [57] in which they incorporated surface polarization effects to favor the biomolecule adsorption to Au(111) interface. Interestingly, the polarization effect facilitated the recognition and adsorption to the layer of Au(111) exposed to the solvent. Nevertheless, it was limited to simulating only an Au (111) surface and could not capture other facets or the restructuring of the surface that can be caused by some adsorbates.

In a further advance, Walsh and coworkers developed the GolP-CHARMM force field [60], which in contrast to GolP, included Au(111) and Au(100) interfaces. Moreover, the same group also developed parameters for Ag(111), Ag(100), and aqueous graphene interfaces, in the same way that was done for GolP-CHARMM [60]. Recently, Hughes and Walsh through first-principles calculations proposed that heteroatomic adsorbates show a higher affinity for the vertex and edge of noble-metal nanoparticles; meanwhile planar facets of nanosurfaces are preferentially a best binding site for hydrocarbon adsorbates [62] (Fig. 2.9).

Given its similarity to organic aromatics, graphene is relatively well represented in standard biomolecular force fields when compared with metals. In their recent work [18], the Comer group has shown that all-atom molecular dynamics simulations can predict the relative adsorption affinities of small

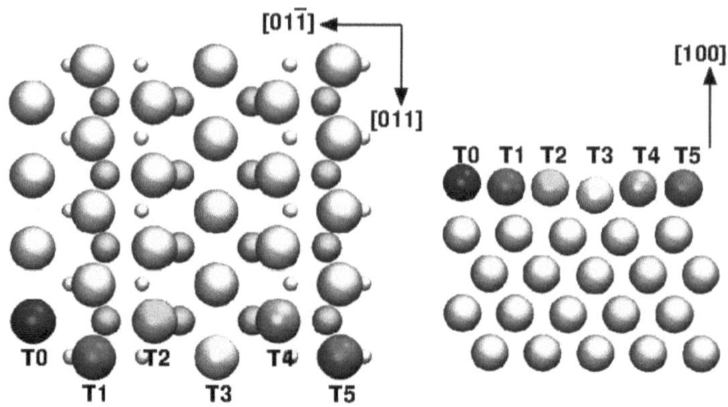

FIGURE 2.9 Au(100)(5 × 1) surface after relaxation using vdW-DF, viewed (left) from above and (right) from the side. For the above view, the surface-layer Au atoms are depicted as largest and gold in color, the second layer of Au atoms are more red and of medium size, and the third layer of Au atoms are the smallest in size and light yellow in color. *Adapted from L.B. Wright, P.M. Rodger, T.R. Walsh, S. Corni, S. First-principles-based force field for the interaction of proteins with Au (100)(5 × 1): an extension of GolP-CHARMM. J. Phys. Chem. C 117 (46) (2013) 24292–24306; T.R. Walsh, M.R. Knecht, Biointerface structural effects on the properties and applications of bioinspired peptide-based nanomaterials. Chem. Rev. 117 (20) (2017) 12641–12704 [63,64].*

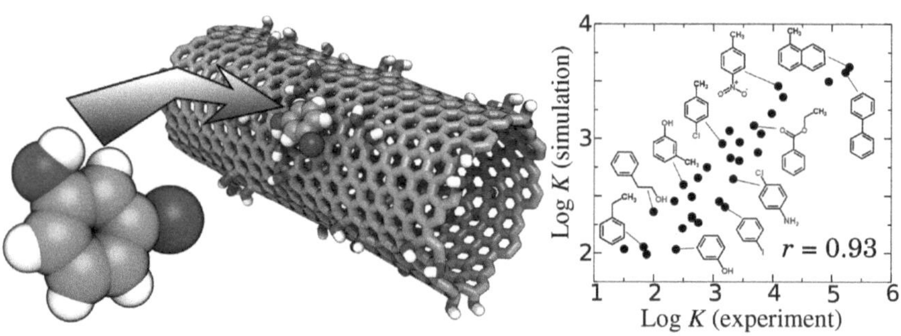

FIGURE 2.10 Comparison of equilibrium constants derived from experiment and simulation for small aromatics adsorbing to large hydroxylated carbon nanotubes [18].

molecules on graphenic materials, comparing with experimental data published previously. The all-atom molecular dynamic model in Fig. 2.10 shows several flat layers of graphene, representing the local environment of small molecules adsorbing to the large-diameter (8–15 nm) carbon nanotubes used in the experiments. Using the CHARMM General Force Field [51] to describe interatomic interactions and the adaptive biasing force technique [49,65] to obtain the binding free-energy, they obtained a high degree of correlation with experiment, yielding Pearson correlation coefficients from 0.90 for naked nanotubes to 0.93 for hydroxylated nanotubes [18]. Notably the simulations revealed structural details that would be difficult to predict from basic chemical principles. Although the surface of carbon nanotubes is considered to be relatively hydrophobic [66], the hydrophobicity of an adsorbate, as characterized by the logarithm of the octanol−water partition coefficient ($\log_{10} P_{OW}$), is a poor predictor of the adsorption affinity with a correlation coefficient of only $r = 0.53$. For example, with its polar nitro group, 4-nitrotoluene is only moderately hydrophobic, with $\log_{10} P_{OW} = 2.4$; however, both experiment and simulation assert that it adsorbs strongly to carbon nanotubes, much more strongly than, for instance, propylbenzene, for which $\log_{10} P_{OW} = 3.7$. This work suggests that experimental data for small molecule adsorption to nanomaterial surfaces could be a robust and convenient means to validate and improve computational models.

With relevance to biomolecules, it has been possible to calculate the free energy of adsorption of amino acids and small protein motifs over nanomaterial surfaces to understand the key interactions driving adsorption. Many authors have made great efforts to increase knowledge in this area. Recently Poblete and Comer studied the interaction of graphene with the common toy model of the protein backbone, alanine dipeptide (Ala-NHMe) [67], which historically has been used to understand the backbone conformational preferences [68−72]. In this work, the authors performed explicit-solvent molecular dynamics and free-energy calculations to reveal how adsorption to carbon nanomaterial surfaces affects the conformational equilibrium of alanine dipeptide, revealing that particular planar conformations (shown in Fig. 2.11) were the most stable structures when the molecule is adsorbed to a pristine graphene surface, due to strong amide−π interactions and these structures were maximizing the area of contact. It was proposed that the hydroxylation degree of the graphene surface changes the balance of the preferred conformations of the peptide. Specifically, it was found that, with more hydroxyl groups, the conformational free-energy landscape of alanine dipeptide becomes more similar to that in an aqueous environment [67]. This study gives insight into how carbon nanotubes, graphene and/or their functionalized derivatives can influence in the secondary structure of proteins.

In recent years, several authors have proposed that the sequence of a given peptide can control the growth of nanomaterials during synthesis, modulating their size and shape [28,30,31,73]. Such peptides might therefore be used to design nanoparticles with specific characteristics by choosing the right peptide sequence. In 2015, using both experiment and molecular dynamics simulations, Poblete et al. tested five pentapeptides of different sequences to determine their influence on the synthesis of silver nanoparticles, revealing that the combination of thiol in Cys and amines in Lys/Arg residues was crucial for providing stable protection for the silver surface [28]. Subsequently, in 2017, Ahumada et al. used the previously identified subsequence (CLK) with additional collagen mimetic sequences to create a superior class of surface stabilizers for silver nanostructures. Fig. 2.12

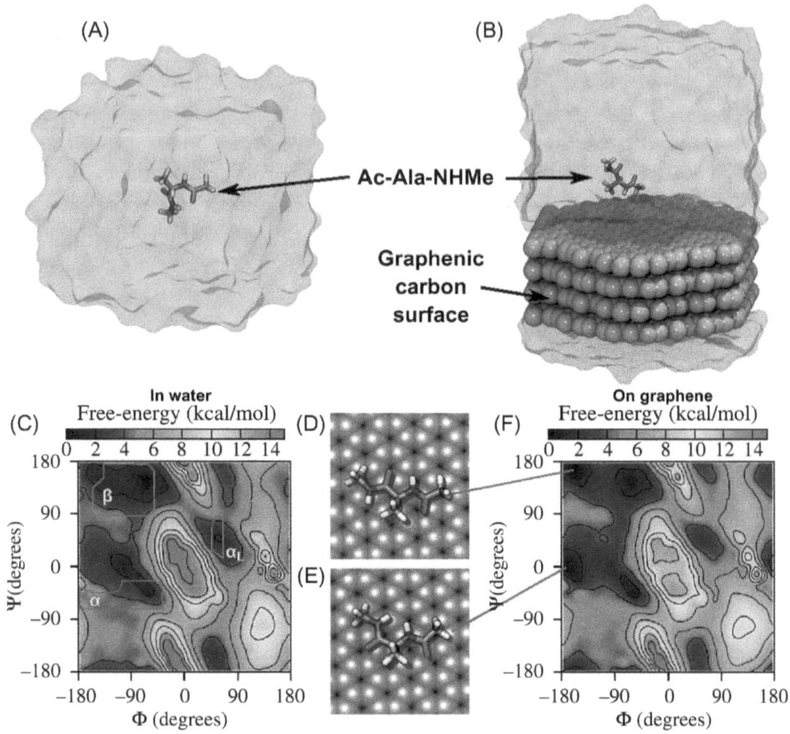

FIGURE 2.11 Conformational free-energy landscapes of Ac-Ala-NHMe in bulk water and at the graphene–water interface. Molecular dynamics models of Ac-Ala-NHMe in water (A) and at the graphene–water interface (B). Both systems are periodic along all three axes; thus the system shown in panel (B) has an effectively infinite graphene surface. The Ac-Ala-NHMe molecules are shown with a ball-and-stick representation. The carbon atoms of the four graphene layers are shown as gray spheres. (C) Ramachandran free-energy landscape for Ac-Ala-NHMe in water. Regions outlined in (magenta in online version) and denoted α, β, and α_L are associated with right-handed α-helices, β-sheet, and left-handed α-helices, respectively. (D) Representative conformation of Ac-Ala-NHMe corresponding to the global free-energy minimum in panel (F) (here called β-planar). (E) Representative conformation of Ac-Ala-NHMe corresponding to the second-lowest local free-energy minimum in panel (F) (here called α-planar). (F) Ramachandran free-energy landscape for Ac-Ala-NHMe at the graphene–water interface. The plots are anchored such that the minimum free energy is zero [67].

demonstrates that peptide length has an important effect on the stability of the resulting nanosilver colloid [31].

Recently it was proposed that this sequence can bind to the growing material and also alter the metal atom deposition on the material surface (see Fig. 2.13), leading to disordered metallic surface structures that deviated from the anticipated crystal structure [74,75]. Such peptides give improved control of the material properties compared with approaches using small molecule stabilizing agents [74,75].

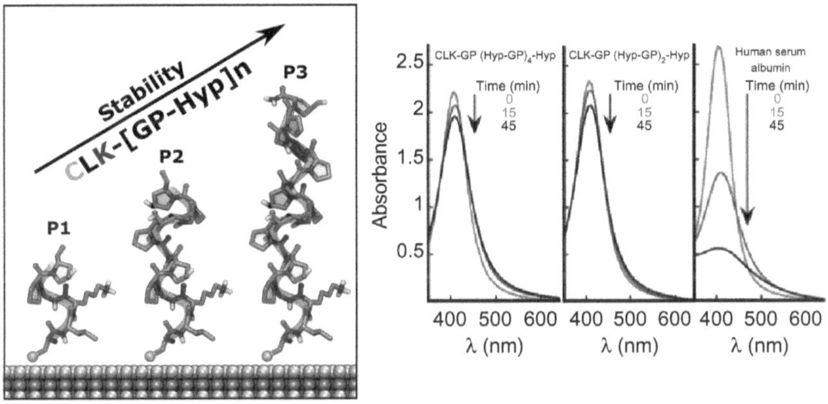

FIGURE 2.12 Changes in nanosilver-surface plasmon band (SPB) seen after heating a sample solution for 45 min at 98°C. Nanosilver samples were prepared using 20 mM of peptides with sequences of CLK-(GP-Hyp)$_3$ or CLK-(GP-Hyp)$_5$ or human serum albumin (HSA). Samples were measured at different time points after heating. All measurements were carried out at room temperature in triplicate [31].

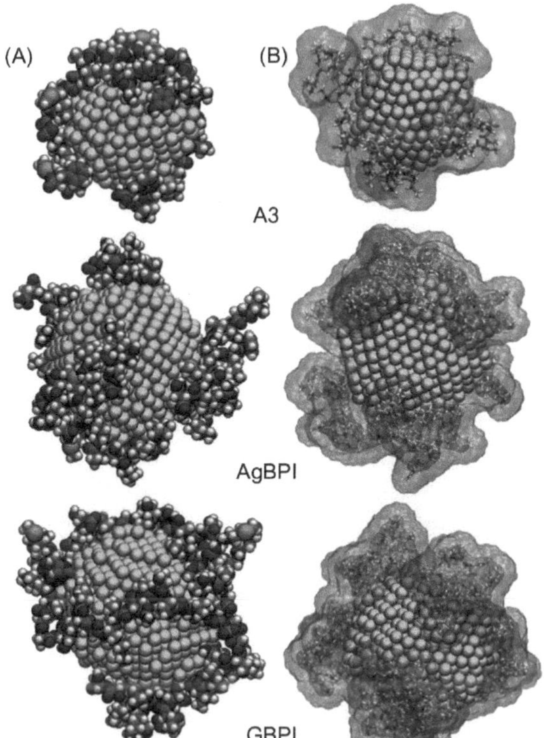

FIGURE 2.13 Structure of exemplary peptide-promoted catalytic Au nanoparticles, A3, AgBP1, and GBP1, obtained from the Boltzmann weighted ensemble of configurations predicted by REST-MD simulations, using NP models originally produced by RMC modeling of experimental PDF data. (A) Typical configuration of the most populated cluster of the peptide overlayer. (B) Indication of the exposed NP surface within the peptide overlayer [74].

In general, this new knowledge allows us to understand the atomic interactions at the interface between capping agents and inorganic nanomaterials in aqueous environment. Further efforts are needed in the development of new techniques and analyses that allow the design of nanoparticles with specific functions in a completely rational way.

2.3 Computational modeling of adsorption: challenges and new methods to understand the adsorption of capping agents to inorganic nanosurfaces

New molecular dynamics force fields maybe needed to accurately describe the interactions between nanomaterials and inorganic surfaces. For instance, the CHARMM General Force Field [54] appears to somewhat underestimate the adsorption affinities of small aromatic molecules on carbon nanotubes as compared with experiment, despite high correlation (>0.9) [18]. Newer polarizable force fields, which explicitly represent atomic polarizability, such as CHARMM Drude [76], promise increased accuracy for interfaces between polar and nonpolar phases, such as the water–graphene oxide interface. However, as a new methodology, it needs more work and refinement in its parameters, and thus reproduce experimental results in a more reliable way. Although this newer force field does not yet cover many of the small compounds in Fig. 2.10, it does have complete coverage of polypeptides. The polarizable force field does increase the computational cost of the simulations (by about four times), both because the models contain additional Drude particles to model polarizability and because a shorter time step (1 vs 2 or 4 fs) is required to integrate the motion of these particles. Recent work has demonstrated that the polarizable CHARMM Drude [77,78] model yields improvement over nonpolarizable CHARMM36 polarizable ions [79]. For instance, a known iodide-binding protein exhibited a strong iodide-binding site when simulated with the CHARMM Drude force field, while a similar calculation with the nonpolarizable CHARMM36 force field apparently erroneously predicted only metastable binding of iodide [80].

In conventional biomolecular molecular dynamics force fields, including CHARMM36 [54] and CHARMM Drude [77,78], covalent bonds between atoms are permanent features of the molecular models and cannot break or form during the simulations. Reactive force fields, exemplified by ReaxFF [82], are relatively recent innovations that model covalent association and dissociation under a classical physics framework, avoiding the computational cost of full quantum mechanical descriptions. Notably, these reactive models promise to allow simulations of chemisorption on nanomaterial surfaces. Fig. 2.14 shows an example of a reactive molecular dynamics simulation of a miniprotein (Trp-cage) [81] on an Au(111) surface; interestingly, using

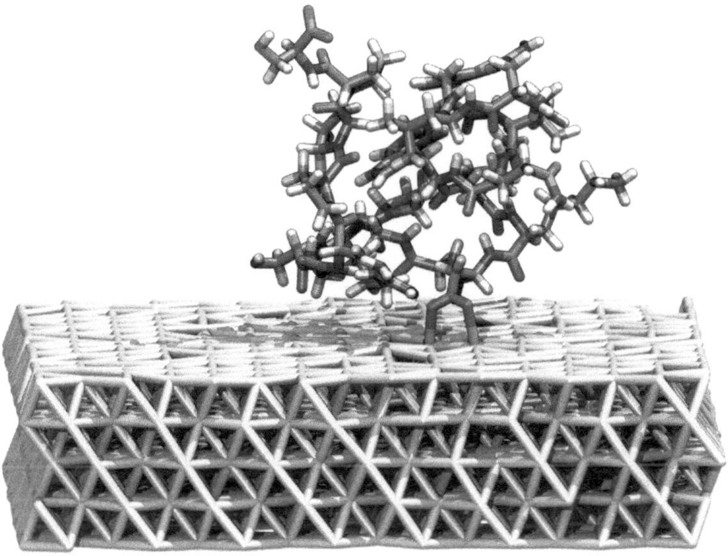

FIGURE 2.14 Reactive molecular dynamics simulation of a miniprotein (Trp-cage) [81] on an Au(111) surface. The formation of a covalent bond between a carboxylate group of the protein and the gold surface is visible.

ReaxFF simulation was possible to see the formation of a covalent bond between a carboxylate group of the protein and the gold surface. Moreover, reactive models allow computational simulations of the fabrication nanomaterials, which may lead to more accurate structural models of reconstructed and disordered surfaces. For instance, graphene oxide can be reduced at high temperatures in ReaxFF simulations, mirroring thermal reduction processes used in experiment to produce reduced graphene oxide [83]. The resulting structures appear similar to those observed in reduced graphene oxide by electron microscopy [84]. A comparison of experimental and computationally created structures is shown in Fig. 2.15.

The chemical space covered by ReaxFF continues to expand, including metals such as gold, silver, or titanium [85], among others. Moreover, ReaxFF has been applied to understand the dynamic deprotonation phenomena at the hydroxylated-titania interfaces [86]. The growing reliability of reactive force fields and the efficiency of their implementation gives an interesting new strategy to address the design of inorganic nanosurfaces. Like conventional force fields, the comparison with experiment that it provides is important for validation of the reactive force fields. Reactive force fields seem to have a bright future in enabling the design of nanotechnology and nanomedicine.

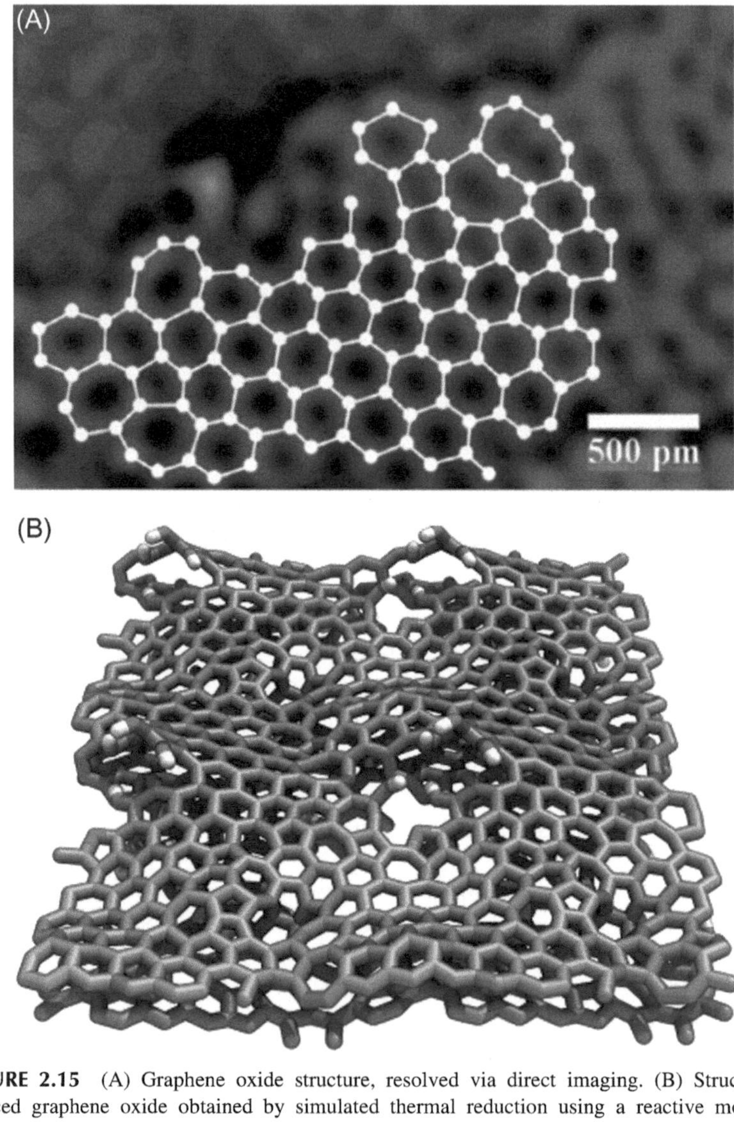

FIGURE 2.15 (A) Graphene oxide structure, resolved via direct imaging. (B) Structure of reduced graphene oxide obtained by simulated thermal reduction using a reactive molecular dynamics force field. *(A) Reproduced with permission from S.H. Dave, C. Gong, A.W. Robertson, J.H. Warner, J.C. Grossman, Chemistry and structure of graphene oxide via direct imaging. ACS Nano. 10 (8) (2016) 7515–7522.*

References

[1] I. Lynch, T. Cedervall, M. Lundqvist, C. Cabaleiro-Lago, S. Linse, K.A. Dawson, The nanoparticle–protein complex as a biological entity; a complex fluids and surface science challenge for the 21st century, Adv. Colloid. Interface Sci. 134 (2007) 167–174.

[2] C. Röcker, M. Pötzl, F. Zhang, W.J. Parak, G.U. Nienhaus, A quantitative fluorescence study of protein monolayer formation on colloidal nanoparticles, Nat Nanotechnol. 4 (9) (2009) 577–580.

[3] S. Tenzer, D. Docter, S. Rosfa, A. Wlodarski, J. Kuharev, A. Rekik, et al., Nanoparticle size is a critical physicochemical determinant of the human blood plasma corona: a comprehensive quantitative proteomicanalysis, ACS Nano. 5 (9) (2011) 7155–7167.

[4] N.A. Monteiro-Riviere, M.E. Samberg, S.J. Oldenburg, J.E. Riviere, Protein binding modulates the cellular uptake of silver nanoparticles into humancells: implications for in vitro to in vivo extrapolations? Toxicol. Lett. 220 (3) (2013) 286–293.

[5] A. Sasidharan, J.E. Riviere, N. Monteiro-Riviere, Gold and silver nanoparticle interactions with humanproteins: impact and implications in biocorona formation, J. Mater. Chem. B 3 (2015) 2075–2082.

[6] M.P. Monopoli, C. AAberg, A. Salvati, K.A. Dawson, Biomolecular coronas provide the biological identity of nanosized materials, Nat Nanotechnol. 7 (12) (2012) 779–786.

[7] J.E. Riviere, Of mice, men and nanoparticle biocoronas: are in vitro to in vivo correlations and interspecies extrapolations realistic? Nanomedicine (Lond) 8 (9) (2013) 1357–1359.

[8] M.A. Dobrovolskaia, S.E. McNeil, Immunological properties of engineered nanomaterials, Nat Nanotechnol. 2 (8) (2007) 469–478.

[9] S. Syed, A. Zubair, M. Frieri, Immune response to nanomaterials: implications for medicine and literature review, Curr. Allergy Asthma Rep. 13 (1) (2013) 50–57.

[10] M.A. Dobrovolskaia, A.K. Patri, J. Zheng, J.D. Clogston, N. Ayub, P. Aggarwal, et al., Interaction of colloidal gold nanoparticles with human blood: effects on particle size and analysis of plasma protein binding profiles, Nanomedicine 5 (2) (2009) 106–117.

[11] M.A. Dobrovolskaia, A.K. Patri, T.M. Potter, J.C. Rodriguez, J.B. Hall, S.E. McNeil, Dendrimer-induced leukocyte procoagulant activity depends on particle size and surface charge, Nanomedicine 7 (2) (2012) 245–256.

[12] S. Bucak, D.A. Jones, P.E. Laibinis, T.A. Hatton, Protein separations using colloidal magnetic nanoparticles, Biotechnol. Prog. 19 (2) (2003) 477–484.

[13] C. Ma, C. Li, N. He, F. Wang, N. Ma, L. Zhang, et al., Preparation and characterization of monodisperse core–shell Fe3O4@SiO2 microspheres and its application for magnetic separation of nucleic acids from *E. coli* BL21, J. Biomed. nanotechnol 8 (6) (2012) 1000–1005.

[14] M. Cano, K. Sbargoud, E. Allard, C. Larpent, Magnetic separation of fatty acids with iron oxide nanoparticles and application to extractive deacidification of vegetable oils, Green Chem. 14 (6) (2012) 1786–1795.

[15] J.R. Cromer, S.J. Wood, K.A. Miller, T. Nguyen, S.A. David, Functionalized dendrimers as endotoxin sponges, Bioorg. Med. Chem. Lett. 15 (5) (2005) 1295–1298.

[16] C.T. Yavuz, J. Mayo, C. Suchecki, J. Wang, A.Z. Ellsworth, H. D'Couto, et al., Pollution magnet: nano-magnetite for arsenic removal from drinking water, Environ. Geochem. Health 32 (4) (2010) 327–334.

[17] B.H.X. Che, S.P. Yeap, A.L. Ahmad, J. Lim, Layer-by-layer assembly of iron oxide magnetic nanoparticles decorated silica colloid for water remediation, Chem. Eng. J. 243 (2014) 68–78.

[18] W. Hu, C. Peng, M. Lv, X. Li, Y. Zhang, N. Chen, et al., Protein corona-mediated mitigation of cytotoxicity of graphene oxide, ACS Nano. 5 (5) (2011) 3693–3700.

[19] J. Comer, R. Chen, H. Poblete, A. Vergara-Jaque, J.E. Riviere, Predicting adsorption affinities of small molecules on carbon nanotubes using molecular dynamics simulation, ACS Nano. 9 (12) (2015) 11761–11774.

[20] M.P. Monopoli, D. Walczyk, A. Campbell, G. Elia, I. Lynch, F. Baldelli Bombelli, et al., Physical – chemical aspects of protein corona: relevance to in vitro and in vivo biological impacts of nanoparticles, J. Am. Chem. Soc. 133 (8) (2011) 2525–2534.

[21] F.D. Sahneh, C. Scoglio, J. Riviere, Dynamics of nanoparticle-protein corona complex formation: analytical results from population balance equations, PLoS. ONE. 8 (5) (2013) e64690.

[22] E.I. Alarcon, C.J. Bueno-Alejo, C.W. Noel, K.G. Stamplecoskie, N.L. Pacioni, H. Poblete, et al., Human serum albumin as protecting agent of silver nanoparticles: role of the protein conformation and amine groups in the nanoparticle stabilization, J. Nanoparticle Res. 15 (1) (2013) 1–14.

[23] E.I. Alarcon, K. Udekwu, M. Skog, N.L. Pacioni, K.G. Stamplecoskie, M. González-Béjar, et al., The biocompatibility and antibacterial properties of collagen-stabilized, photochemically prepared silver nanoparticles, Biomaterials. 33 (19) (2012) 4947–4956.

[24] M. Vignoni, H. de Alwis Weerasekera, M.J. Simpson, J. Phopase, T.-F. Mah, M. Griffith, et al., LL37 peptide@ silver nanoparticles: combining the best of the two worlds for skin infection control, Nanoscale 6 (11) (2014) 5725–5728.

[25] P. Murawala, S.M. Phadnis, R.R. Bhonde, B.L.V. Prasad, In situ synthesis of water dispersible bovine serum albumin capped gold and silver nanoparticles and their cytocompatibility studies, Colloids Surf. B Biointerfaces 73 (2) (2009) 224–228.

[26] R.R. Naik, S.J. Stringer, G. Agarwal, S.E. Jones, M.O. Stone, Biomimetic synthesis and patterning of silver nanoparticles, Nat. Mater. 1 (3) (2002) 169–172.

[27] P. Koley, A. Pramanik, Nanostructures from single amino acid-based molecules: stability, fibrillation, encapsulation, and fabrication of silver nanoparticles, Adv. Funct. Mater. 21 (21) (2011) 4126–4136.

[28] H. Poblete, A. Agarwal, S.S. Thomas, C. Bohne, R. Ravichandran, J. Phopase, et al., New insights on peptide–silver nanoparticle interaction: deciphering the role of cysteine and lysine in the peptide sequence, Langmuir 32 (2015) 265–273.

[29] J.M. Slocik, M.O. Stone, R.R. Naik, Synthesis of gold nanoparticles using multifunctional peptides, Small. 1 (11) (2005) 1048–1052.

[30] Y. Li, Z. Tang, P.N. Prasad, M.R. Knecht, M.T. Swihart, Peptide-mediated synthesis of gold nanoparticles: effects of peptide sequence and nature of binding on physicochemical properties, Nanoscale 6 (6) (2014) 3165–3172.

[31] M. Ahumada, E. Jacques, C. Andronic, J. Comer, H. Poblete, E. Alarcon, Novel specific peptides as superior surface stabilizers for silver nano structures: role of peptide chain length, J. Mater. Chem. B (2017).

[32] R. Jayakumar, D. Menon, K. Manzoor, S.V. Nair, H. Tamura, Biomedical applications of chitin and chitosan basednanomaterials—a short review, Carbohydr. Polym. 82 (2) (2010) 227–232.

[33] T. Douglas, E. Strable, D. Willits, A. Aitouchen, M. Libera, M. Young, Protein engineering of a viral cage for constrained nanomaterials synthesis, Adv. Mater. 14 (6) (2002) 415.

[34] S.T. Stern, D.N. Johnson, Role for nanomaterial-autophagy interaction in neurodegenerative disease, Autophagy. 4 (8) (2008) 1097–1100.

[35] K. Choi, J.E. Riviere, N.A. Monteiro-Riviere, Protein corona modulation of hepatocyte uptake and molecular mechanisms of gold nanoparticle toxicity, Nanotoxicology 11 (1) (2017) 64–75.

[36] R. Braun, M. Sarikaya, K. Schulten, Genetically engineered gold-binding polypeptides: structure prediction and molecular dynamics, J. Biomater. Sci. Polym. Ed. 13 (7) (2002) 747–757.

[37] R.B. Pandey, Z. Kuang, B.L. Farmer, S.S. Kim, R.R. Naik, Stability of peptide (p1 and p2) binding to a graphene sheet via an all-atom to all-residue coarse-grained approach, Soft Matter 8 (35) (2012) 9101–9109.

[38] S. Mitternacht, S. Schnabel, M. Bachmann, W. Janke, A. Irbäck, Differences in solution behavior among four semiconductor-binding peptides, J. Phys. Chem. B 111 (17) (2007) 4355–4360.

[39] A. Swetnam, M.P. Allen, Selective adsorption of lattice peptides on patterned surfaces, Phys. Rev. E 85 (6) (2012) 062901.

[40] L. Darré, M.R. Machado, A.F. Brandner, H.C. González, S. Ferreira, S. Pantano, SIRAH: A structurally unbiased coarse-grained force field for proteins with aqueous solvation and long-range electrostatics, J. Chem. Theory. Comput. 11 (2) (2015) 723–739.

[41] J.C. Gumbart, B. Roux, C. Chipot, Standard binding free energies from computer simulations: what is the best strategy? J. Chem. Theory. Comput. 9 (1) (2012) 794–802.

[42] V. Limongelli, M. Bonomi, M. Parrinello, Funnel metadynamics as accurate binding free-energy method, Proc. Natl. Acad. Sci. USA 110 (16) (2013) 6358–6363.

[43] K.A. Beauchamp, Y.-S. Lin, R. Das, V.S. Pande, Are protein force fields getting better? a systematic benchmark on 524 diverse nmr measurements, J. Chem. Theory. Comput. 8 (4) (2012) 1409–1414.

[44] R. Carr, J. Comer, M.D. Ginsberg, A. Aksimentiev, Microscopic perspective on the adsorption isotherm of a heterogeneous surface, J. Phys. Chem. Lett. 2 (14) (2011) 1804–1807.

[45] C. Chipot, A. Pohorille, *Free Energy Calculations*, Springer, Berlin, Heidelberg, New York, 2007.

[46] J.R. Comer, C. Chipot, F.D. Gonzalez-Nilo, Calculating position-dependent diffusivity in biased molecular dynamics simulations, J. Chem. Theory. Comput. 9 (2) (2013) 876–882.

[47] J. Comer, F. Dehez, W. Cai, C. Chipot, Water Conduction Through a Peptide Nanotube, J. Phys. Chem. C 117 (50) (2013) 26797–26803.

[48] J. Comer, K. Schulten, C. Chipot, Calculation of Lipid-Bilayer Permeabilities Using an Average Force, J. Chem. Theory Comput. 10 (2) (2014) 554–564.

[49] J. Comer, J.C. Gumbart, J. Hénin, T. Lelièvre, A. Pohorille, C. Chipot, The adaptive biasing force method. Everything you always wanted to know, but were afraid to ask, J. Phys. Chem. B 119 (3) (2014) 1129–1151.

[50] J. Comer, K. Schulten, C. Chipot, Diffusive models of membrane permeation with explicit orientational freedom, J. Chem. Theory. Comput. 10 (7) (2014) 2710–2718.

[51] C. Micheletti, G. Bussi, A. Laio, Optimal langevin modeling of out-of-equilibrium molecular dynamics simulations, J. Chem. Phys. 129 (2008) 074105.

[52] S. Türkcan, A. Alexandrou, J.B. Masson, A bayesian inference scheme to extract diffusivity and potential fields from confined single-molecule trajectories, Biophys. J. 102 (10) (2012) 2288–2298.

[53] B.W. Holland, C.G. Gray, B. Tomberli, Calculating diffusion and permeability coefficients with the oscillating forward-reverse method, Phys. Rev. E Stat. Matter. Phys. 86 (3) (2012) 036707.

[54] K. Vanommeslaeghe, E. Hatcher, C. Acharya, S. Kundu, S. Zhong, J. Shim, et al., CHARMM general force field: a force field for drug-like molecules compatible with the charmm all-atom additive biological force fields, J. Comput Chem 31 (4) (2010) 671–690.

[55] H. Heinz, R. Vaia, B. Farmer, R. Naik, Accurate simulation of surfaces and interfaces of face-centered cubic metals using 12–6 and 9–6 lennard-jones potentials, J. Phys. Chem. C 112 (44) (2008) 17281–17290.

[56] H. Heinz, T.-J. Lin, R. Kishore Mishra, F.S. Emami, Thermodynamically consistent force fields for the assembly of inorganic, organic, and biological nanostructures: the interface force field, Langmuir 29 (6) (2013) 1754–1765.

[57] F. Iori, S. Corni, Including image charge effects in the molecular dynamics simulations of molecules on metal surfaces, J. Comput. Chem. 29 (10) (2008) 1656–1666.

[58] J.-W. Park, J.S. Shumaker-Parry, Structural study of citrate layers on gold nanoparticles: role of intermolecular interactions in stabilizing nanoparticles, J. Am. Chem. Soc. 136 (5) (2014) 1907–1921.

[59] M.K. Corbierre, N.S. Cameron, M. Sutton, K. Laaziri, R.B. Lennox, Gold nanoparticle/polymer nanocomposites: dispersion of nanoparticles as a function of capping agent molecular weight and grafting density, Langmuir 21 (13) (2005) 6063–6072.

[60] L.B. Wright, P.M. Rodger, S. Corni, T.R. Walsh, GolP-CHARMM: first-principles based force fields for the interaction of proteins with Au (111) and Au (100), Chem. Theory. Comput. 9 (3) (2013) 1616–1630.

[61] Z.E. Hughes, L.B. Wright, T.R. Walsh, Biomolecular adsorption at aqueous silver interfaces: first-principles calculations, polarizable force-field simulations, and comparisons with gold, Langmuir 29 (43) (2013) 13217–13229.

[62] Z.E. Hughes, T.R. Walsh, Non-covalent adsorption of amino acid analogues on noble-metal nanoparticles: influence of edges and vertices, Phys. Chem. Chem. Phys. 18 (26) (2016) 17525–17533.

[63] L.B. Wright, P.M. Rodger, T.R. Walsh, S. Corni, First-principles-based force field for the interaction of proteins with Au (100)(5 × 1): an extension of GolP-CHARMM, J. Phys. Chem. C 117 (46) (2013) 24292–24306.

[64] T.R. Walsh, M.R. Knecht, Biointerface structural effects on the properties and applications of bioinspired peptide-based nanomaterials, Chem. Rev. 117 (20) (2017) 12641–12704.

[65] E. Darve, A. Pohorille, Calculating free energies using average force, J. Chem. Phys. 115 (2001) 9169–9183.

[66] O. Leenaerts, B. Partoens, F. Peeters, Water on graphene: hydrophobicity and dipole moment using density functional theory, Phys. Rev. B 79 (23) (2009) 235440.

[67] H. Poblete, I. Miranda-Carvajal, J. Comer, Determinants of alanine dipeptide conformational equilibria on graphene and hydroxylated derivatives, J. Phys. Chem. B 121 (2017) 3895–3907.

[68] P.J. Rossky, M. Karplus, Solvation. A molecular dynamics study of a dipeptide in water, J. Am. Chem. Soc. 101 (8) (1979) 1913–1937.

[69] D.J. Tobias, C.L. Brooks III, Conformational equilibrium in the alanine dipeptide in the gas phase and aqueous solution: a comparison of theoretical results, J. Phys. Chem. 96 (9) (1992) 3864–3870.

[70] C. Chipot, A. Pohorille, Conformational equilibria of terminally blocked single amino acids at the water-hexane interface. a molecular dynamics study, J. Phys. Chem. B 102 (1) (1998) 281–290.

[71] P.E. Smith, The alanine dipeptide free energy surface in solution, J. Chem. Phys. 111 (12) (1999) 5568–5579.

[72] V. Parchaňský, J. Kapitán, J. Kaminský, J. Sebestik, P. Bouř, Ramachandran plot for alanine dipeptide as determined from raman optical activity, J. Phys. Chem. Lett. 4 (16) (2013) 2763–2768.

[73] Q. Zeng, X. Jiang, A. Yu, G.M. Lu, Growth mechanisms of silver nanoparticles: a molecular dynamics study, Nanotechnology. 18 (3) (2007) 035708.

[74] N.M. Bedford, Z.E. Hughes, Z. Tang, Y. Li, B.D. Briggs, Y. Ren, et al., Sequence-dependent structure/function relationships of catalytic peptide-enabled gold nanoparticles generated under ambient synthetic conditions, J. Am. Chem. Soc. 138 (2) (2015) 540–548.

[75] N.M. Bedford, A.R. Showalter, T.J. Woehl, Z.E. Hughes, S. Lee, B. Reinhart, et al., Peptide-directed pdau nanoscale surface segregation: toward controlled bimetallic architecture for catalytic materials, ACS Nano. 10 (9) (2016) 8645–8659.

[76] J.A. Lemkul, J. Huang, B. Roux, A.D. MacKerell Jr, An empirical polarizable force field based on the classical drude oscillator model: development history and recent applications, Chem. Rev. 116 (9) (2016) 4983–5013.

[77] G. Lamoureux, A.D. MacKerell Jr, B. Roux, A simple polarizable model of water based on classical drude oscillators, J. Chem. Phys. 119 (10) (2003) 5185–5197.

[78] V.M. Anisimov, I.V. Vorobyov, B. Roux, A.D. MacKerell, Polarizable empirical force field for the primary and secondary alcohol series based on the classical drude model, J. Chem. Theory. Comput. 3 (6) (2007) 1927–1946.

[79] R.B. Best, X. Zhu, J. Shim, P.E. Lopes, J. Mittal, M. Feig, et al., Optimization of the additive CHARMM all-atom protein force field targeting improved sampling of the backbone ϕ, ψ and side-chain X1 and X2 dihedral angles, J. Chem. Theory. Comput. 8 (9) (2012) 3257–3273.

[80] A. Vergara-Jaque, P. Fong, J. Comer, iodide binding in sodium-coupled cotransporters, J. Chem. Inf. Model. (2017).

[81] B. Barua, J.C. Lin, V.D. Williams, P. Kummler, J.W. Neidigh, N.H. Andersen, The trp-cage: optimizing the stability of a globular miniprotein, Protein Eng. Des. Sel. 21 (3) (2008) 171–185.

[82] K. Chenoweth, A.C. Van Duin, W.A. Goddard, ReaxFF reactive force field for molecular dynamics simulations of hydrocarbon oxidation, J. Phys. Chem. A 112 (5) (2008) 1040–1053.

[83] A. Bagri, C. Mattevi, M. Acik, Y.J. Chabal, M. Chhowalla, V.B. Shenoy, Structural evolution during the reduction of chemically derived graphene oxide, Nat. Chem. 2 (7) (2010) 581–587.

[84] S.H. Dave, C. Gong, A.W. Robertson, J.H. Warner, J.C. Grossman, Chemistry and structure of graphene oxide via direct imaging, ACS Nano. 10 (8) (2016) 7515–7522.

[85] T.P. Senftle, S. Hong, M.M. Islam, S.B. Kylasa, Y. Zheng, Y.K. Shin, et al., The ReaxFF reactive force-field: development, applications and future directions, NPJ Comput. Mater. 2 (2016) 15011.

[86] M. Raju, S.-Y. Kim, A.C. Van Duin, K.A. Fichthorn, ReaxFF reactive force field study of the dissociation of water on titania surfaces, J. Phys. Chem. C 117 (20) (2013) 10558–10572.

Chapter 3

Near-infrared photochemistry assisted by upconverting nanoparticles

Alessio Terenzi[a] and Luca Salassa[a,b,*]
[a]*Donostia International Physics Center (DIPC), Donostia, Spain,* [b]*Ikerbasque, Basque Foundation for Science, Bilbao, Spain*
**Corresponding author.*

Chapter Outline

3.1 Introduction

The scientific interest in the use of light to transform chemicals has increased its appeal over the decades. Even today, new reactions are constantly discovered by irradiating all types of molecules with UV–visible light, from organic and biological compounds to coordination and organometallic complexes.

The versatility of excited-state reactions and the accessibility of cheap and reliable light sources have favored applications in virtually all technology fields, including catalysis [1,2], medicine [3,4], bioremediation [5], and sensing [6,7].

Besides, the recent tremendous developments in nanomaterials have enhanced the scope of photochemistry in material sciences. The combination of molecular and material photochemistry has resulted in the design of smart hybrid materials displaying unique light-switchable features. Often, the synergy between molecules and nanomaterials has helped overcoming the flaws

Photoactive Inorganic Nanoparticles. DOI: https://doi.org/10.1016/B978-0-12-814531-9.00003-8
© 2019 Elsevier Inc. All rights reserved.

that affect their photophysics and photochemistry when used as single components [8,9].

The revision of the most important advances in photochemistry would be a titanic mission which is out of the scope of this contribution. Similarly, it is a prohibitive task tracing how molecular photochemistry has been integrated with the plethora of nanomaterials that has been developed in recent years. A number of reviews and books dealing specifically with parts of this vast literature can be found elsewhere [10–12].

Herein, we will focus on a specific type of nanomaterials that have raised great interest for their unique photophysics, that is, upconverting nanoparticles (UCNPs). These materials have been studied extensively in the last decade for uses in light-driven applications. They are characterized by the capability of converting near-infrared (NIR) light into UV–visible via multiphoton processes. Researchers have been inspired by the prospect of using upconverted light to trigger indirectly photochemical reactions under NIR light excitation and implementing new NIR photoassisted chemistry in biology, medicine, and energy-related applications [13–19].

Our contribution will summarize the most fundamental discoveries involving upconverting nanomaterials and photoactivatable organic and inorganic compounds, excluding the numerous and promising applications of UCNPs where their optical and chemical properties have been exploited for sensing or imaging [20–22]. For the sake of clarity, we provide in the following section a brief description of the general properties of UCNPs although these have extensively been reviewed [20–26].

3.1.1 Upconverting nanoparticles

UCNPs are lanthanide-doped nanocrystals, which absorb and convert low-energy NIR photons (e.g., 980 or 808 nm) into high-energy UV–visible light via a nonlinear optical phenomenon. Upconversion is an anti-Stoke emission process and can take place through mechanisms such as excited-state absorption (ESA), energy transfer upconversion (ETU), photon avalanche, cooperative energy transfer and energy migration-mediated upconversion. These can often occur simultaneously in the same nanomaterial; however, in lanthanide-doped nanoparticles ETU emission is the most common process and has been optimized for enhancing the absorption and emission performances of this family of materials [23–25].

Upconverting nanomaterials are constituted by three fundamental components: host matrix, activators, and sensitizers. Among the several types of hosting lattices investigated, fluoride-based lattices of the type $NaREF_4$ (RE = Y, Gd, Lu), where optically active lanthanide ions are introduced to dope the nanocrystals, are the most used in UCNPs. Such matrices are characterized by superior chemical stability, high transparency to NIR irradiation and low phonon energy, which reduce nonradiative decay pathways and maximize the upconverting process [23].

Detailed studies have demonstrated that hexagonal phase $NaYF_4$ (β-$NaYF_4$) nanocrystals display greater upconverting properties than the cubic form (α-$NaYF_4$) [27–29].

The sensitizer and activator are two neighboring lanthanide ions with appropriate alignment of their excited electronic states. In ETU, the sensitizer absorbs a NIR photon to reach the metastable excited level E1, from which energy is transferred to the E1 level of the activator ion. So, the sensitizer goes back to the ground state G, where it can absorb a second photon and excite the activator to the emissive E2 state, relaxing once more to G. These processes are dependent on the distance between the ions which can be controlled by the concentration of the dopants. Conversely, ESA is independent of the dopant concentration as this process occurs by sequential absorption of pump photons by a single ion (Fig. 3.1) [3,25].

The most used sensitizer ions for doping of UCNPs are Yb^{3+} and Nd^{3+}. These are activated at 980 and 800 nm, respectively, being characterized by much higher absorption cross-sections at these wavelengths compared with other lanthanides. For achieving optimal energy transfer to neighboring ions, doping concentrations of Yb^{3+} ions are normally rather high (20–30 mol.%) as these ions have only a single excited state ($^2F_{5/2}$) [23]. Nd^{3+} has instead multiple energy levels, which can lead to adverse cross-relaxation effects with emitter ions. For this reason, the determination of Nd^{3+} ideal doping

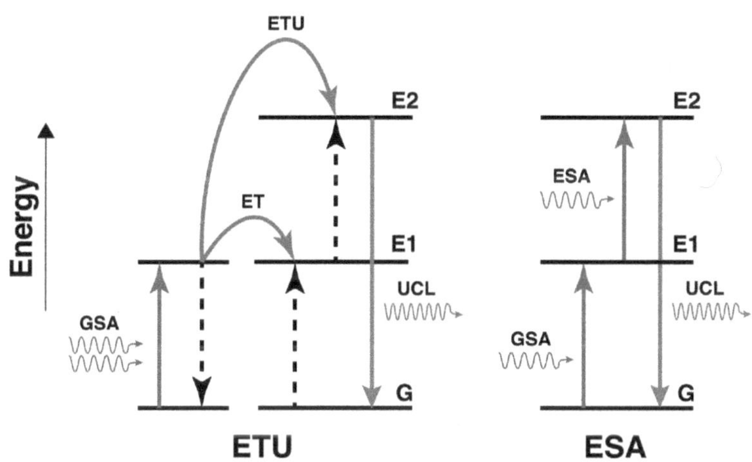

FIGURE 3.1 Schematic representation of selected upconversion mechanisms active in lanthanided-doped UCNPs. *ETU*, Energy transfer upconversion; *ESA*, excited-state absorption; *GSA*, ground state absorption; *ET*, energy transfer; UCL, upconverted luminescence. *Adapted from E.M. Chan, Combinatorial approaches for developing upconverting nanomaterials: high-throughput screening, modeling, and applications, Chem. Soc. Rev. 44 (2015) 1653–1679.*

concentration is more challenging; however, relatively low amounts of Nd^{3+} (1−2 mol.%) are generally employed in the core of core−shell UCNPs [30].

Lanthanides such as Er^{3+}, Tm^{3+}, and Ho^{3+} are the most suitable ions as activators for their ladder-like energy levels. They are employed at low doping concentrations (<2 mol.%) to avoid unfavorable cross-relaxation processes, which cause quenching or disable energy transfer, affecting the upconversion efficiency in a negative manner. For instance, Fig. 3.2 describes the ETU mechanism of the widely studied $NaYF_4:Yb^{3+},Er^{3+}/Tm^{3+}$, and $NaYF_4:Yb^{3+},Nd^{3+},Er^{3+}/Tm^{3+}$ UCNPs [23,25].

Surface passivation is generally employed to increase (up to 30 times) the efficiency of upconverted luminescence [24,31]. Often, the shell is composed of the same material as the host matrix and it allows confining the dopant in the interior of the nanomaterial. In this way, energy transfer to the surface of the UCNP is suppressed and quenching by solvent molecules is reduced [23]. Amorphous shells such as silica have also been extensively employed because they facilitate the introduction of specific chemicals and structural features. However, this type of coating is less effective in enhancing upconverting emission yields and variation of the shell thickness is critical in this case [32].

The features described make UCNPs unique luminescence bio-probes with key advantages with respect to conventional molecular dyes. These include the high photostability, continuous (nonblinking) emissions, long

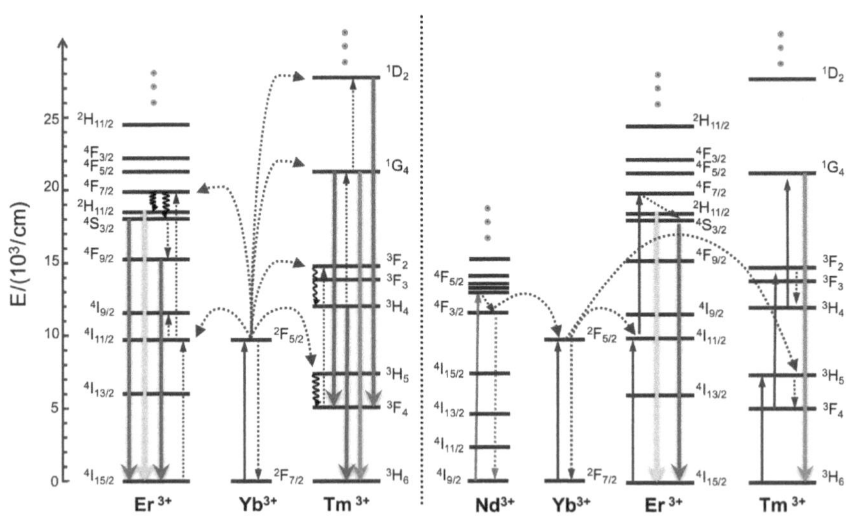

FIGURE 3.2 Upconversion energy transfer mechanisms for (left) $NaYF_4:Yb^{3+},Er^{3+}/Tm^{3+}$ UCNPs under 980-nm diode laser excitation and (right) $NaYF_4:Yb^{3+},Nd^{3+},Er^{3+}/Tm^{3+}$ UCNPs under 800-nm diode laser excitation. *Adapted from A. Bagheri, H. Arandiyan, C. Boyer, M. Lim, Lanthanide-doped upconversion nanoparticles: emerging intelligent light-activated drug delivery systems, Adv. Sci. 3 (2016) 1500437.*

luminescence lifetime, and sharp emission bands. Furthermore, the surface of UCNPs can be modified with biocompatible polymers to increase their biocompatibility, prolong circulation time, and target-specific tissues and cell types.

3.2 Upconverting nanoparticles–assisted photochemistry

The use of light as external stimuli to control chemical processes in space and time is widespread in science and technology. Nevertheless, light as stimulus has its own limitation, principally associated with the capacity of organic and inorganic molecules to absorb photons only in the UV–visible range, where damage to sensitive environments such as tissues or cells and side photoreactions are more likely to occur. UCNPs are intriguing materials because they may help overcoming the intrinsic flaws of light-activated molecular systems.

A key reason for the significant development of UCNPs is related to the possibility of exploiting NIR light excitation. NIR light is attractive for uses in both biology or medicine and technological applications [33,34]. Biomedicine, imaging, and photodynamic therapy (PDT) are leading applications for UCNPs [26], whereas other applications include light harvesting [35], barcoding devices [36], 3D display technologies [37], and molecular sensing [38].

At NIR wavelengths, light tissue penetration is greater and scattering is lower compared with blue–green light. Furthermore, red and NIR light induce less damage to cells and tissues as cell components have low absorptivity above 650 nm [40]. For this reason, the maximum permissible exposure (MPE) of human skin to NIR light is higher (980 nm, 0.73 W/cm^2) when compared with visible light (400–700 nm, 0.20 W/cm^2) [41,42]. Nevertheless, not all the NIR range is ideal for treatment of biological tissues. The optimal window is between 700 and 810 nm. At higher wavelengths (c. 1000 nm) water absorption is predominant and can cause overheating and damaging of cells. Fig. 3.3 shows a schematic representation of the penetration depth of light in skin as function of wavelength [40].

Furthermore, only a fraction of the solar radiation falls in the UV–visible range. A significant portion of the photons generated by the sun is poorly exploited, yet harvesting infrared light has direct impact on fields such as energy conversion and storage, and bioremediation and artificial photosynthesis. Smart spectral conversion schemes using UCNPs promise to make possible the use of the low-energy NIR tail of the solar spectrum to generate photons of sufficient energy for photochemical processes of strategic use [43–46].

The following sections will highlight the most relevant examples in which UCNPs have been used to assist photochemical reactions under NIR light excitation. For the sake of clarity, we have divided photoactivatable systems in organic and inorganic, on the basis of the molecular photochemistry that the upconverting materials trigger upon light irradiation.

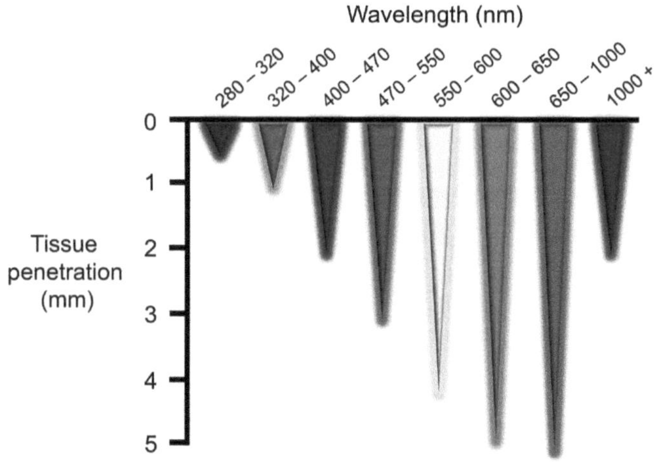

FIGURE 3.3 Wavelength dependence of light penetration into biological tissues (e.g., skin). *Adapted from E. Ruggiero, S. Alonso-deCastro, A. Habtemariam, L. Salassa, Upconverting nanoparticles for the near infrared photoactivation of transition metal complexes: new opportunities and challenges in medicinal inorganic photochemistry, Dalton. Trans. 45 (2016) 13012–13020 [39].*

3.3 Upconverting nanoparticle–assisted organic photochemistry

Organic photocages and photoswitches are the first type of chromophores used in combination with UCNPs to trigger with NIR light photochemical reactions that typically require UV and visible light. Several of these photoresponsive compounds have also been integrated with multiple components to obtain complex supramolecular structures and materials capable of responding to NIR light stimuli.

Early fundamental examples by Branda and coworkers [47–49] comprise the incorporation of diarylethenes (DTEs) with $NaYF_4{:}Yb^{3+},Tm^{3+}$, and $NaYF_4{:}Yb^{3+},Er^{3+}$ UCNPs, which are capable to prompt the ring-closing and ring-opening reactions of this photoswitchable unit by means of their different emission profiles. The closing reaction is favored by the UV emission of Tm-doped UCNPs, while Er-based nanoparticles emit green light which is absorbed by the ring-closed DTE to afford the open form.

The relative intensity of UCNP emission bands can be tuned by the excitation power. So, Tm^{3+} and Er^{3+} doubly doped UCNPs can provide prevalently either UV and blue or green upconverted light depending on the optical density employed. Exploiting such phenomenon (Fig. 3.4A), the Branda group has shown that doubly doped UCNPs can function as selective switches for controlling which photoreaction is taking place on a DTE derivative both in solution and in polymer films (toluene and poly(ethyleneglycol)dimethacrylate, respectively) [47,48].

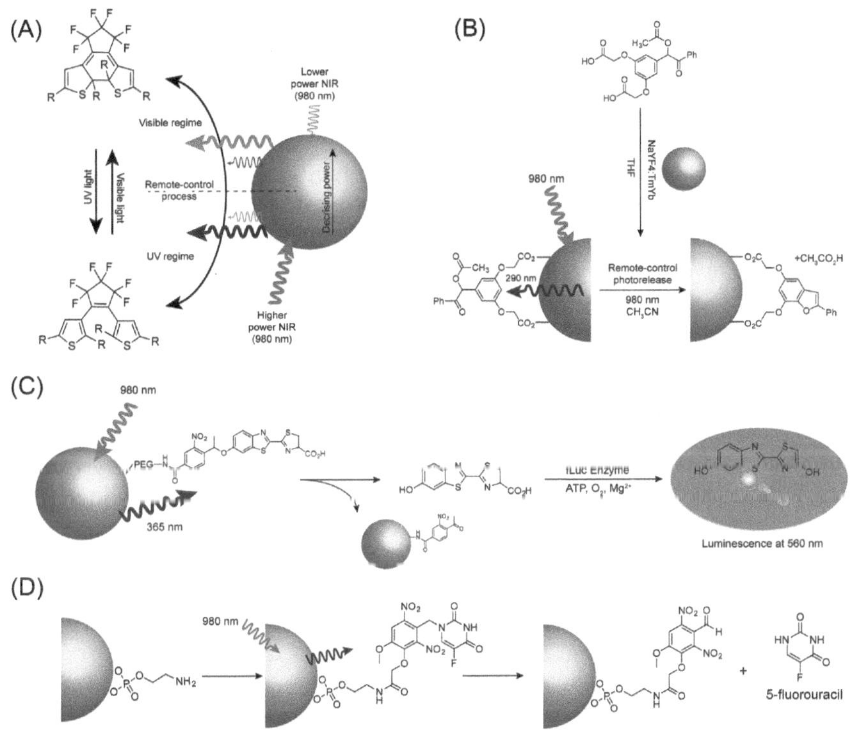

FIGURE 3.4 (A) UCNP-assisted NIR photoactivation of DTE derivatives. In Tm^{3+} and Er^{3+} doubly doped UCNPs the relative intensity of blue and green light can be controlled by the optical density (represented by the sizes of the wavy arrows). (B) UCNP-assisted NIR photorelease of acetic acid using the 3′,5′-di(carboxymethoxy)benzoin cage. (C) NIR uncaging of D-luciferin and subsequent bioluminescence prompted by functionalized core–shell upconversion nanoparticles. (D) Photorelease of 5-fluorouracil upon NIR light excitation of core–shell *O*-phosphorylethanolamine-capped UCNPs. *Adapted from J.-C. Boyer, C.-J. Carling, B.D. Gates, N.R. Branda, Two-way photoswitching using one type of near-infrared light, upconverting nanoparticles, and changing only the light intensity, J. Am. Chem. Soc. 132 (2010) 15766–15772; C.-J. Carling, F. Nourmohammadian, J.-C. Boyer, N.R. Branda, Remote-control photorelease of caged compounds using near-infrared light and upconverting nanoparticles, Angew. Chem. Int. Ed. 49 (2010) 3782–3785; Y. Yang, Q. Shao, R. Deng, C. Wang, X. Teng, K. Cheng, et al., In vitro and in vivo uncaging and bioluminescence imaging by using photocaged upconversion nanoparticles, Angew. Chem. Int. Ed. 51 (2012) 3125–3129; L.L. Fedoryshin, A.J. Tavares, E. Petryayeva, S. Doughan, U.J. Krull, Near-infrared-triggered anticancer drug release from upconverting nanoparticles, ACS Appl. Mater. Interfaces 6 (2014) 13600–13606 [48–51].*

Branda and Xing have also demonstrated that surface functionalization of UCNPs via electrostatic interactions is a suitable synthetic strategy to devise efficient NIR-responsive assemblies. They have reported on the NIR uncaging of active organic compounds from a dialkoxybenzoin [49] and an ortho-nitrobenzyl alcohol derivative [50], two chromophores that are common in photorelease applications (Fig. 3.4B and C).

In particular, Branda employed the 3′,5′-dialkoxybenzoin cage to trap acetic acid as model compound (Fig. 3.4B). The photochemistry of this organic caging group is rather clean in organic solvent and 5,7-dialkoxy-2-phenyl-benzo[b]furan is typically the only photoproduct in addition to the photoreleased acid. Functionalization of $NaYF_4:Yb^{3+},Tm^{3+}$ nanoparticles is straightforwardly achieved by means of the two carboxyl groups of the di (carboxymethoxy)benzoin acetate which undergoes ligand substitution reactions with the oleate capping ligands of the UCNPs. Upon 980-nm light excitation, upconverted UV photons are generated by the promotion of the $Tm^{3+}\,{}^1D_2 \rightarrow {}^3H_6$ transition of the $NaYF_4:Yb^{3+},Tm^{3+}$ nanoparticles and elicit uncaging of the carboxylic acid [49].

In the work of Xing [50], D-luciferin is initially caged with a 1-(2-nitrophenyl)ethyl group modified with a poly(ethylene glycol) linker. Silica-coated UCNPs with reactive terminal thiol groups are then decorated with such linker via maleimide chemistry and the released D-luciferin under NIR light irradiation is monitored in vitro and in vivo by fluorescence and by bioluminescence signals when the fLuc enzyme is coadministered (Fig. 3.4C) [50].

Adopting a similar strategy, Fedoryshin et al. [51] have decorated core–shell UCNPs ($NaYF_4:Yb^{3+},Tm^{3+}@NaYF_4$) with O-phosphorylethanolamine ligands coupled to an *o*-nitrobenzyl derivative of the chemotherapeutic agent 5-fluorouracil (5-FU; Fig. 3.4D). Upconverted UV and blue light from the UCNPs well matches the absorption profile of caged 5-FU, resulting in the photocleavage and release of the anticancer molecule upon low-power NIR light irradiation (<100 mW).

In the development of smart polymeric micelles for the delivery of pharmaceuticals, light has been extensively employed to trigger the release of the payload in irradiated areas. UCNPs allow overcoming limitations associated to the use of short wavelength light and several representative examples have been reported in recent years.

Zhao, Branda (Fig. 3.5A) and collaborators have used block copolymers and core–shell $NaYF_4:Yb^{3+},Tm^{3+}$ to control the destabilization of micelles and liberate hydrophobic molecules. In a proof-of-concept study, they employed a photosensitive diblock copolymer composed of hydrophilic polyethylene oxide and a hydrophobic polymethacrylate functionalized with photolabile *o*-nitrobenzyl groups. The UV light generated upon NIR excitation of the UCNPs photocleaves the nitrobenzyl groups changing the hydrophobic/hydrophilic balance of the micelles and ultimately favoring their destabilization. It is worth highlighting that light-sensitive nitrobenzyl moieties in this system are synthetically modified with two methyl groups to improve the overlap of their absorption profile with the upconverted emission of the nanoparticles. Researchers demonstrate that the nanoconstruct can be used to release payloads incorporating the hydrophobic dye Nile red [52].

Viger et al. have followed a very similar approach designing micelles constituted of cresol monomers highly functionalized with *o*-nitrobenzyl

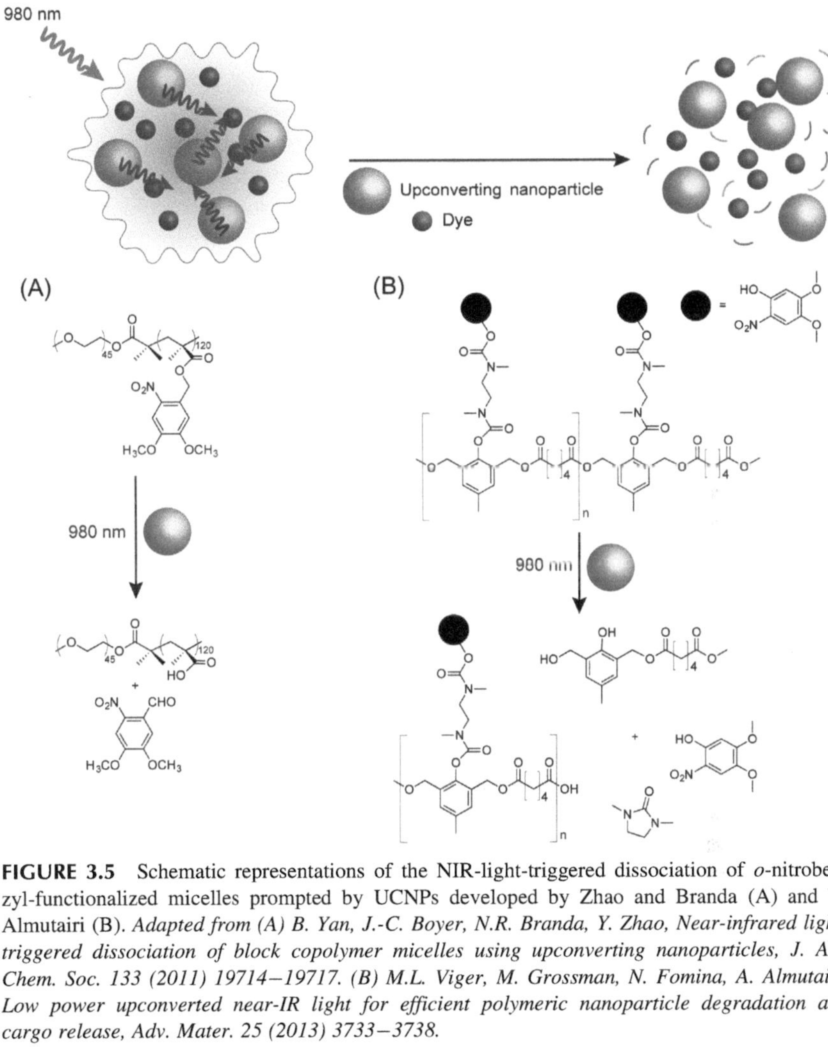

FIGURE 3.5 Schematic representations of the NIR-light-triggered dissociation of *o*-nitrobenzyl-functionalized micelles prompted by UCNPs developed by Zhao and Branda (A) and by Almutairi (B). *Adapted from (A) B. Yan, J.-C. Boyer, N.R. Branda, Y. Zhao, Near-infrared light-triggered dissociation of block copolymer micelles using upconverting nanoparticles, J. Am. Chem. Soc. 133 (2011) 19714–19717. (B) M.L. Viger, M. Grossman, N. Fomina, A. Almutairi, Low power upconverted near-IR light for efficient polymeric nanoparticle degradation and cargo release, Adv. Mater. 25 (2013) 3733–3738.*

moieties. These degrade efficiently through a cascade of cyclization and rearrangement reactions in response to a minimal UV activation by the UCNPs (core–shell $NaYF_4$:Yb^{3+},Tm^{3+}@$NaYF_4$) encapsulated in the polymeric materials. Cleavage of *o*-nitrobenzyl moieties from the polymer unmasks a large number of the secondary amino groups. This process rapidly makes the polymer capsules more permeable to water facilitating the release of the model dye C153 and the UCNPs (Fig. 3.5B). It is notable that the optical density of NIR light employed to release the payload from the micelles in significant amounts is in the range of 0.5–1 W/cm^2 [53].

Besides, coating of UCNPs with amphiphilic polymers (e.g., polyethylene glycol) has served to load photoresponsive molecules onto these nanoparticles for imaging and therapy purposes. The group of Branda has devised UCNP-based systems in which DTE-based photochromic molecules have been encapsulated onto UCNPs by using amphiphilic polymers capable of self-assembling around the oleate ligands capping UCNPs. Typically, the alkyl or aryl chains of these polymers interdigitate around the oleate shell of the nanoparticles, providing a hydrophobic environment where DTEs can be loaded [54]. Related approaches have been exploited using metal complexes as cargos (*vide infra*). In alternative, the amphiphilic polymer backbone can also be directly functionalized with DTE units prior self-assembly, allowing high payloads of chromophores [55].

Incorporation of photoresponsive chromophores and UCNPs into polymeric materials, liquid crystals, and hydrogels has been successfully investigated to induce macroscopic structural and functional changes in such materials upon NIR light stimuli. The key goal of such studies is to minimize the photodegradation that is often associated to the direct and prolonged exposure of these materials to UV and visible light.

Li and Yu introduced $NaYF_4{:}Yb^{3+},Tm^{3+}$ upconverting nanophosphors into cross-linked liquid-crystal polymer films containing azotolane units to promote fast bending upon exposure to continuous-wave NIR light (980 nm). In these systems, UV and visible upconverted light promotes *trans–cis* photoisomerization of the azotolane groups, which causes a continuous stress and an alignment change of the liquid-crystal polymers (Fig. 3.6A) [3,56].

Wang et al. have shown that upconverting $NaGdF_4{:}Yb^{3+},Tm^{3+}$ and NIR laser light can trigger changes in the refractive index of a liquid-crystal phase by promoting the *cis–trans* isomerization of azobenzene mesogens (Fig. 3.6B) [3,57]. Doping of liquid-crystal media with UCNPs and light-switch chiral azobenzenes allows obtaining self-organized helical superstructures and control their organization under NIR light. The initial switching process from the *trans* to *cis* form of the azobenzenes is stimulated in situ by the upconverted UV light generated by UCNPs with high 980-nm excitation power densities. Instead, by employing a low excitation power density, UCNPs prevalently emit photons in the visible and prompt the backward reaction.

Different from azobenzenes, DTE is stable in both isomeric states. For this reason, the same group combined DTE and multiply doped UCNPs with Nd^{3+}, Yb^{3+}, Er^{3+}, and Tm^{3+} to modulate the liquid-crystal phase and organize its helical superstructure. The presence of Yb^{3+} and Nd^{3+} as sensitizer ions allows using 980 and 808 nm light, respectively, for excitation and obtaining different emission profiles that can be exploited to interconvert the photoresponsive DTE reversibly between the two isomeric forms. Upon 808-nm NIR excitation, UV ($^1I_6 \rightarrow {}^3F_4$, $^1D_2 \rightarrow {}^3H_6$) and blue ($^1D_2 \rightarrow {}^3F_4$, $^1G_4 \rightarrow {}^3H_6$) light is generated, whereas at 980-nm green ($^2H_{11/2}$, $^4S_{3/2} \rightarrow {}^4I^{15/2}$) and red ($^4F_{9/2} \rightarrow {}^4I_{15/2}$) light is dominating the upconverted emission. UV

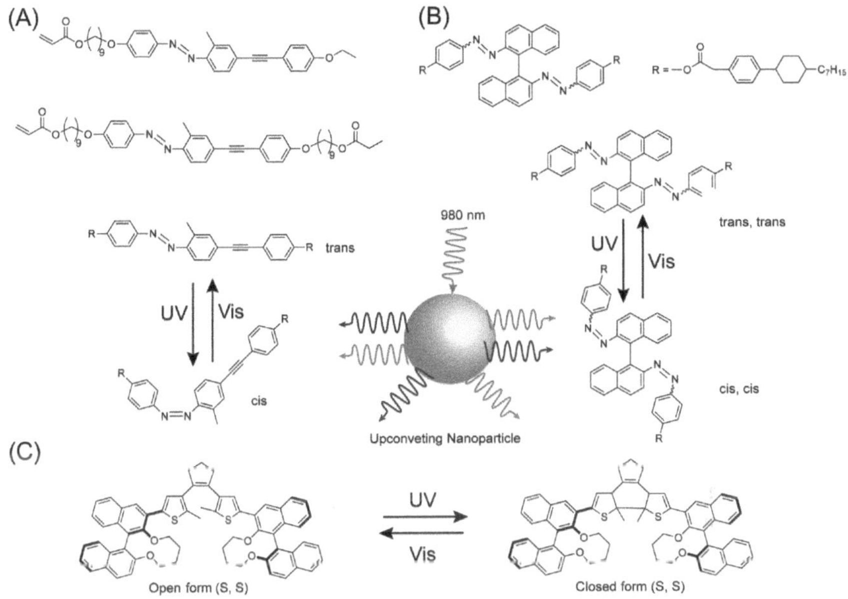

FIGURE 3.6 UCNP-assisted photoisomerization of azotolane-based chromophores used for the NIR-light (A) bending and (B) helical self-organization of liquid crystals; (C) wavelength-selective NIR-light-triggered reversible handedness inversion of the self-organized helical superstructures integrating chiral DTE switch and UCNPs. *Adapted from W. Wu, L. Yao, T. Yang, R. Yin, F. Li, Y. Yu, NIR-light-induced deformation of cross-linked liquid-crystal polymers using upconversion nanophosphors, J. Am. Chem. Soc. 133 (2011) 15810–15813; L. Wang, H. Dong, Y. Li, C. Xue, L.-D. Sun, C.-H. Yan, et al., Reversible near-infrared light directed reflection in a self-organized helical superstructure loaded with upconversion nanoparticles, J. Am. Chem. Soc. 136 (2014) 4480–4483; L. Wang, H. Dong, Y. Li, R. Liu, Y.F. Wang, H.K. Bisoyi, et al., Luminescence-driven reversible handedness inversion of self-organized helical superstructures enabled by a novel near-infrared light nanotransducer, Adv. Mater. 27 (2015) 2065–2069.*

and green emissions obtained at these two excitation wavelengths have good spectral overlaps with the open and close forms of DTE, respectively. Integrating a chiral DTE switch and Yb^{3+}- and Nd^{3+}-doped UCNPs into commercially available achiral liquid crystals (Fig. 3.6C), Yan, Li and coworkers have been able to promote the cholesteric to nematic phase conversion. Further, NIR irradiation affords a new cholesteric phase, however, with inverted handedness. Remarkably, helical superstructures obtained through this NIR-induced change in handedness do not show thermal relaxation [58].

3.4 Upconverting nanoparticles–assisted inorganic photochemistry

The interaction of light with biologically active metal complexes has been extensively explored in the last 20 years to design new prodrugs and chemical biology tools [59,60]. Photoactivation strategies typically take advantage of dark−stable and nontoxic metal complexes, which are converted in biologically active species upon light irradiation. These photochemical processes are often associated to a greater specificity in the biological action of the metal complexes.

Fostered by the success of PDT and the vast background of metal-based anticancer agents, a great number of photoactivatable metal complexes have been developed as anticancer prodrugs in recent years. In 2015, one Ru-polypyridyl complex was approved for clinical trials for the photochemotherapy treatment of nonmuscle invasive bladder cancer [61].

Different approaches have been applied to the design of photoactivatable metal-based prodrugs and for their use in biology and medicine, including light-controlled generation of toxic reactive oxygen species, ligand ejection/exchange followed by metal binding to biomacromolecules (e.g., DNA), or uncaging of bioactive molecules such as CO and NO.

As for the case of organic photoactive molecules, the great majority of metal-based complexes typically require high-energy UV−visible light to trigger a photochemical response, which is not ideal for PDT-like applications. For this reason, UCNPs and NIR light excitation have raised great interest in the bioinorganic and inorganic medicinal chemistry communities [39,42].

Herein we revise key examples of photoactivatable metal complexes integrated into upconverting nanoconstructs and discuss the behavior of such nanomaterials under NIR light irradiation.

Ford and Zhang were the first to report on the use of UCNPs in combination to a transition metal-based compound [62]. They prepared a silica-coated hexagonal $NaYF_4:Yb^{3+}/Er^{3+}@NaYF_4$ UCNPs with a positive outer surface to incorporate the negatively charged iron-nitrosyl Roussin's black salt, $Fe_4S_3(NO)_7^-$ (RBS). In this pioneering work, the iron compound operates as a photoactivatable source of nitric oxide, a well-known small molecule with several biological functions (Fig. 3.7A) [3,63].

In detail, the authors have prepared hexagonal $NaYF_4$:Yb,Er@$NaYF_4$ UCNPs and coated their surface with a 10-nm layer of silica, subsequently modified with primary amines and treated with acids to obtain positively charged groups. These positively charged UCNPs@SiO_2 nanocarriers, having a final diameter of about 60 nm, are electrostatically coupled with the iron salt by mixing and stirring together the two components in aqueous solution.

NO uncaging is eventually produced by 980-nm light irradiation from a NIR diode laser operating in the continuous-wave mode and confirmed by means of a NO-sensitive electrode. Sequential excitation proves that the NO photorelease is power-dependent (1−4.5 W).

The Ford group has also described the preparation of an implantable biocompatible disk capable to simultaneously encapsulate UCNPs and the NO precursor RBS [64]. In particular, the disk is built using poly(dimethylsiloxane) (PDMS), which is NIR transparent and, importantly, approved by FDA for use in vivo. Core−shell $NaGdF_4@NaYF_4$ particles

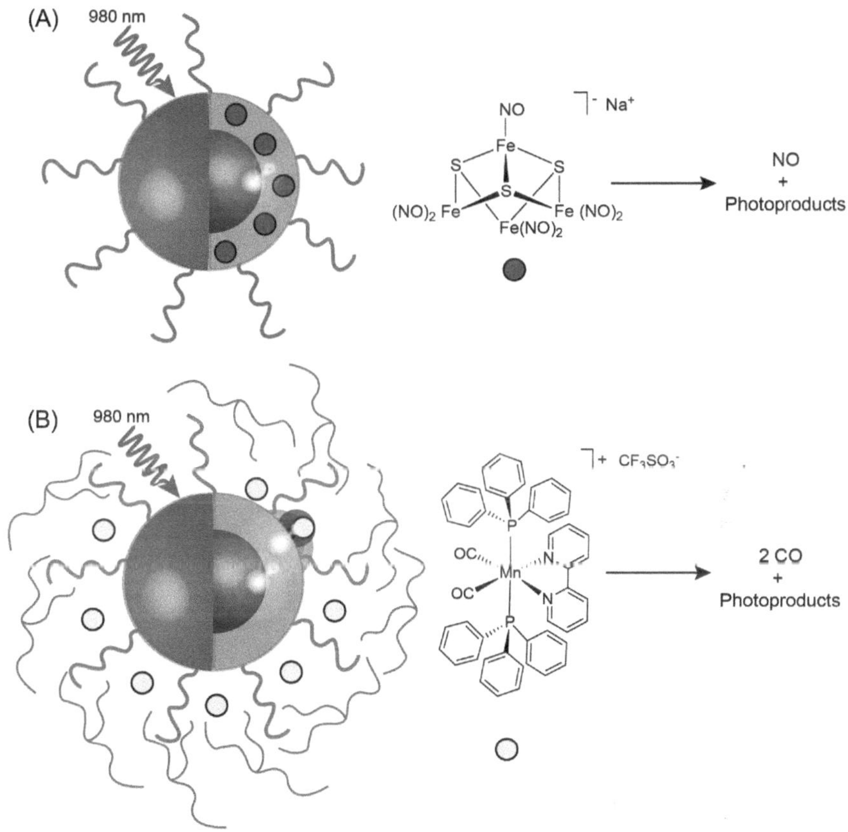

FIGURE 3.7 Schematic representation of UCNP nanocarriers for the NIR light photorelease of (A) NO from RBS and (B) CO from the complex *cis,trans*-[Mn(bpy)$(CO)_2(PPh_3)_2$][CF_3SO_3]. *Adapted from P.T. Burks, J.V. Garcia, R. GonzalezIrias, J.T. Tillman, M. Niu, A.A. Mikhailovsky, et al., Nitric oxide releasing materials triggered by near-infrared excitation through tissue filters, J. Am. Chem. Soc. 135 (2013) 18145–18152. A.E. Pierri, P.-J. Huang, J. V. Garcia, J.G. Stanfill, M. Chui, G. Wu, et al., A photoCORM nanocarrier for CO release using NIR light, Chem. Commun. 51 (2015) 2072–2075.*

doped with Yb^{3+} and either Er^{3+} or Tm^{3+} as light emitters are employed in this system to maximize the overlap between the upconverted visible emissions of the UCNPs and the absorption bands of RBS. Photostimulated NO release from RBS can be achieved by exciting UCNPs with 980-nm light (2.5 W/cm^2). Suspending an aliquot of UCNPs in a viscous solution which contains the PDMS precursor in disk molds affords a polymer disk. NO photorelease capability from the disk is tested using a layer of porcine tissue (skin, muscle, fat) acting as a physical barrier. An oscillating laser beam photolysis apparatus as a light source (300 W/cm^2) is employed to reduce damaging of the biological tissue. NIR light penetrates the tissue protective layer, and generates NO from RBS precursor, though in lower amounts compared with direct excitation of the upconverting disk.

Besides NO, Ford reported an UCNP-based nanocarrier for the photouncaging of another small molecule which possesses biological functions, namely, carbon monoxide [63,65]. CO inhaled as gas and CO-releasing molecules (CORM) have proved to be beneficial for cardiovascular diseases, inflammatory disorders, and organ transplantation in preclinical animal models [66]. In the nanoconstruct designed by Ford and coworkers (Fig. 3.7B), the triflate salt of the manganese carbonyl complex $[Mn(bpy)(CO)_2(PPh_3)_2]^+$, used as photo-CORM, is incorporated into NIR-absorbing $NaGdF_4$ nanoparticles. These are doped with Yb^{3+} and Tm^{3+} ions and coated with the amphiphilic phospholipid-functionalized polyethylene glycol (DSPE-PEG 2000). Such coating allows for the optimal encapsulation of the hydrophobic photo-CORM within the lipid-like core of the nanoparticle ligand shell and facilitates the solubility of the UCNPs in aqueous media due to the ligand hydrophilic PEG layer. $[Mn(bpy)(CO)_2(PPh_3)_2]^+$ slowly infuses into the UCNPs ligand layer by mixing the two entities in a water–acetonitrile solution, with a 80% loading efficiency as proved by ICP. The Mn photo-CORM has a strong absorption band at 400 nm with a tail into the visible which overlaps with the emission bands of the UCNPs upon 980 nm irradiation. As a matter of fact, the nanocarrier elicits photochemical uncaging of CO under NIR light irradiation (2 W), as determined by gas chromatography and myoglobin-binding studies.

In the systems developed by Ford and coworkers, the metal center does not participate actively in the potential therapeutic applications. Our group has employed UCNPs and NIR light for the photoactivation of coordination and organometallic compounds of relevance in chemotherapy, among these are Ru-polypyridyl and arene complexes [67,68] and Pt(IV) complexes [69].

Our first proof-of-concept study describes the UCNP-mediated NIR photoactivation of *cis*-$[Ru(bpy)_2(py)_2][Cl]_2$ (where py = pyridine) [67]. This particular ruthenium compound belongs to the ruthenium polypyridyl complexes class, known for its promising applications in vitro and in vivo [70], spanning from anticancer activity [71–73], to neuronal stimulation [74,75] and protein inhibition [76–79]. The photochemical behavior of *cis*-[Ru $(bpy)_2(py)_2][Cl]_2$ has been studied since the 1980s and is by now well established [80–82]. Upon excitation of its metal-to-ligand charge-transfer band, the complex efficiently photodissociates one pyridine ligand due to the population of accessible metal-centered states of dissociative nature.

We prepared 90-nm oleate-capped hexagon-shaped nanoparticles of the type $NaYF_4:Yb^{3+},Er^{3+}$ to have upconverted green light which matched appropriately the absorption tail of the Ru complex. UCNPs treated with aqueous HCl (pH = 4) show improved water solubility and achieve the photolysis of the complex upon 980-nm excitation (25 W/cm^2). The pyridine photorelease and the subsequent formation of the aqua product *cis*-[Ru $(bpy)_2(py)(H_2O)]^{2+}$ (Fig. 3.8A) is proved by UV–visible (red-shift of the

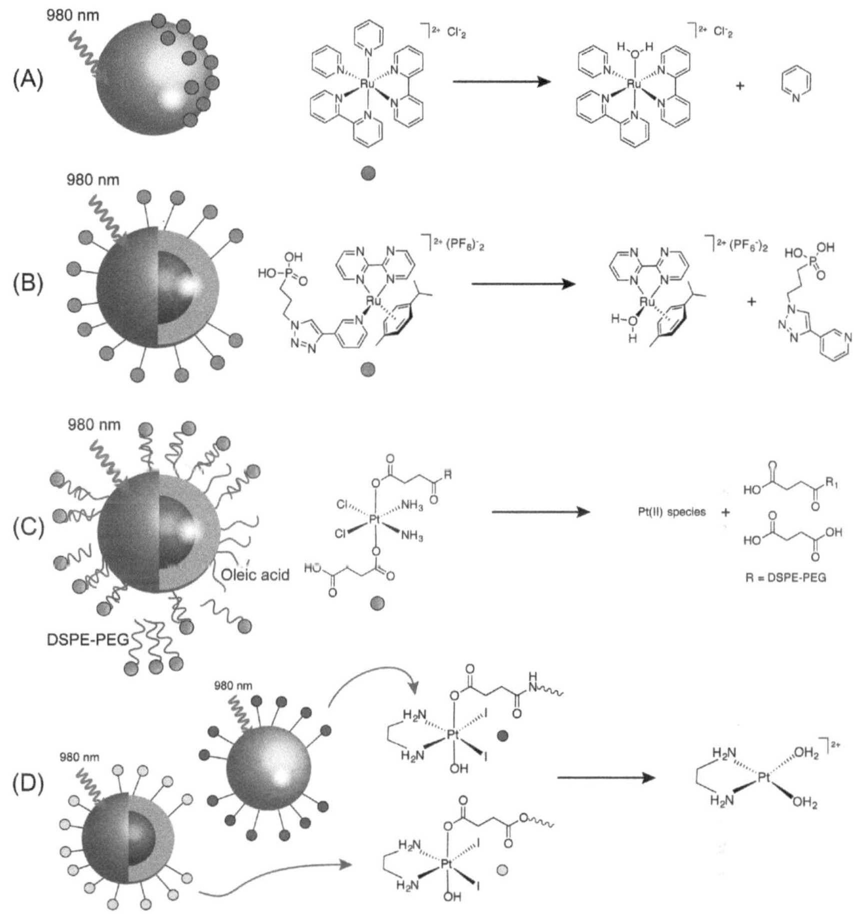

FIGURE 3.8 Schematic representation of UCNP nanoconstructs for the NIR light activation of (A) *cis*-$[Ru(bpy)_2(py)_2][Cl]_2$, (B) $[(\eta^6\text{-p-cym})Ru(bpm)(P\text{-}Trz\text{-}Py)][PF_6]_2$, (C) *cis,cis,trans*-$[Pt(NH_3)_2(Cl)_2(O_2CCH_2CH_2CO_2H)_2]$ functionalized with DSPE-PEG, and (D) *cis,trans*-$[PtI_2(O_2CCH_2CH_2CO_2H)_2(en)]$ complexes. *Adapted from E. Ruggiero, A. Habtemariam, L. Yate, J.C. Mareque-Rivas, L. Salassa, Near infrared photolysis of a Ru polypyridyl complex by upconverting nanoparticles, Chem. Commun. 50 (2014) 1715–1718; E. Ruggiero, C. Garino, J.C. Mareque-Rivas, A. Habtemariam, L. Salassa, Upconverting nanoparticles prompt remote near-infrared photoactivation of Ru(II)–arene complexes, Chem. Eur. J. 22 (2016) 2801–2811; E. Ruggiero, J. Hernandez-Gil, J.C. Mareque-Rivas, L. Salassa, Near infrared activation of an anti-cancer PtIV complex by Tm-doped upconversion nanoparticles, Chem. Commun. 51 (2015) 2091–2094; S. Perfahl, M.M. Natile, H.S. Mohamad, C.A. Helm, C. Schulzke, G. Natile, et al., Photoactivation of diiodido–Pt(IV) complexes coupled to upconverting nanoparticles, Mol. Pharm. 13 (2016) 2346–2362 [87].*

lowest-energy absorption band) and unequivocally, via NMR with the appearance of diagnostic signals for the photoproducts. NMR also allows for a quantitative analysis of the photoconversion, which is of about 70% after 4 h of NIR light irradiation.

Shortly afterward, we proposed a similar approach for the NIR photoactivation of Ru(II) half-sandwich arene complexes [68]. This class of organometallic compounds stands out for its rich structural and electronic diversity, dependent on the ligand choice, which has been exploited to obtain unique anticancer [83] and catalytic features [84]. In the field of chemotherapy, the mechanism of action for most Ru−arene piano-stool complexes is the hydrolysis of a labile ligand to yield aqua products, which are then able to attack biological targets in the cell. A viable anticancer route to tune the activity of this class of derivatives is introducing into the Ru coordination sphere ligands which confer stability toward hydrolysis, but that can undergo photoinduced cleavage. Such a prodrug can be then photoactivated to generate in situ species with antiproliferative properties. In previously reported studies, pyridinato Ru(II) arene complexes proved to be stable in the dark and selectively dissociate the monodentate pyridinato ligand coordinated to the Ru center upon visible light excitation [85,86].

For this work, we anchored the derivative $[(\eta^6\text{-p-cym})\,Ru(bpm)(P\text{-}Trz\text{-}Py)][PF_6]_2$ (where p-cym = p-cymene, bpm = 2,2′-bipyrimidine and P-Trz-Py = [3-(1-pyridin-3-yl-[1−3]triazol-4-yl)-propyl] phosphonic acid diethyl ester) to core−shell $NaYF_4$:Yb,Tm@$NaYF_4$ UCNPs (Fig. 3.8B) [3,68]. The Ru derivative is achieved by using click chemistry and we exploit the high affinity of the $NaYF_4$ matrix for its phosphonate group [88,89].

Upon 980 nm excitation, the Ru complex conjugated to UCNPs selectively generates the aqua derivative $[(\eta^6\text{-p-cym})Ru(bpm)(H_2O)]^{2+}$. Approximately 55% of the starting metal-based complex is photoconverted after 4.5 h (8.1 W/cm^2). NMR and UPLC-MS demonstrated that the aqua photoproduct is able to bind the DNA model base guanosine 5′-monophosphate.

Our group recently demonstrated that the potential of UCNPs can be applied not only to activate anticancer ruthenium-based drug candidates but also to trigger photoreduction of Pt(IV) prodrugs able to release clinically approved chemotherapeutics such as cisplatin [69].

Pt(IV) complexes have attracted a broad interest as anticancer agents, justified by the fact that they are biologically inert in their oxidized state but become active in the tumor tissue where they are reduced to the corresponding active Pt(II) species, either by a light stimuli or by biological reductants [90]. Notably, a number of platinum(IV) produg candidates have entered advanced clinical trials [91].

Nevertheless, light activation of Pt(IV) prodrugs to their Pt(II) counterparts is hardly translatable to clinic applications, due to the extremely poor absorptions in the visible region of Pt(IV) species. In order to demonstrate that this limitation can be overcome, we employed NIR light (7.3 W/cm^2) and core–shell $NaYF_4$:Yb^{3+},Tm^{3+}@$NaYF_4$ UCNPs to promote the photoreduction of *cis,cis,trans*-$[Pt(NH_3)_2(Cl)_2(O_2CCH_2CH_2CO_2H)_2]$ to cisplatin-like Pt(II) species [69]. We selected this particular Pt(IV) complex as it has a high reduction potential compared to other Pt(IV) analogs and is therefore suitable as anticancer prodrug [92].

At first, we demonstrated by NMR and XPS that oleate-capped core–shell UCNPs have the capability of functioning as NIR phototriggers for the activation of Pt(IV) complex in solution (980 nm continuous-wave laser, 4.9 W cm^2). Then we loaded the Pt(IV) complex directly onto the nanoparticles by functionalizing the carboxylic group of its axial ligands. The PEGylated-phospholipid DSPE-PEG(2000)-NH_2 is coupled to the complex via standard bioconjugation techniques and the resulted adduct is introduced into the nanoparticles through the hydrophobic interactions of its phospholipidic arms with the capping oleate chains of the UCNPs. Notably, the PEG units remain exposed toward the solvent, improving the water solubility of the whole system. The NIR-mediated photolysis of the anticancer Pt(IV) prodrug is proven by NMR spectroscopy, demonstrating that the system could be valuable for biological applications.

Bednarski and coworkers investigated the UCNP-assisted NIR activation of Pt(IV) prodrugs by derivation of the complex *cis,trans*-$[PtI_2(OH)_2(en)]$ (where en = ethylenediamine) and its anchoring onto Yb^{3+},Er^{3+}- and Yb^{3+}, Tm^{3+}-doped $NaGdF_4$ [87]. As shown in Fig. 3.8D, they adopted two different approaches: the first strategy uses silica-coated UCNPs functionalized with amino groups and conjugation chemistry to attach diiodido − Pt(IV) carboxylate complexes via amide bond; the second strategy exploits ligand exchange reaction between the capping oleate ligands and the carboxylic groups of the Pt(IV) complexes.

Among the two attachment strategies, the former is more effective for the release of the cytotoxic $[Pt(H_2O)_2(en)]^{2+}$ species under 980-nm light, although control experiments indicate that partial chemical reduction of the diiodido − Pt(IV) prodrugs by the UCNPs occurs in the dark. Authors demonstrate that the released Pt species bind irreversibly to calf thymus DNA and that NIR light irradiation induces a substantial increase in the cytotoxicity of Pt–UCNP conjugates in human leukemia HL60 cell compared to dark controls.

Wu's group has employed photoactive Ru scaffolds and UCNPs with a different strategy. They reported about Ru complexes operating as photoactive molecular valves for the release of the anticancer drug doxorubicin from UCNPs coated with mesoporous silica [93]. Wu and coworkers synthesized core–shell $NaYF_4:Yb^{3+},Tm^{3+}@NaYF_4$ UCNPs which emit blue light upon irradiation with a 974-nm laser at low excitation intensities (0.19–0.81 W/cm^2). These UCNPs are first coated with mesoporous silica (final diameter of ca. 90 nm) and then loaded with the anticancer drug doxorubicin. Last, the ruthenium derivative $[Ru(bpy)_2(PMe_3)((3\text{-amino-propyl})(EtO)_3Si)][PF_6]_2$ is covalently anchored onto UCNPs by means of the silyl group of the aminopropyl ligand, with the aim of trapping doxorubicin molecules inside the silica pores.

Irradiation experiments in solution prove that approximately 42% of doxorubicin is released after 5 h exposure to 974 nm light with an optical density of 0.35/Wcm^2. After a modification of the nanoparticles with PEG and folic acid, the authors show that their system significantly inhibited the growth of HeLa cell, after incubation for 3–6 h followed by NIR irradiation. Lower, yet significant, reduction in cell viability is obtained when irradiation is performed through a thick layer of pork tissue (0.64 W/cm^2 for 5 h), whereas the same effects in solution and in vitro are not observed in a number of control experiments, including using UCNPs where the ruthenium complex is replaced by an azobenzene compound that cannot be cleaved from silica pores. This study is the first demonstrating that intensity lower than the skin MPE dose for NIR light (0.726 W/cm^2) can produce a therapeutic effect.

Si Wu recently reported on an intriguing use of UCNPs and photoactivatable Ru complexes in which these two components are assembled for the NIR light-mediated regulation of cathepsin K enzyme activity in living cells [94]. Cathepsin K is an enzyme with several critical functions in osteoclastic bone resorption, macrophage invasion, and tumor growth [95]. A ruthenium complex is accurately designed in order to contain a photolabile cathepsin K enzyme inhibitor (Fig. 3.9). The selected UCNPs, core–shell $NaYF_4:Yb^{3+},Tm^{3+}@NaYF_4$, are coated with hollow mesoporous silica nanoparticle and the ruthenium enzyme inhibitor is subsequently loaded, obtaining a loading efficiency of 18.3 wt.%. UV–visible and HPLC experiments show that NIR irradiation (974 nm, 1 W/cm^2) of the nanocarrier triggers the photocleavage of the cage ruthenium complex and the release of the bioactive ligand which inhibits cathepsin K. A 30-min irradiation of the Ru-UCNPs reduces the enzyme activity to 13%. Moreover, inhibition of the same enzyme upon irradiation is still possible after placing chicken tissue between the NIR light source and the sample as well as in different cancer cell lines selected for their high cathepsin K expression rate.

The Wang and Jiang research groups adopted a similar strategy and loaded the amphiphilic complex $[Ru(bpy)_2(C_{18}H_{37}CN)_2][Cl]_2$ onto diamond-shaped $LiYF_4:Yb^{3+},Tm^{3+}$ UCNPs taking advantage of the hydrophobic interactions between the long aliphatic chains of the complex and that of the

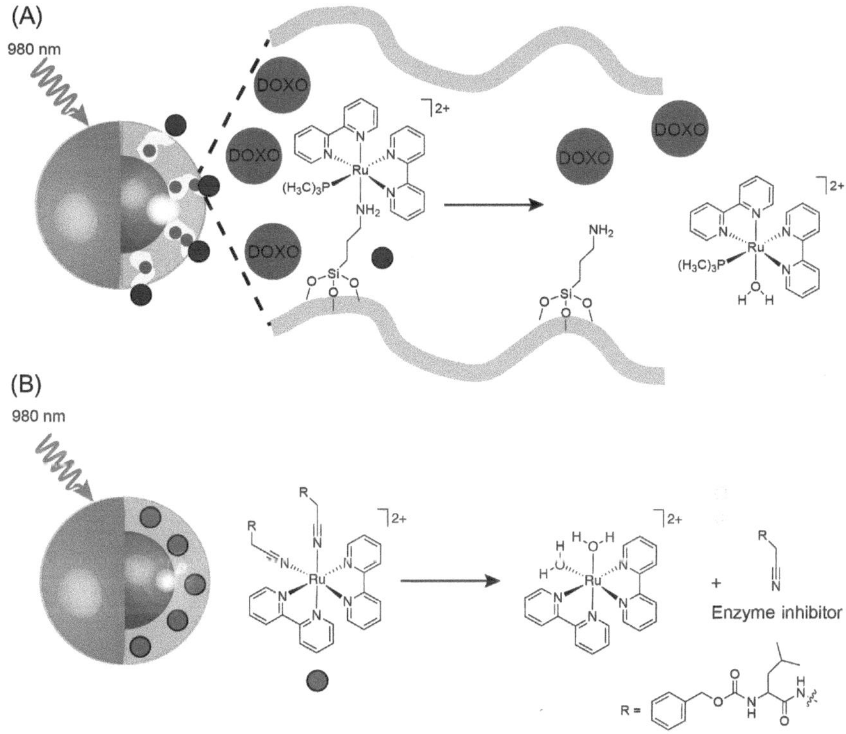

FIGURE 3.9 Schematic representation of the Ru-UCNP-based nanocarriers developed by the group of Wu. (A) Doxorubicin (DOXO) is released by NIR light activation of a Ru-polypyridyl complex acting as a valve. (B) Schematic illustration of photon upconversion enzyme inhibition. *Adapted from S. He, K. Krippes, S. Ritz, Z. Chen, A. Best, H.-J. Butt, et al., Ultralow-intensity near-infrared light induces drug delivery by upconverting nanoparticles, Chem. Commun. 51 (2015) 431–434; Z. Chen, R. Thiramanas, M. Schwendy, C. Xie, S.H. Parekh, V. Mailänder et al., Upconversion nanocarriers encapsulated with photoactivatable Ru complexes for near-infrared light-regulated enzyme activity, Small 13 (2017) 1700997 [93,94].*

oleate layer of the nanoparticles [96]. Additionally, nanoparticles are coated with PEG to enhance their water solubility (Fig. 3.10). The ruthenium complex contains a nitrile monodentate ligand, which is known for its efficient photodissociation rate [97], hence after irradiation by NIR light (980 nm, 3 W/cm^2) for 40 min, 80% of the complex converts to the aqua form *cis*-[Ru(bpy)$_2$(H$_2$O)$_2$]$^{2+}$. Irradiation of the UCNPs and the release of the Ru(II) aqua species from the nanoparticle surface induces the reduction of DNA mobility on agarose gel, accordingly to the binding of the ruthenium aqua complex to the DNA bases.

In another example, Huang, Chen, and Sun coupled UCNPs to ruthenium complexes for the detection of Hg^{2+} ions [98], showing that such systems

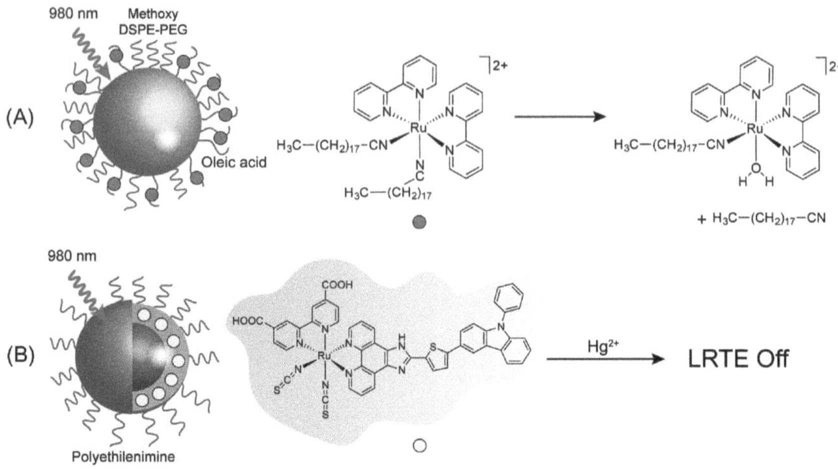

FIGURE 3.10 (A) Schematic illustration of the NIR light activation of $LiYF_4:Yb^{3+},Tm^{3+}$ UCNPs coated with a Ru-polypyridyl complex bearing amphiphilic ligands. (B) Ru-UCNPs hybrid nanomaterial for the detection of Hg^{2+} via luminescence resonance energy transfer. *Adapted from Y. Chen, G. Jiang, Q. Zhou, Y. Zhang, K. Li, Y. Zheng, et al., An upconversion nanoparticle/Ru(II) polypyridyl complex assembly for NIR-activated release of a DNA covalent-binding agent, RSC Adv. 6 (2016) 23804–23808; X. Ge, L. Sun, B. Ma, D. Jin, L. Dong, L. Shi, et al., Simultaneous realization of Hg^{2+} sensing, magnetic resonance imaging and upconversion luminescence in vitro and in vivo bioimaging based on hollow mesoporous silica coated UCNPs and ruthenium complex, Nanoscale 7 (2015) 13877–138870.*

can be used for analytical purposes as well. In detail, the authors synthesized core–shell $NaYF_4:Yb^{3+},Tm^{3+},Er^{3+}@NaGdF_4$ UCNPs and coated them with mesoporous silica. To improve the water solubility of the system, the silica layer is decorated with polyethylenimine. Last, a luminescent ruthenium(II) complex capable of sensing Hg^{2+} ions is physically loaded into the hollow mesoporous silica by solely mixing and stirring. The upconverted green emission band of UCNPs overlaps significantly with the absorption band of the Ru complex and luminescence resonance energy transfer (LRET) between the UCNP and the Ru complex occurs. However, such overlap is reduced when Hg^{2+} ions are added in solution. A blue shift in the absorption spectrum of the ruthenium complex is observed as a result of the formation of an adduct with Hg^{2+}. The "switch off" of the LRET emission after irradiation with a NIR laser is used to quantitatively determine Hg^{2+} ions. Furthermore, UCNPs are doped with Gd^{3+} ions to use the system in vivo for magnetic resonance imaging of mice.

Table 3.1 summarizes key photochemical features for UCNP-based photoresponsive nanoconstructs used in the activation of transition metal complexes.

TABLE 3.1 Upconverting nanoparticles-triggered photochemistry of metal complexes.

Compound	UCNP	Power	Application	Reference
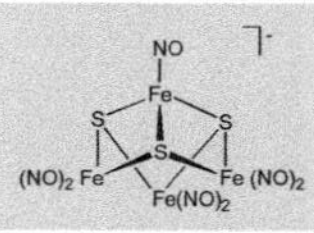	$NaYF_4$:Yb,Er@$NaYF_4$ and $NaYF_4$:Yb,Tm@$NaYF_4$	400 W/cm^2	Ligand photorelease	[64]
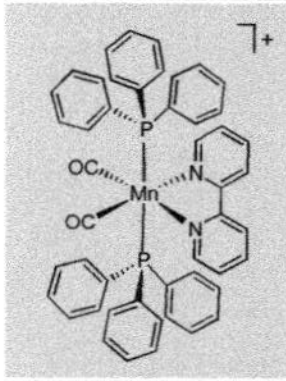	$NaYF_4$:Gd,Tm@$NaGdF_4$	NA	Ligand photorelease	[65]
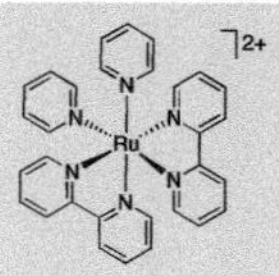	$NaYF_4$:Yb,Er	25 W/cm^2	Ligand photorelease	[67]
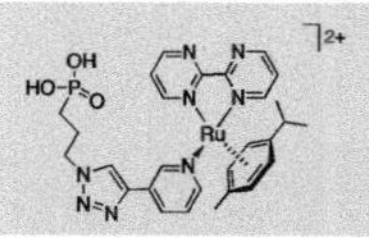	$NaYF_4$:Yb,Tm@$NaYF_4$	8.8 W/cm^2	Prodrug photoactivation	[68]
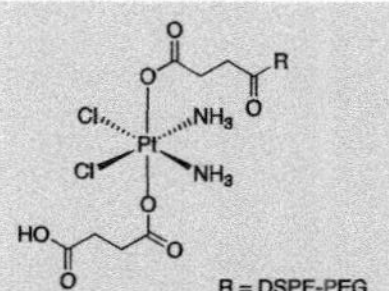	$NaYF_4$:Yb^{3+},Tm^{3+}@$NaYF_4$	4.9/7.3 W/cm^2	Prodrug photoactivation	[69]

(*Continued*)

TABLE 3.1 (Continued)

Compound	UCNP	Power	Application	Reference
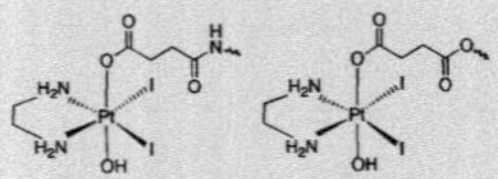	$NaGdF_4$:Yb^{3+},Tm^{3+} and $NaGdF_4$:Yb^{3+},Er^{3+}@SiO_2	1.2 W/cm^2	Prodrug photoactivation	[87]
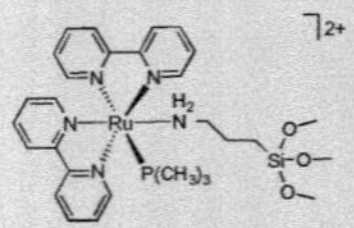	$NaYF_4$:Yb^{3+},Tm^{3+}@$NaYF_4$@SiO_2	0.35/0.64 W/cm^2	Drug photorelease	[93]
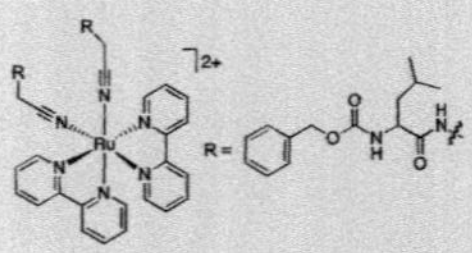	$NaYF_4$:Yb^{3+},Tm^{3+}@$NaYF_4$@SiO_2	1 W/cm^2	Inhibitor photorelease	[94]
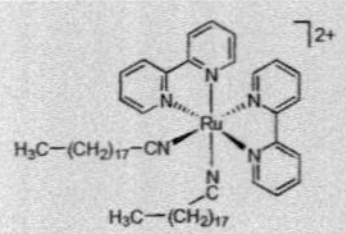	$LiYF_4$:Yb^{3+},Tm^{3+}	3 W/cm^2	Prodrug photoactivation	[96]
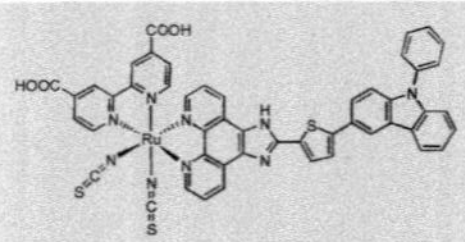	$NaYF_4$:Yb^{3+},Tm^{3+},Er^{3+}@$NaGdF_4$ @SiO_2	NA	Sensing	[98]

All systems were irradiated at 974–980 nm.

3.5 Conclusions

These encouraging results demonstrate that UCNPs and photoreactive molecules can be integrated to provide hybrid nanomaterials whose action can be controlled with NIR light. This is a fine prospect particularly for applications in biology and medicine. NIR light excitation and the localized nature of the upconverted emission of UCNPs can avoid some of the damage produced by UV or visible photons, as well as limit unwanted side photoreactions that direct UV and visible irradiation of biological components can elicit.

Nevertheless, some important challenges are still awaiting solution before UCNP-based strategies become the blue print in photochemistry.

The major drawback of these materials is the low efficiency of the upconverting process. This is an intrinsic limitation, linked to the nature of the lanthanides 4f orbitals and their excited-state photophysics. For this reason, high intensities of excitation lasers are required to be above a threshold for UCNPs. This can cause damage to the surrounding of the laser focus. Such issue is dramatic in biological samples and especially at 980 nm, a wavelength where water strongly absorbs and is overheated. As established by the American National Standards for the Use of Lasers, only few hundreds of mW /cm^2 is the maximum permissible dose for NIR light and the great majority of UCNP systems so far reported require much higher optical power densities.

Several research groups have demonstrated that upconverting efficiency can be raised by improving lattice design and surface passivation approaches in UCNPs, as well as by surface plasmon enhancement [99]. Nevertheless, major breakthroughs in these directions are needed to really push forward the use of UCNP-based photoresponsive nanomaterials.

Acknowledgments

This work was supported by the Spanish Ministry of Economy and Competitiveness (Grant CTQ2016−80844-R). L.S. thank members of the European COST 1403 action for stimulating discussions.

References

[1] C.K. Prier, D.A. Rankic, D.W.C. MacMillan, Visible light photoredox catalysis with transition metal complexes: applications in organic synthesis, Chem. Rev. 113 (2013) 5322−5363.

[2] D.M. Schultz, T.P. Yoon, Solar synthesis: prospects in visible light photocatalysis, Science 343 (2014).

[3] K. Szaciłowski, W. Macyk, A. Drzewiecka-Matuszek, M. Brindell, G. Stochel, Bioinorganic photochemistry: frontiers and mechanisms, Chem. Rev. 105 (2005) 2647−2694.

[4] C.W. Brian, S.P. Michael, The physics, biophysics and technology of photodynamic therapy, Phys. Med. Biol. 53 (2008) R61.

[5] M.N. Chong, B. Jin, C.W.K. Chow, C. Saint, Recent developments in photocatalytic water treatment technology: a review, Water Res. 44 (2010) 2997–3027.
[6] A.P. de Silva, D.B. Fox, A.J.M. Huxley, T.S. Moody, Combining luminescence, coordination and electron transfer for signalling purposes, Coord. Chem. Rev. 205 (2000) 41–57.
[7] X. Fan, I.M. White, S.I. Shopova, H. Zhu, J.D. Suter, Y. Sun, Sensitive optical biosensors for unlabeled targets: a review, Anal. Chim. Acta 620 (2008) 8–26.
[8] T.P.A. Ruberu, Y. Dong, A. Das, R. Eisenberg, Photoelectrochemical generation of hydrogen from water using a cdse quantum dot-sensitized photocathode, ACS Catal. 5 (2015) 2255–2259.
[9] R.D. Harris, S. Bettis Homan, M. Kodaimati, C. He, A.B. Nepomnyashchii, N.K. Swenson, et al., Electronic processes within quantum dot-molecule complexes, Chem. Rev. 116 (2016) 12865–12919.
[10] K. Watanabe, D. Menzel, N. Nilius, H.-J. Freund, Photochemistry on metal nanoparticles, Chem. Rev. 106 (2006) 4301–4320.
[11] Y. Zhang, S. He, W. Guo, Y. Hu, J. Huang, J.R. Mulcahy, et al., Surface-plasmon-driven hot electron photochemistry, Chem. Rev. 118 (2018) 2927–2954.
[12] M.L. Brongersma, N.J. Halas, P. Nordlander, Plasmon-induced hot carrier science and technology, Nat. Nanotechnol. 10 (2015) 25.
[13] M. González-Béjar, L. Francés-Soriano, J. Pérez-Prieto, Upconversion nanoparticles for bioimaging and regenerative medicine, Front. Bioeng. Biotechnol. 4 (2016) 47.
[14] G. Chen, H. Qiu, P.N. Prasad, X. Chen, Upconversion nanoparticles: design, nanochemistry, and applications in theranostics, Chem. Rev. 114 (2014) 5161–5214.
[15] X. Huang, S. Han, W. Huang, X. Liu, Enhancing solar cell efficiency: the search for luminescent materials as spectral converters, Chem. Soc. Rev. 42 (2013) 173–201.
[16] L. Prodi, E. Rampazzo, F. Rastrelli, A. Speghini, N. Zaccheroni, Imaging agents based on lanthanide doped nanoparticles, Chem. Soc. Rev. 44 (2015) 4922–4952.
[17] Y.I. Park, K.T. Lee, Y.D. Suh, T. Hyeon, Upconverting nanoparticles: a versatile platform for wide-field two-photon microscopy and multi-modal in vivo imaging, Chem. Soc. Rev. 44 (2015) 1302–1317.
[18] A. Sedlmeier, H.H. Gorris, Surface modification and characterization of photon-upconverting nanoparticles for bioanalytical applications, Chem. Soc. Rev. 44 (2015) 1526–1560.
[19] O.S. Wolfbeis, An overview of nanoparticles commonly used in fluorescent bioimaging, Chem. Soc. Rev. 44 (2015) 4743–4768.
[20] S. Shikha, T. Salafi, J. Cheng, Y. Zhang, Versatile design and synthesis of nano-barcodes, Chem. Soc. Rev. 46 (2017) 7054–7093.
[21] J. Zhou, Z. Liu, F. Li, Upconversion nanophosphors for small-animal imaging, Chem. Soc. Rev. 41 (2012) 1323–1349.
[22] W. Zheng, P. Huang, D. Tu, E. Ma, H. Zhu, X. Chen, Lanthanide-doped upconversion nano-bioprobes: electronic structures, optical properties, and biodetection, Chem. Soc. Rev. 44 (2015) 1379–1415.
[23] A. Bagheri, H. Arandiyan, C. Boyer, M. Lim, Lanthanide-doped upconversion nanoparticles: emerging intelligent light-activated drug delivery systems, Adv. Sci. 3 (2016) 1500437.
[24] F. Wang, X. Liu, Recent advances in the chemistry of lanthanide-doped upconversion nanocrystals, Chem. Soc. Rev. 38 (2009) 976–989.
[25] E.M. Chan, Combinatorial approaches for developing upconverting nanomaterials: high-throughput screening, modeling, and applications, Chem. Soc. Rev. 44 (2015) 1653–1679.

[26] J. Zhou, Q. Liu, W. Feng, Y. Sun, F. Li, Upconversion luminescent materials: advances and applications, Chem. Rev 115 (2015) 395–465.

[27] T. Rinkel, J. Nordmann, A.N. Raj, M. Haase, Ostwald-ripening and particle size focussing of sub-10 nm $NaYF_4$ upconversion nanocrystals, Nanoscale 6 (2014) 14523–14530.

[28] B. Voss, M. Haase, Intrinsic focusing of the particle size distribution in colloids containing nanocrystals of two different crystal phases, ACS Nano 7 (2013) 11242–11254.

[29] T. Yang, Y. Sun, Q. Liu, W. Feng, P. Yang, F. Li, Cubic sub-20 nm $NaLuF_4$-based upconversion nanophosphors for high-contrast bioimaging in different animal species, Biomaterials. 33 (2012) 3733–3742.

[30] S. Guo, M.-K. Tsang, W.-S. Lo, J. Hao, W.-T. Wong, 808 nm excited energy migration upconversion nanoparticles driven by a Nd^{3+}-trinity system with color-tunability and superior luminescence properties, Nanoscale 10 (2018) 2790–2803.

[31] F. Vetrone, R. Naccache, V. Mahalingam, C.G. Morgan, J.A. Capobianco, The active-core/active-shell approach: a strategy to enhance the upconversion luminescence in lanthanide-doped nanoparticles, Adv. Funct. Mater. 19 (2009) 2924–2929.

[32] J.-N. Liu, W.-B. Bu, J.-L. Shi, Silica coated upconversion nanoparticles: a versatile platform for the development of efficient theranostics, Acc. Chem. Res. 48 (2015) 1797–1805.

[33] E. Hemmer, P. Acosta-Mora, J. Mendez-Ramos, S. Fischer, Optical nanoprobes for biomedical applications: shining a light on upconverting and near-infrared emitting nanoparticles for imaging, thermal sensing, and photodynamic therapy, J. Mater. Chem. B 5 (2017) 4365–4392.

[34] B. Zhou, B. Shi, D. Jin, X. Liu, Controlling upconversion nanocrystals for emerging applications, Nat. Nanotechnol. 10 (2015) 924.

[35] W.G. van Sark, J. de Wild, J.K. Rath, A. Meijerink, R.E. Schropp, Upconversion in solar cells, Nanoscale Res. Lett. 8 (2013) 81.

[36] J. Lee, P.W. Bisso, R.L. Srinivas, J.J. Kim, A.J. Swiston, P.S. Doyle, Universal process-inert encoding architecture for polymer microparticles, Nat. Mater. 13 (2014) 524.

[37] T.R. Hinklin, S.C. Rand, R.M. Laine, Transparent, polycrystalline upconverting nanoceramics: towards 3-D displays, Adv. Mater. 20 (2008) 1270–1273.

[38] S. Hao, G. Chen, C. Yang, Sensing using rare-earth-doped upconversion nanoparticles, Theranostics 3 (2013) 331–345.

[39] E. Ruggiero, S. Alonso-deCastro, A. Habtemariam, L. Salassa, Upconverting nanoparticles for the near infrared photoactivation of transition metal complexes: new opportunities and challenges in medicinal inorganic photochemistry, Dalton. Trans. 45 (2016) 13012–13020.

[40] S. Mallidi, S. Anbil, A.-L. Bulin, G. Obaid, M. Ichikawa, T. Hasan, Beyond the barriers of light penetration: strategies, perspectives and possibilities for photodynamic therapy, Theranostics 6 (2016) 2458–2487.

[41] *Laser Safety Handbook*, Northwestern University, Evanston, 2011.

[42] S. Wu, H.J. Butt, Near-infrared-sensitive materials based on upconverting nanoparticles, Adv. Mater. 28 (2016) 1208–1226.

[43] L. Wondraczek, E. Tyystjärvi, J. Méndez-Ramos, F.A. Müller, Q. Zhang, Shifting the sun: solar spectral conversion and extrinsic sensitization in natural and artificial photosynthesis, Adv. Sci. 2 (2015) 1500218.

[44] W. Wang, W. Huang, Y. Ni, C. Lu, Z. Xu, Different upconversion properties of β-$NaYF_4$: Yb^{3+}, Tm^{3+}/Er^{3+} in affecting the near-infrared-driven photocatalytic activity of high-reactive TiO_2, ACS Appl. Mater. Interfaces 6 (2014) 340–348.

[45] S. Obregon, G. Colon, Evidence of upconversion luminescence contribution to the improved photoactivity of erbium doped TiO_2 systems, Chem. Commun. 48 (2012) 7865–7867.

[46] M.E. Borges, M. Sierra, J. Méndez-Ramos, P. Acosta-Mora, J.C. Ruiz-Morales, P. Esparza, Solar degradation of contaminants in water: TiO_2 solar photocatalysis assisted by up-conversion luminescent materials, Sol. Energy Mater. Sol. Cells 155 (2016) 194–201.

[47] C.-J. Carling, J.-C. Boyer, N.R. Branda, Remote-control photoswitching using nir light, J. Am. Chem. Soc. 131 (2009) 10838–10839.

[48] J.-C. Boyer, C.-J. Carling, B.D. Gates, N.R. Branda, Two-way photoswitching using one type of near-infrared light, upconverting nanoparticles, and changing only the light intensity, J. Am. Chem. Soc. 132 (2010) 15766–15772.

[49] C.-J. Carling, F. Nourmohammadian, J.-C. Boyer, N.R. Branda, Remote-control photorelease of caged compounds using near-infrared light and upconverting nanoparticles, Angew. Chem. Int. Ed. 49 (2010) 3782–3785.

[50] Y. Yang, Q. Shao, R. Deng, C. Wang, X. Teng, K. Cheng, et al., In vitro and in vivo uncaging and bioluminescence imaging by using photocaged upconversion nanoparticles, Angew. Chem. Int. Ed. 51 (2012) 3125–3129.

[51] L.L. Fedoryshin, A.J. Tavares, E. Petryayeva, S. Doughan, U.J. Krull, Near-infrared-triggered anticancer drug release from upconverting nanoparticles, ACS Appl. Mater. Interfaces 6 (2014) 13600–13606.

[52] B. Yan, J.-C. Boyer, N.R. Branda, Y. Zhao, Near-infrared light-triggered dissociation of block copolymer micelles using upconverting nanoparticles, J. Am. Chem. Soc. 133 (2011) 19714–19717.

[53] M.L. Viger, M. Grossman, N. Fomina, A. Almutairi, Low power upconverted near-IR light for efficient polymeric nanoparticle degradation and cargo release, Adv. Mater. 25 (2013) 3733–3738.

[54] T. Wu, J.-C. Boyer, M. Barker, D. Wilson, N.R. Branda, A. Plug-and-Play, Method to prepare water-soluble photoresponsive encapsulated upconverting nanoparticles containing hydrophobic molecular switches, Chem. Mater. 25 (2013) 2495–2502.

[55] T. Wu, D. Wilson, N.R. Branda, Fluorescent quenching of lanthanide-doped upconverting nanoparticles by photoresponsive polymer shells, Chem. Mater. 26 (2014) 4313–4320.

[56] W. Wu, L. Yao, T. Yang, R. Yin, F. Li, Y. Yu, NIR-light-induced deformation of cross-linked liquid-crystal polymers using upconversion nanophosphors, J. Am. Chem. Soc. 133 (2011) 15810–15813.

[57] L. Wang, H. Dong, Y. Li, C. Xue, L.-D. Sun, C.-H. Yan, et al., Reversible near-infrared light directed reflection in a self-organized helical superstructure loaded with upconversion nanoparticles, J. Am. Chem. Soc. 136 (2014) 4480–4483.

[58] L. Wang, H. Dong, Y. Li, R. Liu, Y.F. Wang, H.K. Bisoyi, et al., Luminescence-driven reversible handedness inversion of self-organized helical superstructures enabled by a novel near-infrared light nanotransducer, Adv. Mater. 27 (2015) 2065–2069.

[59] N.J. Farrer, L. Salassa, P.J. Sadler, Photoactivated chemotherapy (PACT): the potential of excited-state d-block metals in medicine, Dalton. Trans. (2009) 10690–10701.

[60] N.A. Smith, P.J. Sadler, Photoactivatable metal complexes: from theory to applications in biotechnology and medicine, Philos. Trans. A Math. Phys. Eng. Sci. 371 (2013) 20120519.

[61] G. Shi, S. Monro, R. Hennigar, J. Colpitts, J. Fong, K. Kasimova, et al., Ru(II) dyads derived from α-oligothiophenes: a new class of potent and versatile photosensitizers for PDT, Coord. Chem. Rev. 282–283 (2015) 127–138.

[62] J.V. Garcia, J. Yang, D. Shen, C. Yao, X. Li, R. Wang, et al., NIR-triggered release of caged nitric oxide using upconverting nanostructured materials, Small. 8 (2012) 3800–3805.

[63] V. Somasundaram, D. Basudhar, M. Greer, R.Y.S. Cheng, L.A. Ridnour, D.A. Wink, NO in cancer—carcinogenesis, metastasis, and therapy, in: B.A. Freeman (Ed.), Nitric Oxide, third ed., Academic Press, 2017.

[64] P.T. Burks, J.V. Garcia, R. GonzalezIrias, J.T. Tillman, M. Niu, A.A. Mikhailovsky, et al., Nitric oxide releasing materials triggered by near-infrared excitation through tissue filters, J. Am. Chem. Soc. 135 (2013) 18145–18152.

[65] A.E. Pierri, P.-J. Huang, J.V. Garcia, J.G. Stanfill, M. Chui, G. Wu, et al., A photoCORM nanocarrier for CO release using NIR light, Chem. Commun. 51 (2015) 2072–2075.

[66] R. Motterlini, L.E. Otterbein, The therapeutic potential of carbon monoxide, Nat. Rev. Drug. Discov. 9 (2010) 728.

[67] E. Ruggiero, A. Habtemariam, L. Yate, J.C. Mareque-Rivas, L. Salassa, Near infrared photolysis of a Ru polypyridyl complex by upconverting nanoparticles, Chem. Commun. 50 (2014) 1715–1718.

[68] E. Ruggiero, C. Garino, J.C. Mareque-Rivas, A. Habtemariam, L. Salassa, Upconverting nanoparticles prompt remote near-infrared photoactivation of Ru(II)–arene complexes, Chem. Eur. J. 22 (2016) 2801–2811.

[69] E. Ruggiero, J. Hernandez-Gil, J.C. Mareque-Rivas, L. Salassa, Near infrared activation of an anticancer Pt^{IV} complex by Tm-doped upconversion nanoparticles, Chem. Commun. 51 (2015) 2091–2094.

[70] F.E. Poynton, S.A. Bright, S. Blasco, D.C. Williams, J.M. Kelly, T. Gunnlaugsson, The development of ruthenium(II) polypyridyl complexes and conjugates for in vitro cellular and in vivo applications, Chem. Soc. Rev. 46 (2017) 7706–7756.

[71] B.S. Howerton, D.K. Heidary, E.C. Glazer, Strained ruthenium complexes are potent light-activated anticancer agents, J. Am. Chem. Soc. 134 (2012) 8324–8327.

[72] M. Dickerson, B. Howerton, Y. Bae, E.C. Glazer, Light-sensitive ruthenium complex-loaded cross-linked polymeric nanoassemblies for the treatment of cancer, J. Mater. Chem. B 4 (2016) 394–408.

[73] E. Wachter, D.K. Heidary, B.S. Howerton, S. Parkin, E.C. Glazer, Light-activated ruthenium complexes photobind DNA and are cytotoxic in the photodynamic therapy window, Chem. Commun. 48 (2012) 9649–9651.

[74] E. Fino, R. Araya, D. Peterka, M. Salierno, R. Etchenique, R. Yuste, RuBi-glutamate: two-photon and visible-light photoactivation of neurons and dendritic spines, Front. Neural Circuits 3 (2009).

[75] R. Araya, V. Andino-Pavlovsky, R. Yuste, R. Etchenique, Two-photon optical interrogation of individual dendritic spines with caged dopamine, ACS Chem. Neurosci. 4 (2013) 1163–1167.

[76] T. Respondek, R.N. Garner, M.K. Herroon, I. Podgorski, C. Turro, J.J. Kodanko, Light activation of a cysteine protease inhibitor: caging of a peptidomimetic nitrile with $Ru^{II}(bpy)_2$, J. Am. Chem. Soc. 133 (2011) 17164–17167.

[77] R. Sharma, J.D. Knoll, P.D. Martin, I. Podgorski, C. Turro, J.J. Kodanko, Ruthenium *tris* (2-pyridylmethyl)amine as an effective photocaging group for nitriles, Inorg. Chem. 53 (2014) 3272–3274.

[78] T. Respondek, R. Sharma, M.K. Herroon, R.N. Garner, J.D. Knoll, E. Cueny, et al., Inhibition of cathepsin activity in a cell-based assay by a light-activated ruthenium compound, ChemMedChem. 9 (2014) 1306–1315.

[79] S.D. Ramalho, R. Sharma, J.K. White, N. Aggarwal, A. Chalasani, M. Sameni, et al., Imaging sites of inhibition of proteolysis in pathomimetic human breast cancer cultures by light-activated ruthenium compound, PLoS. ONE. 10 (2015) e0142527.

[80] E. Borfecchia, C. Garino, L. Salassa, T. Ruiu, D. Gianolio, X. Zhang, et al., X-ray transient absorption structural characterization of the ^{3}MLCT triplet excited state of cis-[Ru $(bpy)_2(py)_2]^{2+}$, Dalton. Trans. 42 (2013) 6564–6571.

[81] L. Salassa, C. Garino, G. Salassa, R. Gobetto, C. Nervi, Mechanism of ligand photodissociation in photoactivable $[Ru(bpy)_2L_2]^{2+}$ complexes: a density functional theory study, J. Am. Chem. Soc. 130 (2008) 9590–9597.

[82] L. Salassa, E. Borfecchia, T. Ruiu, C. Garino, D. Gianolio, R. Gobetto, et al., Photo-induced pyridine substitution in cis-$[Ru(bpy)_2(py)_2]Cl_2$: a snapshot by time-resolved x-ray solution scattering, Inorg. Chem. 49 (2010) 11240–11248.

[83] Y.K. Yan, M. Melchart, A. Habtemariam, P.J. Sadler, Organometallic chemistry, biology and medicine: ruthenium arene anticancer complexes, Chem. Commun. (2005) 4764–4776.

[84] P. Kumar, R.K. Gupta, D.S. Pandey, Half-sandwich arene ruthenium complexes: synthetic strategies and relevance in catalysis, Chem. Soc. Rev. 43 (2014) 707–733.

[85] S. Betanzos-Lara, L. Salassa, A. Habtemariam, P.J. Sadler, Photocontrolled nucleobase binding to an organometallic Ru^{II} arene complex, Chem. Commun. (2009) 6622–6624.

[86] S. Betanzos-Lara, L. Salassa, A. Habtemariam, O. Novakova, A.M. Pizarro, G.J. Clarkson, et al., Photoactivatable organometallic pyridyl ruthenium(II) arene complexes, Organometallics. 31 (2012) 3466–3479.

[87] S. Perfahl, M.M. Natile, H.S. Mohamad, C.A. Helm, C. Schulzke, G. Natile, et al., Photoactivation of diiodido–Pt(IV) complexes coupled to upconverting nanoparticles, Mol. Pharm. 13 (2016) 2346–2362.

[88] J.-C. Boyer, M.-P. Manseau, J.I. Murray, F.C.J.M. van Veggel, Surface modification of upconverting $NaYF_4$ nanoparticles with PEG – phosphate ligands for NIR (800 nm) biolabeling within the biological window, Langmuir 26 (2010) 1157–1164.

[89] L.-L. Li, P. Wu, K. Hwang, Y. Lu, An exceptionally simple strategy for DNA-functionalized up-conversion nanoparticles as biocompatible agents for nanoassembly, DNA delivery, and imaging, J. Am. Chem. Soc. 135 (2013) 2411–2414.

[90] E. Ruggiero, S. Alonso-de Castro, A. Habtemariam, L. Salassa, The photochemistry of transition metal complexes and its application in biology and medicine, in: K.K.-W. Lo (Ed.), Luminescent and Photoactive Transition Metal Complexes as Biomolecular Probes and Cellular Reagents, Springer Berlin Heidelberg, Berlin, Heidelberg, 2015, pp. 69–107.

[91] T.C. Johnstone, K. Suntharalingam, S.J. Lippard, The next generation of platinum drugs: targeted Pt(II) agents, nanoparticle delivery, and Pt(IV) prodrugs, Chem. Rev. 116 (2016) 3436–3486.

[92] P. Gramatica, E. Papa, M. Luini, E. Monti, M.B. Gariboldi, M. Ravera, et al., Pt(IV) complexes: synthesis, biological activity, and quantitative structure–activity relationship modeling, JBIC 15 (2010) 1157–1169.

[93] S. He, K. Krippes, S. Ritz, Z. Chen, A. Best, H.-J. Butt, et al., Ultralow-intensity near-infrared light induces drug delivery by upconverting nanoparticles, Chem. Commun. 51 (2015) 431–434.

[94] Z. Chen, R. Thiramanas, M. Schwendy, C. Xie, S.H. Parekh, V. Mailänder, et al., Upconversion nanocarriers encapsulated with photoactivatable Ru complexes for near-infrared light-regulated enzyme activity, Small. 13 (2017) 1700997.

[95] M.M. Mohamed, B.F. Sloane, Multifunctional enzymes in cancer, Nat. Rev. Cancer 6 (2006) 764.

[96] Y. Chen, G. Jiang, Q. Zhou, Y. Zhang, K. Li, Y. Zheng, et al., An upconversion nanoparticle/Ru(II) polypyridyl complex assembly for NIR-activated release of a DNA covalent-binding agent, RSC Adv. 6 (2016) 23804–23808.

[97] Y. Liu, D.B. Turner, T.N. Singh, A.M. Angeles-Boza, A. Chouai, K.R. Dunbar, et al., Ultrafast ligand exchange: detection of a pentacoordinate Ru(II) intermediate and product formation, J. Am. Chem. Soc. 131 (2009) 26–27.

[98] X. Ge, L. Sun, B. Ma, D. Jin, L. Dong, L. Shi, et al., Simultaneous realization of Hg^{2+} sensing, magnetic resonance imaging and upconversion luminescence in vitro and in vivo bioimaging based on hollow mesoporous silica coated UCNPs and ruthenium complex, Nanoscale 7 (2015) 13877–13887.

[99] D.M. Wu, A. García-Etxarri, A. Salleo, J.A. Dionne, Plasmon-enhanced upconversion, J. Phys. Chem. Lett. 5 (2014) 4020–4031.

Chapter 4

Dressing up for the occasion: the many faces of decorated titanium dioxide in photocatalysis

Ayda Ali Elhage, Juan C. Scaiano and Anabel Estela Lanterna
Department of Chemistry and Biomolecular Sciences and Centre for Advanced Materials Research (CAMaR), University of Ottawa, Ottawa, ON, Canada

Chapter Outline

4.1 Introduction

This chapter rationalizes the use of modified titanium dioxide (TiO_2) in different applications involving UV and visible light harvesting, with emphasis on photochemical transformations. TiO_2 is a well-known semiconductor, and its more common form, anatase, has a large band gap ($\sim$3.1 eV) that can absorb light below $\sim$400 nm. It has been used in a variety of different applications such as base color of paints [1], addition to cement and tiles surfaces conferring the material sterilizing and antifouling properties [2,3], self-cleaning glass [4], the Grätzel solar cell [5], cosmetics, particularly sunscreens [6–7], and water disinfection [8–10], among others. However, many

Photoactive Inorganic Nanoparticles. DOI: https://doi.org/10.1016/B978-0-12-814531-9.00004-X
© 2019 Elsevier Inc. All rights reserved.

of these applications benefit from the modification of TiO_2, usually increasing its photochemical properties in the visible region of the spectrum.

To understand the way in which the surface modifications can confer TiO_2 new or enhanced properties, it is useful to compare this semiconductor with the case of typical organic molecules where the consequences of photoexcitation are readily explained by the paradigms of organic photochemistry. When a diamagnetic molecule is irradiated with visible or UV light, the consequence is the promotion of a highest occupied molecular orbital (HOMO) electron to the lowest unoccupied molecular orbital (LUMO) or a higher molecular orbital. If excitation leads to higher orbital levels (e.g., LUMO + 1 and LUMO + 2), most molecules undergo rapid relaxation to the LUMO, leading to the first singlet (S_1) excited state. This electronic state with a vacancy in the HOMO and an electron in the LUMO is a better electron donor and a better electron acceptor than the ground state, S_0 [11]. This is illustrated in Fig. 4.1.

Although both donor and acceptor redox properties are enhanced in S_1 relative to S_0, these characteristics can be modified by substitution of the molecule under study. Furthermore, the actual efficiency of charge transfer reactions from S_1 will be controlled by its lifetime, as this will determine the probability that S_1 will encounter electron donors or acceptors before its inevitable return to S_0.

As we deal with semiconductors, such as TiO_2, the use of discrete orbital energy levels is no longer appropriate; instead we can distinguish two regions where there is a non-zero probability of finding an electron or a hole: the valence band (VB) and conduction band (CB) (Fig. 4.2). Thus

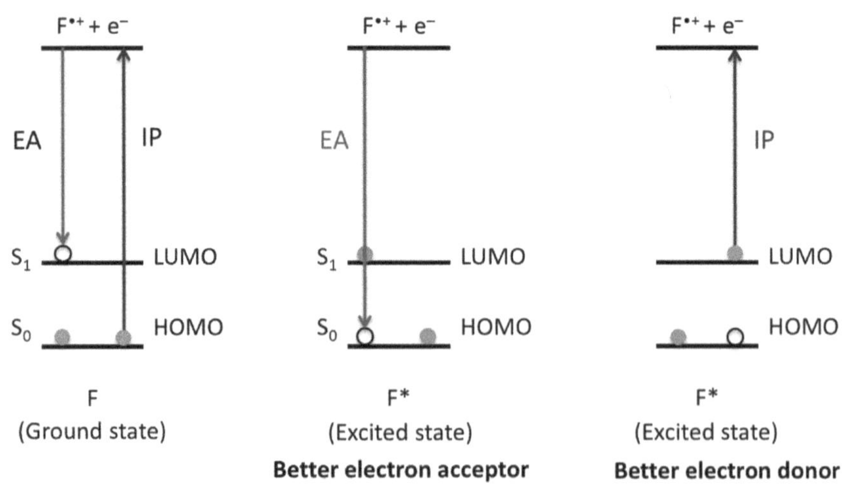

FIGURE 4.1 Changes in the electron affinity (EA) and the ionization potential (IP) of a fluorophore (F) upon excitation.

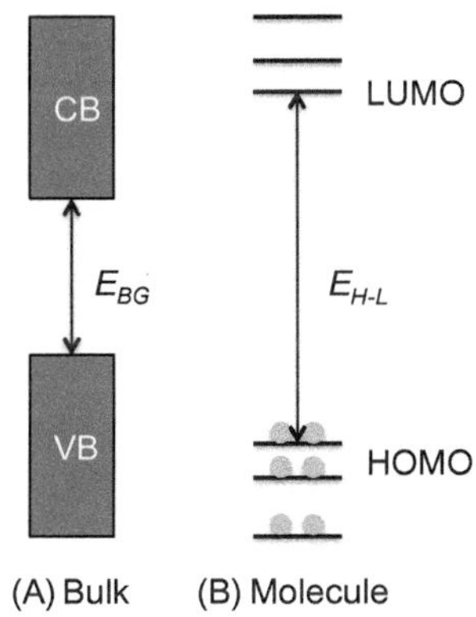

FIGURE 4.2 Energy level structure of a bulk semiconductor material comparing to a molecular system. E_{BG}, Band gap energy; $E_{H\text{-}L}$, energy difference HOMO–LUMO.

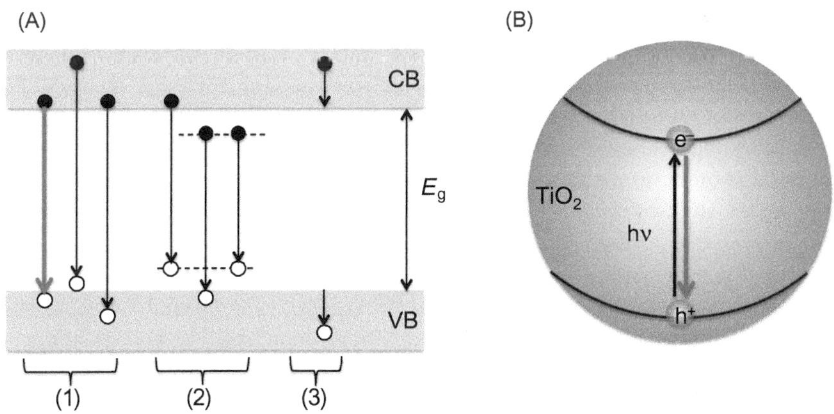

FIGURE 4.3 (A) Excitation of TiO_2 leads to the promotion of an electron from the VB to the CB: (1) Interband transitions, (2) band to localized defects transitions, (3) intraband transitions (non-radiative transition: phonon). (B) Simplification of the diagram in (A) used along this chapter. The *blue (gray in print version) arrow* represents all interband transitions. *Adapted from S.M. Sze, K.N. Kwok, Physics and Semiconductor Devices, John Wiley & Sons, Inc., New Jersey, 2007 [12].*

photoexcitation leads to the promotion of VB electrons into the CB forming the so-called electron–hole (e–h) pair (or exciton). The "hole" in the VB becomes an excellent electron acceptor, while the electron now populating the CB is an excellent reducing agent. The electron and the hole can eventually recombine through different pathways as described in Fig. 4.3A [12]. For the sake of simplicity, the process of electron excitation and e–h recombination in a semiconductor will be represented as shown in Fig. 4.3B. Just as in the case of organic molecules, energy levels (S_0, S_1, or VB and CB)

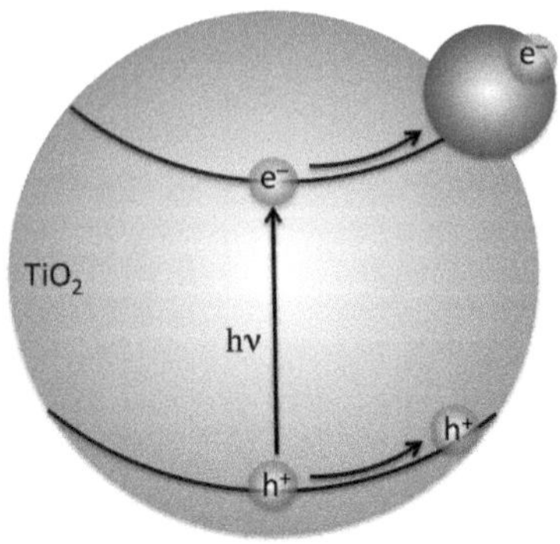

FIGURE 4.4 Electron transfer to a MNP induces charge separation delaying the e–h recombination process.

control the thermodynamics and thus feasibility of the process, the practical efficiencies are determined by the dynamics of the system, or in other words, how long the charge separation persists. In the case of photoexcited TiO_2 rapid charge recombination or back electron transfer severely restricts the efficiency of any process that tries to harvest electrons or holes.

Decorating TiO_2 with metal or metal oxide nanoparticles (MNP@TiO_2) is a common strategy to increase the lifetime of the charge-separated states in semiconductors, as this provides electronic and spatial barriers to e–h recombination. It is appropriate to define, at this stage, the characteristics of the materials we will be discussing in this contribution. TiO_2 has three common crystalline structures, anatase, brookite, and the denser rutile (densities 3.78 and 4.23 g/mL for anatase and rutile, respectively). Unless otherwise indicated, we will be referring to nanometric anatase, typically with particle size in the 10–100 nm range. Some of the anatase-rich commercial materials incorporate trace amounts of rutile. The decorating materials will be nanoparticles of smaller dimensions than the TiO_2 itself, frequently in the 1–10 nm range. It is easy to produce hybrid nanomaterials where the transfer of the electron from the CB to the metal nanoparticle (MNP) is energetically favorable, as illustrated in Fig. 4.4.

Under normal conditions, numerous MNPs decorate each TiO_2 particle. Such modification affects both the VB and the CB behavior, although these changes are dominated by effects of the CB electron, rapidly transferred to the MNP that now determines the fate of the electron. New opportunities created for this electron are always in competition with relaxation via back electron transfer, a process that is usually delayed significantly once the electron sits in the MNP rather than at the CB. While generally discussed as MNP, some of these decorating materials may be metallic or oxidized metals, partially or totally, depending on their synthesis, media, and history. Furthermore, electron transfer from the CB can lead to localized changes in the excitation state of the MNP. Although the common goal of adding MNP is to increase the charge recombination time, making both the electron and the hole readily available for chemical reactions, these hybrid materials are actually new materials with their own characteristic properties. The properties of both hole and electron are modified in these new hybrids; however, upon UV excitation the effect is generally more pronounced on the electron, as the donor is no longer TiO_2, but rather the decorating nanoparticle. These MNPs confer the new material properties characteristic of each metal or oxide. Furthermore, these new or enhanced characteristics are generally dependent not only on the metal constituents of MNP, but also on their abundance, dimensions, and morphology, as well as their history; thus, reused materials may or may not retain the properties they exhibited in their first use. This is extremely important when the application involves catalysis. Furthermore, MNP polydispersity, or changes in polydispersity resulting from usage, can also alter the properties of these hybrid materials [13].

Beyond the catalytic processes discussed here the electron relay properties of TiO_2 are also a key feature of some solar cells, such as the Grätzel or dye-sensitized solar cell that takes advantage of dye-decorated TiO_2, in a system where the dye, rather than TiO_2, is the visible light absorber [5].

This contribution summarizes the different uses of TiO_2 involving light-harvesting processes with emphasis on applications of MNP@TiO_2 in catalysis. It is not a comprehensive review of the literature and it is not the intent to cover all the metals or metal oxides that have been reported. It is our goal to illustrate the types of photocatalysis that decorated TiO_2, endowed with new properties, can perform. Sections 4.2 and 4.3 deal with the first uses of TiO_2 as a photocatalyst and its modifications with MNPs, metal oxide nanoparticles, and organic dyes. There is also inclusion of a brief section on organic molecules, such as indoles [14], that when they decorate TiO_2 can function as visible light photocatalysts reflecting the properties of a charge transfer complex with the semiconductor.

4.2 Photocatalytic activity of TiO_2

As mentioned previously TiO_2 can only absorb light below 400 nm (Fig. 4.5) and therefore much of the work using the semiconductor as a photocatalyst was done under UV irradiation [15], frequently in the UVA region. Although TiO_2 has been extensively used as a pollutant remover, particularly for decomposition, or just bleaching, of dyes [16,17], the focus of this chapter centers on photocatalytic activity that leads to organic synthesis. However, it is important to know the basic mechanism involved in the decomposition of dyes to understand further applications of TiO_2. It is well known that TiO_2 can generate reactive oxygen species (ROS) when irradiated with UV light in aqueous media (Fig. 4.6A) [15]. Although the mechanism has been accepted for many years, there are still some controversial comments related

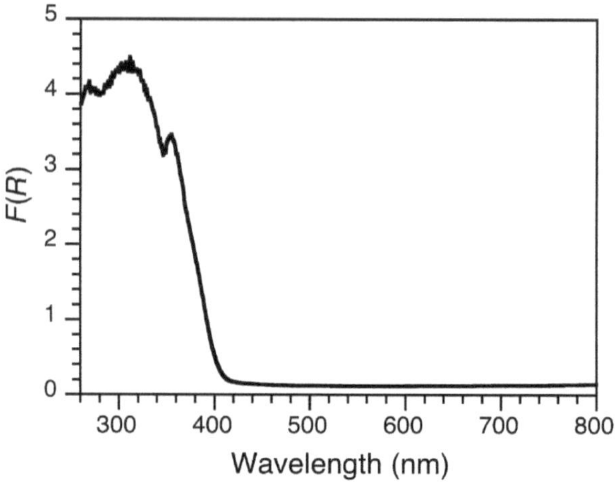

FIGURE 4.5 Diffuse reflectance spectrum of TiO_2 showing absorption below 400 nm.

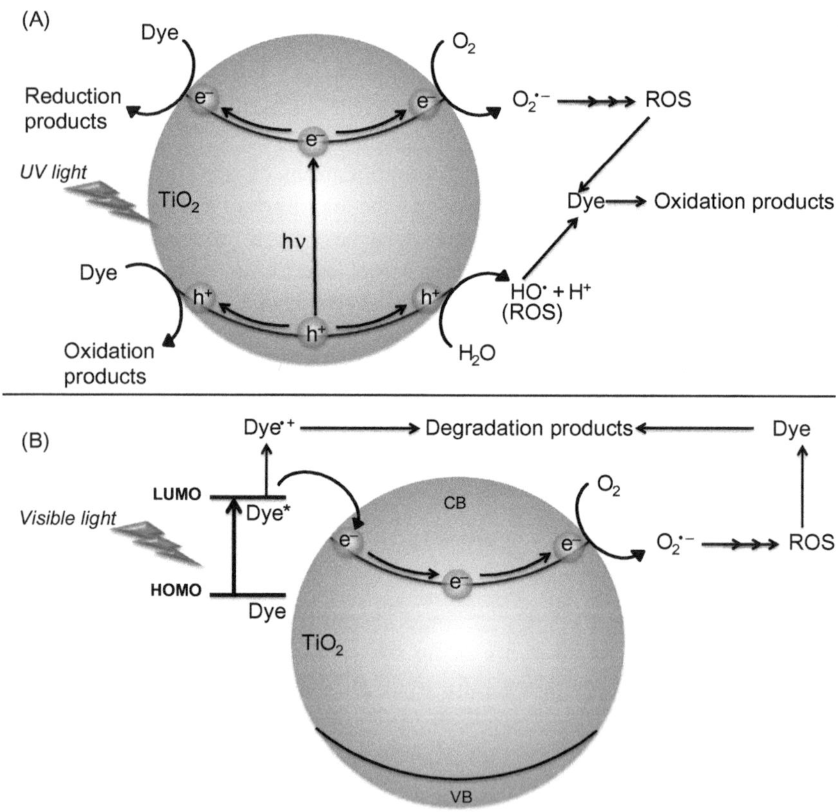

FIGURE 4.6 Dye degradation processes taking place on the surface of TiO_2: (A) indirect degradation upon UV light irradiation, (B) direct degradation upon visible light irradiation absorbed by the dye. *Adapted from A. Ajmal, I. Majeed, R.N. Malik, H. Idriss, M.A. Nadeem, Principles and mechanisms of photocatalytic dye degradation on TiO_2 based photocatalysts: a comparative overview, RSC Adv. 4 (2014) 37003–37026 [16].*

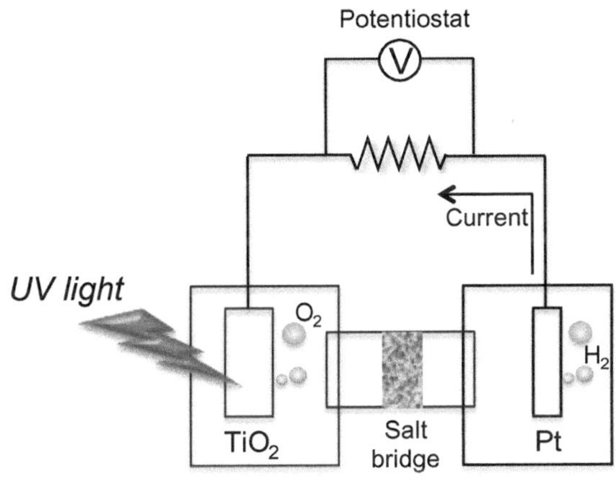

FIGURE 4.7 Electrochemical cell showing how O_2 and H_2 can be generated without the application of an external potential, but upon irradiation with UV light. *Adapted from A. Fujishima, K. Honda, Electrochemical photolysis of water at a semiconductor electrode, Nature 238 (1972) 37–38 [19].*

to the role of the dye during this process, as direct dye-sensitization cannot be ruled out (*vide infra*) (Fig. 4.6B) [18].

In a very groundbreaking publication dated on 1972, Fujishima and Honda first described the use of TiO_2 for the electrochemical photolysis of water at the surface of the semiconductor electrode [Eqs. (4.1–4.3)] [19].

$$TiO_2 + 2\,h\nu \rightarrow 2\,e^- + 2\,h^+ \tag{4.1}$$

$$4\,h^+ + 2\,H_2O \rightarrow O_2 + 4\,H^+ \tag{4.2}$$

$$2\,e^- + 2\,H^+ \rightarrow H_2 \tag{4.3}$$

where the overall reaction is

$$2\,H_2O + 2\,h\nu \rightarrow O_2 + 2\,H_2 \tag{4.4}$$

Fig. 4.7 shows the electrochemical cell used demonstrating that O_2 and H_2 can be generated without the application of an external potential, but upon irradiation with UV light. Thus the anode (the TiO_2 electrode) and the cathode (the platinum electrode) are placed in different compartments but linked through a salt bridge and an external circuit. When the TiO_2 electrode is irradiated with UV light, the current flows from the platinum electrode toward the TiO_2 electrode and there is a concomitant formation of H_2 and O_2, respectively.

Although this process was discovered more than 40 years ago, the difficulties to increase the efficiency of this reaction have pushed back its application in the field of H_2 energy technologies. In contrast, H_2 production using sacrificial molecules can generate H_2 efficiently but is neither "green" nor economical, as valuable chemicals are consumed in its production (e.g., as alcohols, organic acids, and hydrocarbons). This approach has been criticized as the generation of H_2 gas using such sacrificial system does not generate any mechanistic information about water splitting, but rather the sacrificial donor degradation [Eqs. (4.1) and (4.5)–(4.7)] [20]. However, the sacrificial donor approach has helped to understand the H_2 formation, half of the water-splitting reaction [Eq. (4.3)].

$$CH_3OH + h^+ \rightarrow {}^{\bullet}CH_2OH + H^+ \tag{4.5}$$

$$^{\bullet}CH_2OH \rightarrow HCHO + H^+ + e^- \tag{4.6}$$

where the overall reaction is shown in Eq. (4.7) and its effectively methanol splitting, not water splitting.

$$CH_3OH + h\nu \rightarrow CH_2O + H_2 \tag{4.7}$$

This approach could be more cost-effective replacing methanol for non-valuable molecules such as common contaminants found in water streams [21–23]. Still, the use of solar light for water splitting has been the Holy Grail of H_2 fuel generation, but finding good catalysts is challenging. Therefore many efforts are concentrating on the modification of the semiconductor optical properties in order to increase the efficiency of the process under visible irradiation [24].

4.3 Modification of the photocatalytic activity of TiO_2

Understanding how the modifications on the TiO_2 change its photocatalytic properties is the starting point to design new materials based on TiO_2. In order to have a material that can be photoresponsive under visible light irradiation, the band gap of a semiconductor should be higher than 1.6 eV ($\sim$700 nm) and less than 3 eV ($\sim$400 nm). Several strategies have been applied to modify the band-gap energy including doping, creating defects (inserting impurities), or decorating TiO_2 surface. At this point, it is important to distinguish between intrinsic modifications of TiO_2 crystal lattice (here referred as "doping") and superficial modifications of the material by adding metal or metal oxide nanoparticles as well as organic dyes (here referred as "decoration"). As mentioned previously two main factors are taken into consideration when the modification of the semiconductor is designed:

1. The possibility of creating visible light–induced e–h pairs.
2. The increase on the charge carriers separation to delay the e–h recombination [25].

These characteristics can be achieved by two different strategies: (1) generation of typical defects such as vacancies or surface-related sites (doping)

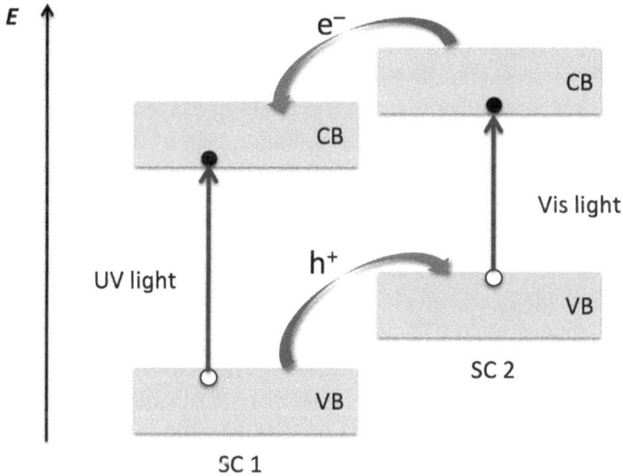

FIGURE 4.8 Electron transfer in a semiconductor heterojunction. The semiconductor 2 (SC 2) with the higher conduction or higher VB energy edges is prone to trap the holes while the electrons are trapped by the lower energy of the conduction or VBs of the semiconductor 1 (SC 1).

where the defect energy levels reside within the band gap [Fig. 4.3A(2)], or (2) the formation of semiconductor heterojunctions (decoration) in which the presence of staggered band-gap alignment leads to the electron (or the hole) confinement into one of the two materials (Fig. 4.8). Nevertheless, light harvesting could be hampered if the exposed surface area is low, therefore, the material morphology plays an important role on its catalytic efficiency. For good photocatalytic efficiency, the reagent diffusion along the material is needed as much as the light harvesting [26,27]. Various TiO_2 morphologies, such as nanotubes, nanowires, nanosheets, and hollow nanostructures, have been successfully prepared to date, although these materials are beyond the scope of this book chapter [28–31].

4.3.1 TiO_2 doping

Examples on the use of doped TiO_2 show high catalytic activity compared to bare TiO_2. In general, doped TiO_2 is synthesized utilizing $TiCl_4$ or $Ti(OEt)_4$ as precursors [32,33]. The use of nonmetal elements (E) was explored for many including B, C, N, F, P, S, Se, Cl, and Br [33–40]. Doping can be either substitutional or interstitial, in both cases the resulting material is expected to show activity in the visible region. Substitutional doping involves replacement of oxygen in the TiO_2 lattice ($TiO_{2-x}E_x$) by elements with similar atomic p level and ionic radius. Thus N doping has been found to be the most promising as its p states contribute to narrowing the band gap [41–43]. Interstitial doping is based on incorporating elements into the TiO_2 lattice (TiO_2E_x), leading to the formation of localized mid-gap states above the VB of TiO_2, and as a result the band gap of the new material can be efficiently narrowed, that is ~2.5 eV for N doping [44], leading to a blue-absorbing material. Thus the nature of the increase photocatalytic activity in the visible region is due to the decrease of the band gap as a result of a formation of new energy levels in the forbidden band of TiO_2 and the decrease on the recombination rate of the photogenerated e−h pair [44]. The ease synthesis of N-doped TiO_2, low cost, high chemical stability and intense visible light absorption make this material a suitable candidate for practical applications. Different routes of N-doped TiO_2 synthesis using variable conditions were reported to date, that is, solvothermal [45], hydrothermal [46,47], mixing [48], chemical vapor deposition [49,50], sol-gel [46,51,52], atomic layer deposition [53], microwave [54], electrospinning [55], vapor-thermal [50], and wet methods; [56] and their applications as photocatalyst for visible light−driven degradation of various types of organic pollutants (colored and noncolored compounds) such as methylene blue (MB), methyl orange, 2-propanol, 2-chlorophenol, benzamide, stearic acid Rhodamine B, and ciprofloxacin [44].

There are some reports that demonstrate phosphorous doping of TiO_2 can narrow the semiconductor band gap increasing its photocatalytic activity [57,58]. S-doped TiO_2 is also reported as a good candidate for visible light−driven catalysis with higher efficiency compared to undoped TiO_2 under UV irradiation. Hamadanian et al. [59] found 0.05% S-doped TiO_2 exhibit higher photocatalytic activity for methyl orange degradation compared to undoped TiO_2 under both visible and UV light. Furthermore, degradation of 98% of MB under sunlight was achieved on S-doped hollow TiO_2 nanoparticles with good reusability (up to five catalytic cycles) compared to only 30% achieved using standard commercial Degussa P25 as photocatalyst [60].

Additional energy levels introduced into the TiO_2 band gap can also be obtained by utilizing metal ions. Examples involve iron (Fe^{3+}/Fe^{2+}) [39], vanadium (V^{4+}/V^{3+}) [61], and tungsten (W^{6+}/W^{4+}) [32,62]. Piskunov et al. [39] reported the use of Fe-doped TiO_2 nanotubes for visible light−driven photocatalytic water splitting, where Fe^{2+}/Fe^{3+} serves as electron trap centers, while Fe^{3+}/Fe^{4+} serve as hole-trap centers (Fig. 4.9) [63]. Iron-doped

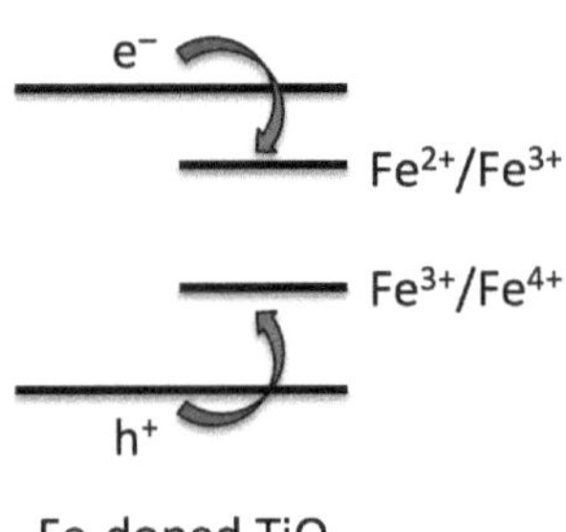

FIGURE 4.9 Electron and hole trapping mechanism on Fe-doped TiO_2 nanotubes. *Adapted from V.J. Babu, S. Vempati, T. Uyar, S. Ramakrishna, Review of one-dimensional and two-dimensional nanostructured materials for hydrogen generation, Phys. Chem. Chem. Phys. 17 (2015) 2960−2986 [63].*

TiO_2 prepared by hydrothermal method [64] has also shown enhanced catalytic activity for dye degradation in aqueous solution under UV and visible light. The V-doped TiO_2 showed intense absorption in the visible region where the V^{4+}/V^{5+} pair acts as electron and hole traps for photocatalytic water splitting [61]. Tungsten-doped TiO_2 nanowires (diameter ~50 nm) synthesized by electrospinning showed high performance in dye-sensitized solar cells (*vide infra*) where 2% W-TiO_2 gave 90% higher short-circuit current density compared with the undoped TiO_2 nanowires [62].

4.3.2 TiO_2 decoration

Since the first report by Grätzel et al. [65] extensive investigations have been carried out into development of thermally, photochemically stable, and visible light–responsive photocatalysts based on TiO_2 by incorporating small amount of different components. As mentioned before, the decoration of TiO_2 can be divided into three categories based on the material used for this purpose: semiconductors (including metal oxide nanoparticles), MNPs, and organic dyes, the latter also referred to as sensitized TiO_2.

The decoration using semiconductors can yield long-lived e–h pairs on TiO_2 due to the formation of a semiconductor–semiconductor heterojunction (Fig. 4.8). There is a plethora of semiconductor combinations that can improve the photocatalytic properties of the material. For instance, the proper band alignment between host–guest photoanode system $TiO_2/BiVO_4$ allows the charge transfer from one to another. Due to the small band gap of $BiVO_4$ (2.5 eV) and the good electron affinity of TiO_2, $BiVO_4$ serves as a visible light harvester and TiO_2 acts as electron acceptor. The VB energy level of $BiVO_4$ is higher than TiO_2 while the CB energy level of the TiO_2 is lower than the other semiconductor, resulting in charge separation and long-lived e–h pairs [66]. Additionally, the use of transition metal oxide nanoparticles such as copper oxides was recently explored as one method to inject electrons into the metal oxide nanoparticle upon excitation of the TiO_2 band gap and hence form Cu(I) species on demand [13]. This strategy has been very successful as a method to photocatalyze "click" chemistry with excellent spatial and temporal control (*vide infra*) [13].

Noble MNP are also used for decorating TiO_2 surfaces. The noble MNPs can be classified into two categories depending on their optical properties: plasmonic and nonplasmonic MNPs. The first group of MNPs, represents gold, silver, and copper NPs; while the second group can include, but is not restricted to, palladium, platinum, ruthenium, and iridium NPs, among others [25].

Plasmonic MNPs are known for their high efficiency of visible light harvesting mediated by the localized surface plasmon resonance (LSPR). LSPR is a collective oscillation of free CB electrons at the metal–dielectric interface upon light excitation [67,68]. LSPR is observed when the photon frequency of incident light matches the frequency of the free electrons in the

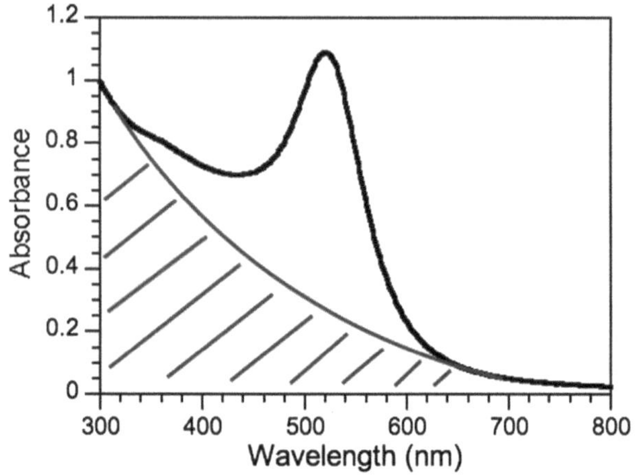

FIGURE 4.10 Extinction spectrum of AuNP showing (*black*) intraband absorption (LSPR) and (*striped area*) interband absorption.

metal, typically in the visible region (e.g., <500 nm for gold). LSPR depends not only on the size of the NP but also on the shape, chemical, and dielectric nature of both the metal and its surrounding [68]. As the particles grow bigger the absorption band broadens and shifts to longer wavelengths [69,70]. Also, the shape of NPs affect the plasmon band, for example, in the case of AgNPs the absorption band is shifted from 400 to 670 nm as the particle varies from a sphere to a cube [71], while AuNPs (that typically absorb around 530 nm) exhibit strong absorption band in the red or NIR region when they are in nanorod-like shape compared to the spherical form [72]. Thus, the response to light excitation varies from metal to metal; upon light excitation, excited electrons in the VB move to high energy levels through interband transitions; whereas, free electrons in the CB (sea or gas electrons) of plasmonic NPs react to incident light via LSPR effect (Fig. 4.10) [68]. Therefore, the contribution from both LSPR effect and interband transitions varies from metal to metal. In the case of metal NPs supported on the TiO_2 surface, the absorption of the new material can now reflect the absorption of the LSPR in the visible region and the contributions in the UV regions of both, MNP and the semiconductor. In some cases this is beneficial for the photocatalytic enhancement either in the UV or visible region.

The electromagnetic field arising from plasmon generation enhances the charge transfer process from TiO_2 to metal or vice versa. The presence of MNP on the surface of TiO_2 contributes to either direct or indirect photocatalysis by enhancing the TiO_2 catalytic performance. On the other hand, metal NPs supported on insulator particles such as SiO_2 (9.0 eV), Al_2O_3 (8.7 eV), or ZrO_2 (5.8 eV), prevent any photocatalytic interference from the support itself, and therefore, electron transfer between MNPs and the support is unlikely to take place. Since the support itself cannot absorb visible light, the MNPs function as light absorbers and photocatalytic active centers, by either charge injection into the reactants or electromagnetic filed enhancement [73]. The photocatalytic activity of supported MNPs is characterized by a relationship between quantum efficiency [11] and photon flux (light intensity of incident light) [71,74,75]. On the other hand, semiconductors like TiO_2 exhibit low quantum efficiency at higher light intensities [76]. In general, semiconductor photocatalysts show lower reaction rates at higher temperature, compared with semiconductors decorated with plasmonic MNPs, where rate and quantum efficiency of photocatalytic reactions increase with temperature [75,77]. For instance, AuNPs supported on TiO_2 (or CeO_2) showed good catalytic activity toward *azo*-compounds synthesis at 100°C or higher temperatures. The mechanism involves the reduction of nitrobenzene to aniline on the gold surface under H_2 atmosphere ($\sim$9 bar) followed by oxidation in the presence of O_2 (5 bar) at 100°C to produce the aromatic azo-compounds, widely used in the production of dyes, food additives and pharmaceutical products [78]. Another clear example is shown with the use of commercial TiO_2 for the ethylene epoxidation reaction, which showed maximum oxidation between 80°C and 300°C. A dramatic decrease of reaction yields was found at higher temperatures attributed to the nonradiative, multiphonon recombination of photogenerated charge carriers [79]. On the other hand, when the same reaction is catalyzed by silver ($Ag@Al_2O_3$) the rate of ethylene epoxidation increases with increasing light intensity at constant temperature (Fig. 4.11). The same behavior was observed with the increase of the temperature at constant light intensity [75]. Thus, plasmonic NPs, unlike semiconductor photocatalysts, are able to couple thermal energy and photonic energy sources to drive chemical reactions with high rates. The effect of both temperature and photons on the photocatalysis rate is explained by two main processes: (1) at high temperature the population of excited vibrational states of the adsorbate is larger, which means less energy from the conduction electrons on metal NPs is required to overcome the activation barrier and (2) it is known that when the molecule is in an excited vibrational state is more prone to gain photons [73,75].

The application of nonplasmonic MNP doped TiO_2 was first observed for photo-generation of H_2 using $Pt@TiO_2$ (*vide infra*) [80]. Upon UV excitation an electron migrates to the metal site (PtNP); thus e−h recombination process slows down, and as result, reduction reactions are favored by trapping the metal-supported electron, while photogenerated holes are free to diffuse to the semiconductor surface to induce oxidation processes [81]. The resulting material does not show strong absorbance in the visible region via the LSPR effect, but the interband transition in the visible and UV range contributes to the high efficiency toward catalyzing light-mediated organic reactions at ambient or moderate temperatures [82]. Sarina et al. proposed a full description of the mechanism of the photocatalytic reactions with

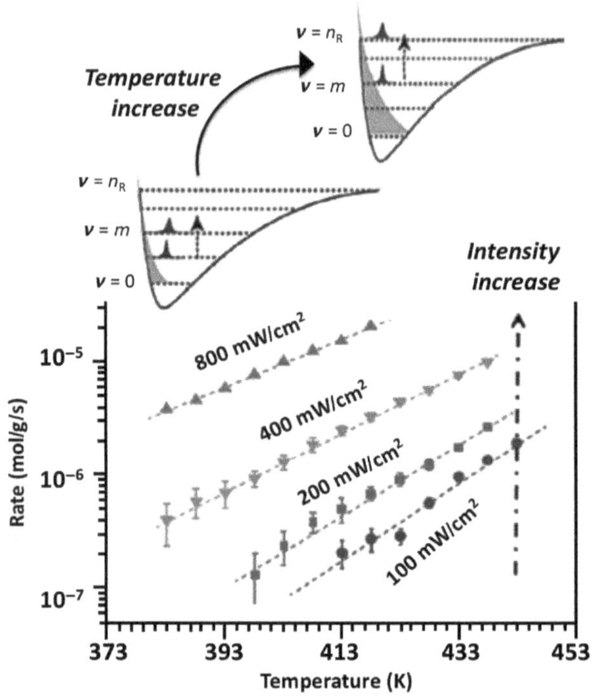

FIGURE 4.11 Reaction rate versus temperature for different intensities. At high temperature there is a higher occupancy of the excited vibrational states of the substrate and therefore higher probability for the reaction. *Reproduced from S. Sarina, E.R. Waclawik, H. Zhu, Photocatalysis on supported gold and silver nanoparticles under ultraviolet and visible light irradiation. Green. Chem. 15 (2013) 1814–1833, [73] with permission of The Royal Society of Chemistry.*

nonplasmonic MNP. Thus no electron transfer can occur in the dark at low temperatures for high physisorption or weak chemisorption, but in the case of strong chemisorption on the MNP, the interaction of reactant molecules with the metal *d* electrons influences the bonding and antibonding orbitals of the adsorbed molecules. The electrons can populate the antibonding orbitals via thermal or photoexcitation processes, facilitating the reaction of the adsorbed molecules. The light intensity has a direct impact on the number of excited electrons, whereas the light wavelength selectively excites electrons to different energy levels. A direct consequence of this implies that different reagents can show different reactivity toward the same reaction conditions [83].

Despite the very well-known reactivity of the MNP-based catalysts, the goal now is to improve the catalytic activity of TiO_2 minimizing the use of expensive and rare metals, such as Ru, Ir, Pd, or Au, by developing bimetallic systems using earth-abundant and less expensive elements such as, Ni, Co, and Cu [84]. Although not covered in great detail in this contribution, it is important to note that particle size and bonding of surface atoms play an important role in determining catalytic activity [85]. It has been shown that bimetallic catalysts can exhibit higher catalytic activity and selectivity than their monometallic counterparts. Thus size, shape, and composition of a bimetallic system can play a critical role into designing catalytic systems showing high activity and selectivity [86].

Additionally, as an approach for developing visible light–responsive TiO_2 photocatalysts, wide selections of photosensitive dyes were loaded (adsorbed) on the TiO_2 surface and their catalytic activity investigated. The major drawbacks of the resulting material are either thermal and/or photochemical instability [65,87].

4.4 TiO_2 in solar cells

As discussed previously, changes on the absorption profile of TiO_2 are desirable in order to facilitate efficient visible light harvesting. Indeed, this strategy has been used in the development of the dye-sensitized solar cells (DSSC), or Grätzel solar cell [5], in which the dye absorbs the light while the semiconductor (TiO_2) is involved in the charge carrier transport. Thus the system works as a conventional alkaline battery where there is an anode electrode (TiO_2), a counter-electrode and an electrolyte solution. Arguably, stability is one of the main challenges that the use of DSSCs faces nowadays. There is a plethora of different combinations of dyes [88], transition metal complexes, and TiO_2 morphologies that have been explored since the first DSSC was designed by Grätzel and coworkers [5]. The principle of operation of the Grätzel cell is described in Fig. 4.12. The schematic representation shows the four main components of the cell: (1) photoanode deposited on a transparent conductive glass substrate, (2) a monolayer of the dye, (3) an electrolyte containing a redox couple, and (4) a counter-electrode usually made of a platinum-coated conductive glass. The light can be absorbed through the transparent conducting oxide, exciting the dye. The sensitizer can then inject electrons into the TiO_2 CB. The dye is regenerated by the electrolyte solution, which in turn can be regenerated in the counter-electrode by the electrons passing through the load.

Besides solar energy harvesting the use of sensitizers to drive chemical transformations under visible light has been used for more than three decades [89].

4.5 Photocatalytic activity of decorated TiO_2 as an aid for organic transformations

This section deals with representative organic transformations and is subdivided accordingly. In many cases the use of decorated TiO_2 reflects replacing unrecoverable homogeneous catalysts with heterogeneous ones that can routinely be recovered and frequently reused. Although many applications of TiO_2 involve photocatalysis and photoelectrochemical reactions, there are

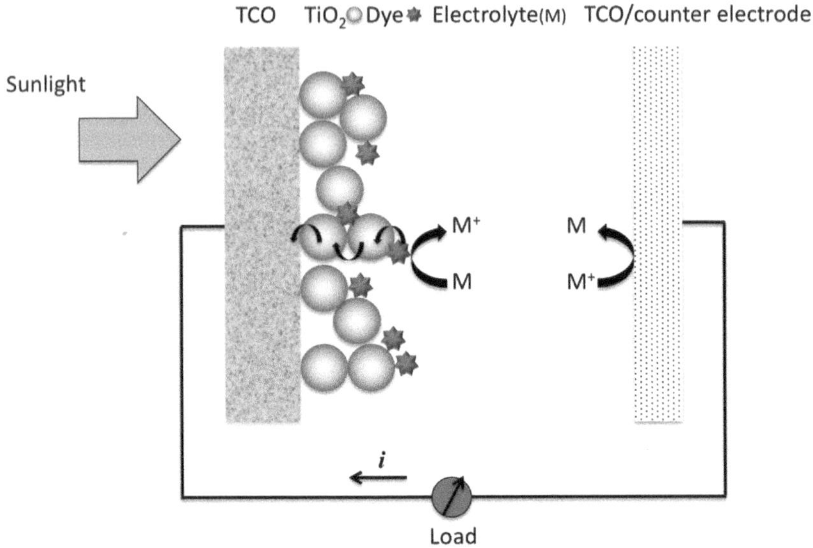

FIGURE 4.12 Schematic illustration of the Grätzel cell mechanism.

numerous interesting discoveries of TiO_2 for other reactions. Recent development of TiO_2 as a catalyst in green organic synthesis includes hydrodeoxygenation, hydrogenation, esterification/transesterification, the water–gas shift reaction, and visible light–induced organic transformation. All this is possible due to its strong metal-support interaction (SMSI), high chemical stability, acidity, and high redox reaction at low temperature. The relationship between the catalytic performance and different metal or metal oxide dopants, and different polymorphs of TiO_2 are discussed in detail. It is interesting to note that the reduction temperature and addition of promoters have a significant effect on the catalytic performance of TiO_2 [90].

4.5.1 TiO_2-mediated photoreductions

The reduction of nitro-aromatic compounds has been widely studied under both thermal and photocatalytic conditions. The generally accepted reaction mechanism for the catalytic reduction of aromatic nitro compounds is based on the electrochemical model presented by Haber (Scheme 4.1) [91]. Thus the direct route (left pathway in Scheme 4.1) proposes the reduction of nitro compound to the nitroso compound and finally the hydroxylamine compound. These two first steps are believed to be fast and the final reduction to aniline is the slowest reaction step. The condensation route involves the condensation of the nitroso compound with a molecule of hydroxylamine to form the azoxy compound, which is subsequently reduced to azo, hydrazo, and aniline compounds (right pathway in Scheme 4.1). There are many examples of different reaction conditions that have been explored in order to either favor the formation of the aniline or the azo compounds during the reduction of nitro compounds. Thermal and photoinduced catalysis of the reduction of nitro compounds involving supported gold nanostructures on different semiconductors have been subject of many studies [92–94], showing formation of aniline or azo compounds can easily be driven. The photochemical approach was also used to synthesize azo polymers under mild conditions [93].

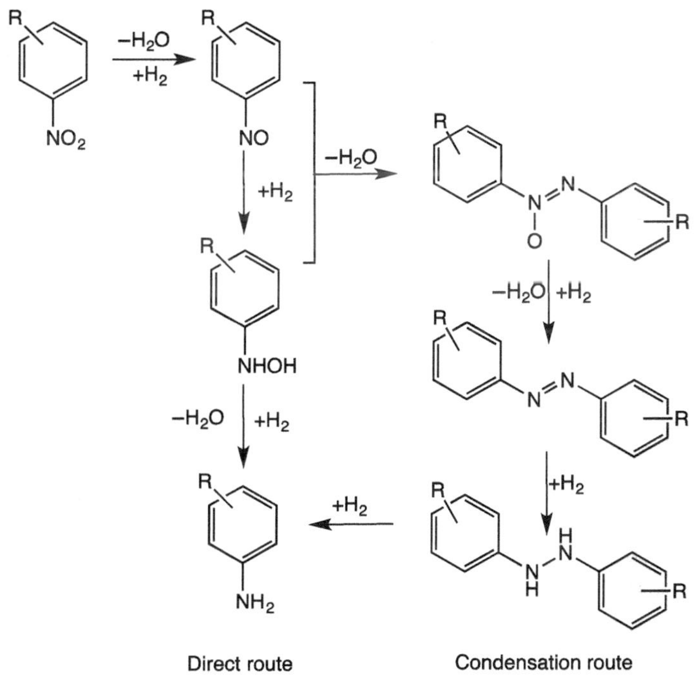

SCHEME 4.1 Generally accepted reaction mechanism for the catalytic reduction of aromatic nitro compounds based on the electrochemical model presented by Haber. [91].

Interestingly, Au@TiO_2 shows remarkable chemoselectivity, and in the reduction of *m*-nitrostyrene, only the reduction of the nitro group is strongly favored (Scheme 4.2) [95]. Other studies [96] have noted the importance of particle size and morphology in controlling the reaction outcome [97].

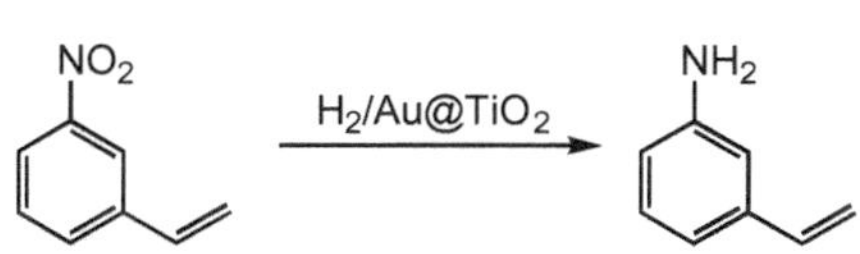

SCHEME 4.2 Selective reduction of m-nitrostyrene to the corresponding aniline using Au@TiO_2.

Additionally, photocatalytic hydrogenation of alkenes to alkanes can be easily promoted in the absence of H_2 utilizing alcoholic suspensions of Pd@TiO_2 as shown in Scheme 4.3 [98]. The suggested mechanism involves the solvent oxidation and the in situ formation of H_2, which is believed to form Pd-hydride species on the nanoparticle's surface [99].

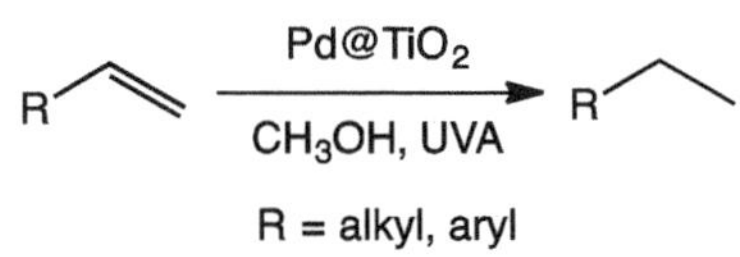

SCHEME 4.3 Photocatalytic hydrogenation of alkenes to alkanes using alcoholic suspensions as H-sources and Pd@TiO_2 photocatalyst.

Another interesting redox reaction involves the reduction of carbon dioxide, which due to its great stability, its conversion to based fuels usually requires high-energy input. The photochemical reduction of CO_2 using solar light as a means of alternative fuel production is an effective way of reducing carbon dioxide in the atmosphere. There are several reports utilizing different semiconductors, particularly TiO_2. An interesting report by Ishitani et al. [100] suggests that TiO_2 can reduce CO_2 mainly to CO (with small amounts of CH_4), a 2-electron process. In contrast the use of Pd-decorated TiO_2 can generate CH_4 (8-electron process) as a major product and a considerable amount of C_2H_6.

There are many factors that can affect CO_2 photoreduction including type of reductant, temperature, pressure, and particle size. The most popular reducing agents tested are H_2O, NaOH, and C_3H_7OH. To use CO_2 photoreduction as an economically feasible technology, it is very important to utilize abundant hydrogen sources, such as water. The main problem of using water as reducing agent is the low solubility of CO_2 in water and the competition between CO_2 reduction and hydrogen generation processes (*vide infra*). CO_2 photoreduction is usually carried out at room temperature as solubility decreases with increasing temperature. Additionally, product selectivity is usually favored at high CO_2 concentrations (high pressure) in aqueous media [101].

Additionally, engineered combinations of different semiconductors having different band-edge potentials are an effective means of improving charge separation efficiency and photocatalytic activity, and proved to be efficient for CO generation [102].

4.5.2 TiO_2-mediated photooxidations

A review by Augugliaro et al. [103] provides the reader with a general overview on heterogeneous photocatalytic oxidation mechanisms in the presence of TiO_2, specially dedicated to aliphatic and aromatic organic species. As already mentioned, TiO_2 can form radical species upon excitation is shown in the following Eqs. (4.8)–(4.13):

$$TiO_2 + h\nu \rightarrow TiO_2(e^-_{(CB)} + h^+_{(VB)}) \tag{4.8}$$

$$HO^- + h^+_{(VB)} \rightarrow HO^\bullet \tag{4.9}$$

$$O_2 + e^-_{(CB)} \rightarrow O_2^{\bullet -} \tag{4.10}$$

$$O_2^{\bullet -} + H^+ \rightarrow HO_2^\bullet \tag{4.11}$$

$$2HO_2^\bullet \rightarrow O_2 + H_2O_2 \tag{4.12}$$

$$H_2O_2 + O_2^{\bullet -} \rightarrow HO^- + HO^\bullet + O_2 \tag{4.13}$$

The oxygenation of alkanes is believed to initiate after activation of the photocatalyst via reaction (4.14a) or (4.14b):

$$RCH_3 + h^+_{(VB)} \rightarrow {}^\bullet RCH_2 + H^+ \tag{4.14a}$$

$$RCH_3 + HO^\bullet \rightarrow {}^\bullet RCH_2 + H_2O \tag{4.14b}$$

There are several paths of reaction for the alkyl radical, among them reactions (4.15)–(4.19) are possible:

$${}^\bullet RCH_2 + HO^\bullet \rightarrow RCH_2OH \tag{4.15}$$

$${}^\bullet RCH_2 + O_2 \rightarrow RCH_2OO^\bullet \tag{4.16}$$

$$RCH_2OO^\bullet + RCH_3 \rightarrow RCH_2OOH + {}^\bullet RCH_2 \tag{4.17}$$

$$RCH_2OOH + e^-_{(CB)} \rightarrow RCH_2O^\bullet + HO^- \tag{4.18}$$

$$RCH_2O^\bullet + RCH_3 \rightarrow RCH_2OH + {}^\bullet RCH_2 \tag{4.19}$$

According to the authors from Ref. [103], subsequent oxidation of the alcohol to the corresponding aldehyde and acid are also possible through the following reactions:

$$RCH_2OH + HO^\bullet \rightarrow RCH_2O^\bullet + H_2O \tag{4.20}$$

$$RCH_2O^\bullet + HO^\bullet \rightarrow RCHO + H_2O \tag{4.21}$$

$$RCHO + HO^\bullet \rightarrow R^\bullet CHOOH \tag{4.22}$$

$$R^\bullet CHOOH + HO^\bullet \rightarrow RCOOH + H_2O \tag{4.23}$$

Considering for instance the simplest hydrocarbon, methane, this can generate methanol, which is the precursor of other high-value molecule such as formaldehyde and formic acid. The synthesis of methanol from CH_4 can be performed on the surface of different semiconductors following the reactions described previously.

This mechanism has been studied by Gondal et al. [104] on the surface of different semiconductors including rutile TiO_2. Since all the reactions were performed at room temperature, the methanol produced remains in the solution and as the reaction proceeds, the concentration of methanol increases competing with water for VB holes. In other words, methanol can be oxidized under the conditions is being formed leading to the formation of formaldehyde and additional hydrogen.

It is clear that the basic requirement for a semiconductor catalyst to convert methane into methanol is its ability to oxidize the solvent (i.e., water) to produce hydroxyl radicals [104]. Examples of using better oxidants such as H_2O_2 can also be found in literature [105]. Hydrogen peroxide is known to decompose on the TiO_2 surface; some of the proposed [106] decomposition pathways are shown in reactions (4.24) and (4.25).

$$H_2O_2 + {}^{\bullet}OH/h^+ \rightarrow HO_2^{\bullet} + H_2O/H^+ \tag{4.24}$$

$$HO_2^{\bullet} + h^+ \rightarrow O_2 + H^+ \tag{4.25}$$

According to Masik et al. [107] for reactions requiring a higher selectivity, the use of rutile TiO_2 should be considered. The oxygen adsorption at the surface of rutile TiO_2 is significantly better than in anatase TiO_2, and consequently reactions that require singlet oxygen and superoxide radical formation are favored. Indeed, higher selectivity toward aldehyde formation during alcohol oxidations has been reported [108,109].

TiO_2 and its modified derivatives have been widely used in work aimed at the oxidative destruction of water contaminants. The details of this extensive work are beyond the scope of this chapter and the reader is referred to the many reviews and feature articles on the topic [110,111]. Beyond the oxidative aspects that were illustrated in Fig. 4.6B, recent isotopic studies show the direct chemical involvement of oxygen atoms (and related vacancies) in TiO_2-mediated photooxidations [112]. Thus the incorporation of ^{18}O into CO_2 molecules, formed during the photocatalytic mineralization process of benzene, derives from the $Ti^{18}O_2$ surface lattice oxygen ions.

An interesting approach combining both oxidation and reduction properties of Pd@TiO_2 has been studied for the synthesis of secondary amines utilizing the in situ reduction of azobenzenes and oxidation of alcohols (Fig. 4.13) [113]. Thus aldehydes can be formed by oxidation of the corresponding alcohol, whereas the azobenzene is concurrently reduced to aniline on the TiO_2 surface. By condensation of these two precursors an imine is formed, to undergo photocatalytic hydrogenation and furnish the secondary amine formation.

4.5.3 C–C couplings

C–C coupling reactions are among the most important strategies in synthetic organic chemistry, and heterogeneous catalysts are preferred choices for

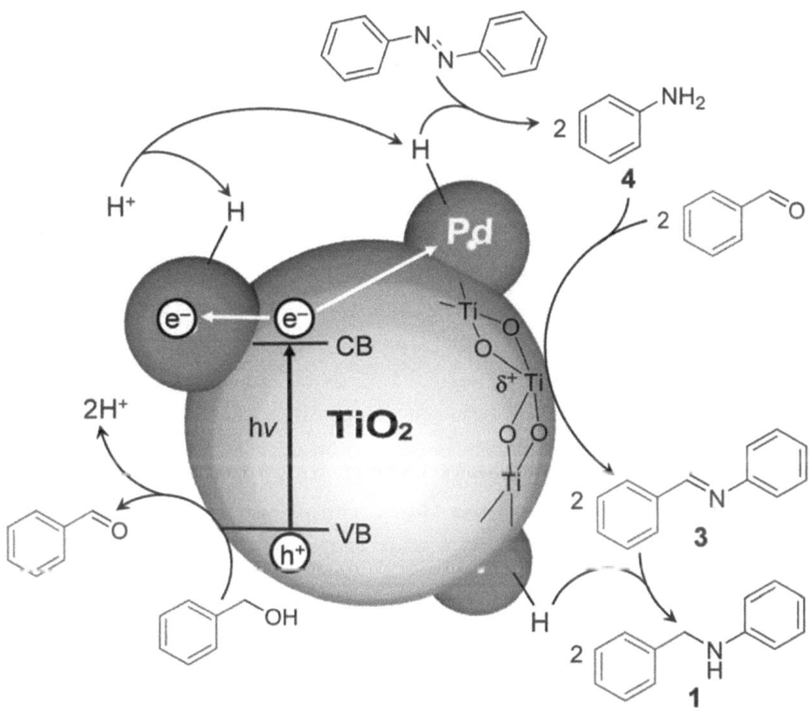

FIGURE 4.13 Proposed mechanism for secondary amine production from alcohol and azobenzene on a Pd@TiO_2 catalyst under UV irradiation. *Reproduced from J.F. Montoya, I. Ivanova, R. Dillert, D.W. Bahnemann, P. Salvador, J. Peral, Catalytic role of surface oxygens in TiO_2 photooxidation reactions: aqueous benzene photooxidation with $Ti^{18}O_2$ under anaerobic conditions, J. Phys. Chem. Lett. 4 (2013) 1415–1422 [112] with permission of The Royal Society of Chemistry.*

thermal and photoinduced processes. There are examples utilizing TiO_2 as a catalyst support for thermally-catalyzed C–C couplings that are not covered in this chapter.

The formation of carbon-centered radicals from alkanes on the surface of a semiconductor was described in the previous section. These radicals are versatile as reactive intermediates, especially for C–C bond formation. Compelling reports dealing with the use of photocatalysis for the formation of the C–C bond can be found in Refs. [114–117].

The following examples we use to illustrate C–C couplings include one where the reaction is dominated by free radical recombination and others with a more surface-specific coupling. Gold nanoparticles supported on TiO_2 (Au@TiO_2) are excellent photocatalysts for photoinduced reductive C–C couplings. Thus surface plasmon excitation of supported AuNPs in the presence of an amine leads to the C–C coupling of a variety of substituted benzyl bromides at room temperature with good yields in a free radical–mediated reaction. The overall efficiency of the C–C coupling is largely dependent on the nature of the amine used and requires solvents that are rather inert toward free radicals such as dichloromethane [118]. As many radical reactions the yield of the dimerization product depends on the fast accumulation of the carbon-centered radical, benzyl radical in this case, which can be modulated by changing the irradiation photon flux (Scheme 4.4).

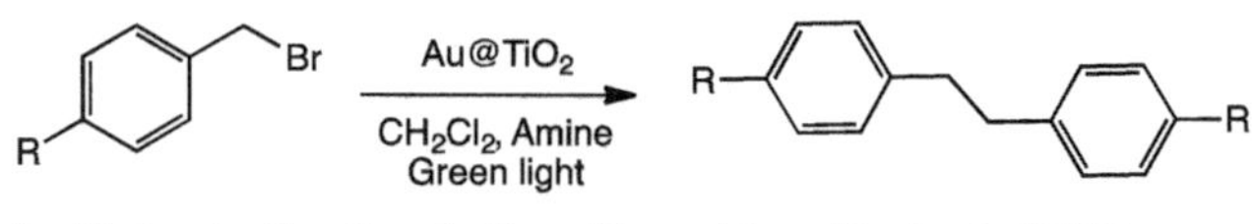

SCHEME 4.4 Photoreductive dimerization of benzyl bromides by Au@TiO_2.

The Sonogashira reaction can easily be photocatalyzed by supported palladium nanoparticles. Thus direct photoexcitation of PdNPs can catalyze the C–C coupling between different aryl iodides and acetylenes under very mild conditions in short reaction times [119]. The catalyst is air- and moisture-tolerant and can be supported on a wide range of materials, including inert ones such as nanodiamonds, as well as the central subject of this chapter, TiO_2. Study of the action spectrum demonstrates that direct excitation of the PdNPs is required, such as in the case of Pd@TiO_2, where visible excitation works well whereas UVA (368 nm) irradiation is ineffective because of TiO_2 blocks light absorption by the PdNP. The catalyst can be reused a couple of times, and if it loses activity, it can be readily reactivated by a simple reductive photochemical strategy (Scheme 4.5).

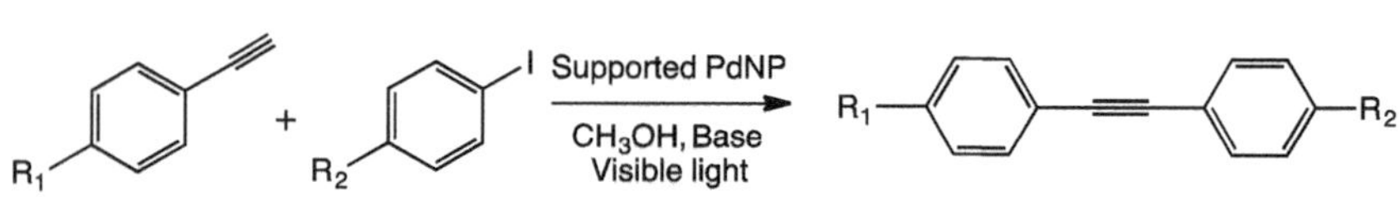

SCHEME 4.5 Sonogashira reaction photocatalyzed by Pd@TiO_2.

More interestingly the formation of methyl formate from the photocatalytic coupling of chemisorbed formaldehyde with chemisorbed methoxy species contributes to the photocatalytic oxidation of methanol [120]. The suggested mechanism involves reactions 4.26–4.28 (O_{br}: bridging-bonded O sites of the TiO_2 surface):

$$CH_3OH + O_{br} + h\nu \rightarrow CH_3O^{\bullet} + HO^{\bullet}_{br} \quad (4.26)$$

$$CH_3O^{\bullet} + O_{br} + h\nu \rightarrow HCHO + HO^{\bullet}_{br} \quad (4.27)$$

$$CH_3O^{\bullet} + HCHO + O_{br} + h\nu \rightarrow HCOOCH_3 + HO^{\bullet}_{br} \quad (4.28)$$

More recently the synthesis of 3-butanediol was described through TiO_2-photocatalytic C–C coupling of bioethanol concurrently with the formation of

H_2 [121]. More complex reactions can be driven by TiO_2 under irradiation including multicomponent reaction for the formation of α-amino amides [122].

4.5.4 Cyclizations

Among the cyclizations catalyzed by TiO_2 the click reaction where an alkyne and an azide react to form a heterocyclic five-membered ring has been of particular interest as building blocks for pharmaceutical compounds. Illustrated in Scheme 4.6, the process requires Cu(I) and thus it is not catalyzed by the CuO nanoparticles supported on TiO_2. However, upon photoexcitation of the material an electron from the CB of TiO_2 is donated to CuO, thus generating a Cu(I) site. This results in efficient photoinduced TiO_2 catalysis [13].

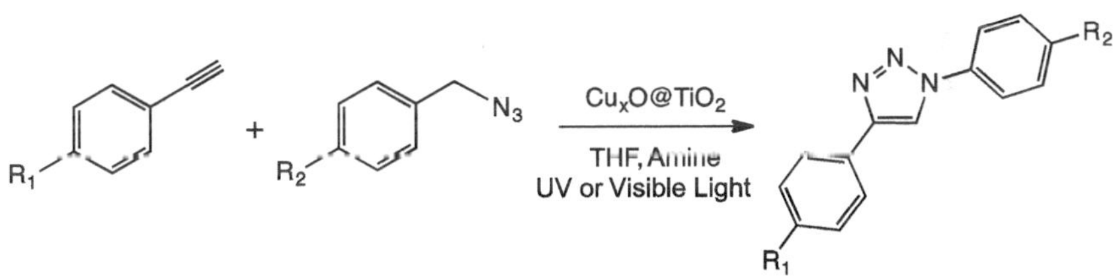

SCHEME 4.6 Photocatalytic heterogeneous click chemistry using Cu-decorated TiO_2.

TiO_2-photocatalytic cyclization reactions can also achieve the formation of strained carbon rings for the syntheses of aryltetralones. Wang et al. have demonstrated that conventional over oxidation can be overcome utilizing radical cations with sufficient reactivity toward nucleophilic addition [123].

Even more complex cyclization, usually performed through transition metal complexes photocatalysis, can be attained using nickel(II) oxide-modified TiO_2 catalyst (Scheme 4.7) [124].

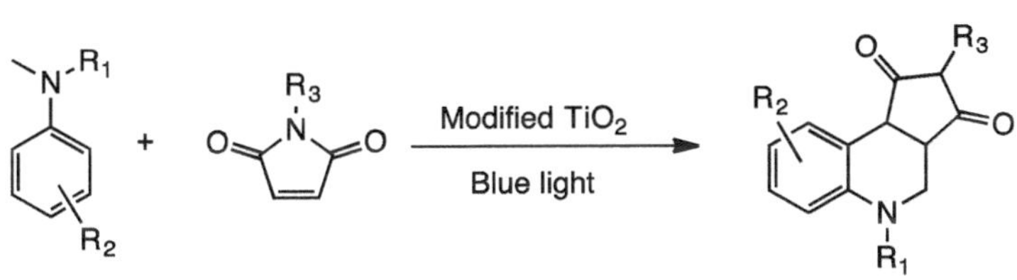

SCHEME 4.7 Photocatalytic cyclization of tertiary anilines with maleimides using Ni-decorated TiO_2.

4.5.5 TiO_2 catalysts for hydrogen generation

Perhaps the most obvious way to generate hydrogen fuel is simple water splitting into H_2 and O_2 [125]. Yet, over four decades since Fujishima and

Honda [19] showed the feasibility of this process, it remains very inefficient and under most conditions requiring use of the UV component of the solar spectrum. Numerous researchers have employed sacrificial electron donors (SED), such as alcohols, amines, or formic acid to trap the hole produced upon TiO_2 excitation in the hope of understanding the hydrogen generating path, formally half of the water split reaction. Unfortunately, electron scavenging and hole trapping are really inseparable processes and artificially trapping the hole teaches little about the overall process, even if hydrogen yields are dramatically increased [23]. The approach has been severely criticized as one that provides no mechanistic information [20]. The question is whether there is value in the SED approach, even if it is not true water splitting. Other than mechanistic criticisms (because it does not help understand true water splitting), the only valid objection to the use of SED is that valuable chemicals are consumed as raw materials to generate H_2. Usually the efficiency of the H_2 generation is correlated to the presence of hydroxyl groups in the SED, which are sometimes regarded as essential [126]. Other SED have been tested over the past 30 years, among them glucose as model for biomass [126]. Different reactivity was found when α-D- and β-D-glucose were used as SED for the photocatalytic formation of H_2 over Pt@TiO_2. According to Li et al. [127], the low reactivity of β-D-glucose is connected to its lower adsorption on the surface of TiO_2, which emphasizes the importance of the interaction of the substrates with the photocatalyst surface. Therefore the presence of functionalities, such as –OH or C=C are believed to help the adsorption on the surface. Furthermore, the use of cellulose as SED has been recently explored upon immobilization onto a Pt@TiO_2 photocatalyst [128].

The use of glycerol as sacrificial agent for H_2 production was also explored. Glycerol is a by-product of the biodiesel industry and considered as a waste, therefore its use as a SED may reduce the cost of its disposal [129]. The idea of using waste or organic pollutants as SED was tested many years ago by Sakata et al. using hydrocarbons and fossil fuels as SED. Particularly, phenol presented 12 times more efficiency toward H_2 generation compared to pure water utilizing Pt@TiO_2 [130]. The mineralization of other organic pollutants—4-chlorophenol—concomitant to the formation of hydrogen was also studied over fluoride modified Pt@TiO_2 photocatalysts. The treatment of industrial wastes was also explored although the high content of anions tends to inhibit the H_2 formation via the competitive adsorption on the photocatalyst surface [131]. Degradation of dyes as common contaminants was extensively explored over TiO_2-based photocatalysts [132], although it is hard to find examples with the associated formation of H_2. This approach has been recently criticized [18,133] as it is an inappropriate method to evaluate the photocatalytic activity of novel photocatalysts. As mentioned before, dye-sensitization can increase the visible light-harvesting properties of the new material and therefore can no longer be considered a model contaminant to be used for testing a novel photocatalyst. Despite all this, it has been proposed that normal organic contaminants in water (such as river water) may serve as SED capable of generating hydrogen. Beyond using SED with zero value, there is value added in their removal. This strategy is capable of killing bacteria while reducing the level or organic contaminants, and simultaneously generating hydrogen [22,23].

Beside the selection of a good SED, the role of the cocatalyst is also important. Photocatalysts need to be stable and harvest sunlight efficiently. As mentioned earlier, the challenge is to replace scarce and precious materials, such as Pt or Pd, for earth-abundant materials in order to facilitate scale-up synthesis and industrial applications.

4.6 Sunscreens based on TiO_2

Although sometimes described as a societal experiment [134], TiO_2 has a niche in the world of cosmetics, where it is widely used as a sun block [135], frequently replacing organic active ingredients. The main advantages of using inorganic sun blockers rely on their high UV absorption and great photostability, comparing to organic sunscreens. Particularly, TiO_2 shows great absorption in the UVB region whereas ZnO is better absorber in the UVA region. Therefore combinations of TiO_2 and ZnO are frequently used, in order to increase the UVB and UVA protection. However, microsized TiO_2 and ZnO absorption properties are accompanied by high refraction indexes [136,137] across the UV and visible spectral region. This produces an unwanted effect in the final appearance of the cosmetic products, known as whitening effect. In order to solve this, the size of semiconductors' particles has been decreased to the nanoscale level (<100 nm). This approach has raised many concerns as not only the UV protection but also the toxicity of the material can drastically change at the nanoscale level [138].

There are few reports on the toxicity of inorganic ingredients such as TiO_2 and ZnO for sunscreen applications [135,139,140]. Although there are some concerns about skin penetration of particles, particularly for nanosized TiO_2 [141], this is not regarded as a significant health issue [142]. According to Pavel and Smijs [138] cyto- and genotoxicity of TiO_2 and ZnO NPs is often associated with their photocatalytic activity, and the concomitant formation of free radical species, such as ROS [143]. ROS could penetrate the skin layers easily [144] and produce cellular damage [145], they also diminish the lifetime of the cosmetic formulation due to photodegradation of other emulsion components, including organic sun blockers [146]. Attempts to stop TiO_2 from generating free radicals under conditions of exposure to sunlight, oxygen, and water have been generally unsuccessful. This is hardly surprising, as we reach the final items in this contribution, it should be clear that the band gap and absorption properties of TiO_2 are ideally set up for it to generate free radicals. The use of passivation coatings has been implemented in order to reduce the photocatalytic activity of TiO_2, usually involving silica or alumina coatings, with different degrees of success [147–149]. More recently, Gao et al. showed that hierarchical composites formed with TiO_2 and ZnO can show higher solar protection factors whereas lowering the photodegradation of the formulation [150].

In a very pragmatic approach we decided to coat TiO_2 with a very thin ($\sim$3 nm) layer of lignin, a tree-derived natural material with antioxidant properties. While TiO_2 no doubt continues to make free radicals, they are quantitatively trapped by the lignin shell, and none escape to produce damage to biomolecules [143]. Few more examples utilizing biomolecules were also explored, such as the use of chitosan (natural polymer) [151] or morin (plant flavonoid) to encapsulate TiO_2 [152]. Notice that the cytotoxicity of the particles should not only be evaluated versus human skin cells, but also against cutaneous bacteria, which is the first skin barrier. Cosmetic ingredients have already demonstrated a negative impact in the development of the human microbiota inducing skin disorders [148,153,154].

Additionally, it is important to notice that using inorganic particles as sunscreens also challenges the formulation of the final product. Prior to commercialization the cosmetic emulsion must show robustness in terms of UV protection, appearance, and particle stability. Therefore interaction of TiO_2 particles with the emulsifiers present in the final formulation have to be tuned, usually adding a second coating that improves dispersion properties [155].

Lastly, the effect of the sunscreens in the marine life has raised many concerns in the last few years, as it has been proof that some organic sun blockers can kill coral reefs [156]. With the insertion of inorganic nanoparticles, new bioassays should be developed taking into account that not only the inorganic but also the organic compounds present in the sunscreen formulation may contribute to the formation of ROS, and their hazards for marine life [157].

Overall the development of new sunscreens needs to be accompanied by conscious analyses of potential risks including stabilization of the NPs in the cosmetic formulation, chronic exposures, and the reduction of the photoproduction of free radicals. There is a plethora of material combinations that are yet to be explored, and very little is known about their hazards upon commercialization. Regulations are still in their infancy, with no clear recommendations but enough to suggest developers "*... to continue monitoring and collecting new safety data as it emerges*" [158].

Acknowledgments

The various citations of work at the University of Ottawa have been supported for many years by the Natural Sciences and Engineering Research Council of Canada, by the Canada Research Chairs program and by the Canada Foundation for Innovation. The work cited would not have been possible without the contributions from present and past coworkers who are gratefully thanked for their experimental and intellectual contribution to our group's research.

References

[1] J.H. Braun, A. Baidins, R.E. Marganski, TiO_2 pigment technology: a review, Prog. Org. Coat. 20 (1992) 105–138.

[2] H.T. Yu, L. Song, Y.F. Hao, N. Lu, X. Quan, S. Chen, et al., Fabrication of pilot-scale photocatalytic disinfection device by installing TiO_2 coated helical support into UV annular reactor for strengthening sterilization, Chem. Eng. J. 283 (2016) 1506–1513.

[3] S.Y. Kwak, S.H. Kim, S.S. Kim, Hybrid organic/inorganic reverse osmosis (RO) membrane for bactericidal anti-fouling. 1. Preparation and characterization of TiO_2 nanoparticle self-assembled aromatic polyamide thin-film-composite (TFC) membrane, Environ. Sci. Technol. 35 (2001) 2388–2394.

[4] E.I. Cedillo-Gonzalez, R. Ricco, M. Montorsi, M. Montorsi, P. Falcaro, C. Siligardi, Self-cleaning glass prepared from a commercial TiO_2 nano-dispersion and its photocatalytic performance under common anthropogenic and atmospheric factors, Build Environ. 71 (2014) 7–14.

[5] B. O'Regan, M. Grätzel, A low-cost, high-efficiency solar cell based on dye-sensitized colloidal TiO2 films, Nature 353 (1991) 737–740.

[6] C. Berkey, K. Biniek, R.H. Dauskardt, Screening sunscreens: protecting the biomechanical barrier function of skin from solar ultraviolet radiation damage, Int. J. Cosmetic Sci. 39 (2017) 269–274.

[7] I. de la Calle, M. Menta, M. Klein, F. Seby, Screening of TiO_2 and Au nanoparticles in cosmetics and determination of elemental impurities by multiple techniques (DLS, SP-ICP-MS, ICP-MS and ICP-OES), Talanta 171 (2017) 291–306.

[8] I. Garcia-Fernandez, I. Fernandez-Calderero, M.I. Polo-Lopez, P. Fernandez-Ibanez, Disinfection of urban effluents using solar TiO_2 photocatalysis: a study of significance of dissolved oxygen, temperature, type of microorganism and water matrix, Catal. Today 240 (2015) 30–38.

[9] S. Sreeja, K.V. Shetty, Photocatalytic water disinfection under solar irradiation by Ag@TiO_2 core-shell structured nanoparticles, Sol Energy 157 (2017) 236–243.

[10] D. Venieri, A. Fraggedaki, M. Kostadima, E. Chatzisymeon, V. Binas, A. Zachopoulos, et al., Solar light and metal-doped TiO_2 to eliminate water-transmitted bacterial pathogens: photocatalyst characterization and disinfection performance, Appl. Catal. B 154 (2014) 93–101.

[11] N.J. Turro, V. Ramamurthy, J.C. Scaiano, Modern Molecular Photochemistry of Organic Molecules, University Science Publishers, New York, 2010, p. 1100.

[12] S.M. Sze, K.N. Kwok, Physics and Semiconductor Devices, John Wiley & Sons, Inc, New Jersey, 2007.

[13] B.W. Wang, J. Durantini, J. Nie, A.E. Lanterna, J.C. Scaiano, Heterogeneous photocatalytic click chemistry, J. Am. Chem. Soc. 138 (2016) 13127–13130.

[14] S.P. Pitre, T.P. Yoon, J.C. Scaiano, Titanium dioxide visible light photocatalysis: surface association enables photocatalysis with visible light irradiation, Chem. Comm. 53 (2017) 4335–4338.

[15] K. Hashimoto, H. Irie, A. Fujishima, TiO_2 photocatalysis: a historical overview and future prospects, Jpn. J. Appl. Phys., Part 1 44 (2005) 8269–8285.

[16] A. Ajmal, I. Majeed, R.N. Malik, H. Idriss, M.A. Nadeem, Principles and mechanisms of photocatalytic dye degradation on TiO_2 based photocatalysts: a comparative overview, RSC Adv. 4 (2014) 37003–37026.

[17] S. Naskar, S.A. Pillay, M. Chanda, Photocatalytic degradation of organic dyes in aqueous solution with TiO_2 nanoparticles immobilized on foamed polyethylene sheet, J. Photochem. Photobiol. A: Chem. 113 (1998) 257–264.

[18] N. Barbero, D. Vione, Why dyes should not be used to test the photocatalytic activity of semiconductor oxides, Environ. Sci. Technol. 50 (2016) 2130–2131.

[19] A. Fujishima, K. Honda, Electrochemical photolysis of water at a semiconductor electrode, Nature 238 (1972) 37–38.

[20] J. Schneider, D.W. Bahnemann, Undesired role of sacrificial reagents in photocatalysis, J. Phys. Chem. Lett. 4 (2013) 3479–3483.

[21] M.V. Pavliuk, A.B. Fernandes, M. Abdellah, D.L.A. Fernandes, C.O. Machado, I. Rocha, et al., Nano-hybrid plasmonic photocatalyst for hydrogen production at 20% efficiency, Sci. Rep. 7 (2017) 8670.

[22] A.E. Lanterna, J.C. Scaiano, Photoinduced hydrogen fuel production and water decontamination technologies. Orthogonal strategies with a parallel future? ACS Energy Lett. 2 (2017) 1909–1910.

[23] A.S. Hainer, J.S. Hodgins, V. Sandre, M. Vallieres, A.E. Lanterna, J.C. Scaiano, Photocatalytic hydrogen generation using metal-decorated TiO_2: sacrificial donors vs true water splitting, ACS Energy Lett. (2018) 542–545.

[24] V. Etacheri, C. Di Valentin, J. Schneider, D. Bahnemann, S.C. Pillai, Visible-light activation of TiO_2 photocatalysts: advances in theory and experiments, J. Photochem. Photobiol., C: Photochem. Rev. 25 (2015) 1–29.

[25] J. Schneider, M. Matsuoka, M. Takeuchi, J. Zhang, Y. Horiuchi, M. Anpo, et al., Understanding TiO_2 photocatalysis: mechanisms and materials, Chem. Rev. 114 (2014) 9919–9986.

[26] L. Song, H. Bin Yang, X. Wang, S.Y. Khoo, C.C. Wong, X.W. Liu, et al., Improved utilization of photogenerated charge using fluorine-doped TiO_2 hollow spheres scattering layer in dye-sensitized solar cells, ACS Appl. Mater. Inter. 4 (2012) 3712–3717.

[27] X. Wu, G.Q. Lu, L.Z. Wang, Dual-functional upconverter-doped TiO_2 hollow shells for light scattering and near-infrared sunlight harvesting in dye-sensitized solar cells, Adv. Energy Mater. 3 (2013) 704–707.

[28] S.P. Albu, A. Ghicov, J.M. Macak, R. Hahn, P. Schmuki, Self-organized, free-standing TiO_2 nanotube membrane for flow-through photocatalytic applications, Nano Lett. 7 (2007) 1286–1289.

[29] H.B. Wu, H.H. Hng, X.W. Lou, Direct synthesis of anatase TiO_2 nanowires with enhanced photocatalytic activity, Adv. Mater. 24 (2012) 2567–2571.

[30] X.G. Han, Q. Kuang, M.S. Jin, Z.X. Xie, L.S. Zheng, Synthesis of titania nanosheets with a high percentage of exposed (001) facets and related photocatalytic properties, J. Am. Chem. Soc. 131 (2009) 3152–3153.

[31] Q.D. Truong, T.S. Le, H.T. Hoa, Fluoride-free self-templated synthesis of hollow TiO_2 nanostructures for hydrogen evolution, CrystEngComm. 14 (2012) 4274–4278.

[32] S. Sathasivam, D.S. Bhachu, Y. Lu, N. Chadwick, S.A. Althabaiti, A.O. Alyoubi, et al., Tungsten doped TiO_2 with enhanced photocatalytic and optoelectrical properties via aerosol assisted chemical vapor deposition, Sci. Rep. 5 (2015) 10952.

[33] H.M. Luo, T. Takata, Y.G. Lee, J.F. Zhao, K. Domen, Y.S. Yan, Photocatalytic activity enhancing for titanium dioxide by co-doping with bromine and chlorine, Chem. Mater. 16 (2004) 846–849.

[34] A. Zaleska, J.W. Sobczak, E. Grabowska, J. Hupka, Preparation and photocatalytic activity of boron-modified TiO_2 under UV and visible light, Appl. Catal. B. 78 (2008) 92–100.

[35] S. Sakthivel, H. Kisch, Daylight photocatalysis by carbon-modified titanium dioxide, Angew. Chem. Int. Ed. 42 (2003) 4908–4911.

[36] I. Hiroshi, W. Yuka, H. Kazuhito, Carbon-doped anatase TiO_2 powders as a visible-light sensitive photocatalyst, Chem. Lett. 32 (2003) 772–773.

[37] S. Sakthivel, M. Janczarek, H. Kisch, Visible light activity and photoelectrochemical properties of nitrogen-doped TiO_2, J. Phys. Chem. B 108 (2004) 19384–19387.

[38] K. Takeshita, A. Yamakata, T. Ishibashi, H. Onishi, K. Nishijima, T. Ohno, Transient IR absorption study of charge carriers photogenerated in sulfur-doped TiO_2, J. Photochem. Photobiol. A Chem. 177 (2006) 269–275.

[39] S. Piskunov, O. Lisovski, J. Begens, D. Bocharov, Y.F. Zhukovskii, M. Wessel, et al., C-, N-, S-, and Fe-doped TiO_2 and $SrTiO_3$ nanotubes for visible-light-driven photocatalytic water splitting: prediction from first principles, J. Phys. Chem. C 119 (2015) 18686–18696.

[40] C. Di Valentin, G. Pacchioni, Trends in non-metal doping of anatase TiO_2: B, C, N and F, Catal. Today 206 (2013) 12–18.

[41] R. Asahi, T. Morikawa, H. Irie, T. Ohwaki, Nitrogen-doped titanium dioxide as visible-light-sensitive photocatalyst: designs, developments, and prospects, Chem. Rev. 114 (2014) 9824–9852.

[42] R. Asahi, T. Morikawa, T. Ohwaki, K. Aoki, Y. Taga, Visible-light photocatalysis in nitrogen-doped titanium oxides, Science 293 (2001) 269–271.

[43] C.W. Dunnill, I.P. Parkin, Nitrogen-doped TiO_2 thin films: photocatalytic applications for healthcare environments, Dalton Trans. 40 (2011) 1635–1640.

[44] S.A. Ansari, M.M. Khan, M.O. Ansaric, M.H. Cho, Nitrogen-doped titanium dioxide (N-doped TiO_2) for visible light photocatalysis, New J. Chem. 40 (2016) 3000–3009.

[45] G.D. Yang, Z. Jiang, H.H. Shi, T.C. Xiao, Z.F. Yan, Preparation of highly visible-light active N-doped TiO_2 photocatalyst, J. Mater. Chem. 20 (2010) 5301–5309.

[46] X.F. Zhou, J. Lu, J.J. Jiang, X.B. Li, M.N. Lu, G.T. Yuan, et al., Simple fabrication of N-doped mesoporous TiO_2 nanorods with the enhanced visible light photocatalytic activity, Nanosc. Res. Lett. 9 (2014) 34.

[47] J. Wang, C.Y. Fan, Z.M. Ren, X.X. Fu, G.D. Qian, Z.Y. Wang, N-doped TiO_2/C nanocomposites and N-doped TiO_2 synthesised at different thermal treatment temperatures with the same hydrothermal precursor, Dalton Trans. 43 (2014) 13783–13791.

[48] Y. Nosaka, M. Matsushita, J. Nishino, A.Y. Nosaka, Nitrogen-doped titanium dioxide photocatalysts for visible response prepared by using organic compounds, Sci. Tech. Adv. Mater. 6 (2005) 143–148.

[49] S. Buzby, M.A. Barakat, H. Lin, C. Ni, S.A. Rykov, J.G. Chen, et al., Visible light photocatalysis with nitrogen-doped titanium dioxide nanoparticles prepared by plasma assisted chemical vapor deposition, J. Vac. Sci. Technol. B. 24 (2006) 1210–1214.

[50] Y.J. Yao, J.C. Qin, H. Chen, F.Y. Wei, X.T. Liu, J.L. Wang, et al., One-pot approach for synthesis of N-doped TiO_2/$ZnFe_2O_4$ hybrid as an efficient photocatalyst for degradation of aqueous organic pollutants, J. Hazard. Mater. 291 (2015) 28–37.

[51] X.D. Wang, K. Zhang, X.L. Guo, G.D. Shenb, J.Y. Xiang, Synthesis and characterization of N-doped TiO_2 loaded onto activated carbon fiber with enhanced visible-light photocatalytic activity, New J. Chem. 38 (2014) 6139–6146.

[52] X. Zhang, J. Zhou, Y. Gu, D. Fan, Visible-light photocatalytic activity of N-doped TiO_2 nanotube arrays on acephate degradation, J. Nanomater. 2015 (2015) 6. Article ID 527070.

[53] K.I. Liu, C.Y. Su, T.P. Perng, Highly porous N-doped TiO_2 hollow fibers with internal three-dimensional interconnected nanotubes for photocatalytic hydrogen production, RSC Adv. 5 (2015) 88367–88374.

[54] A.N. Kadam, R.S. Dhabbe, M.R. Kokate, Y.B. Gaikwad, K.M. Garadkar, Preparation of N doped TiO_2 via microwave-assisted method and its photocatalytic activity for degradation of Malathion, Spectrochim. Acta A 133 (2014) 669–676.

[55] X.P. Liu, Y.Y. Chen, C.L. Cao, J. Xu, Q.R. Qian, Y.J. Luo, et al., Electrospun nitrogen and carbon co-doped porous TiO_2 nanofibers with high visible light photocatalytic activity, New J. Chem. 39 (2015) 6944–6950.

[56] Z.F. Jiang, W. Wei, D.J. Mao, C. Chen, Y.F. Shi, X.M. Lv, et al., Silver-loaded nitrogen-doped yolk-shell mesoporous TiO_2 hollow microspheres with enhanced visible light photocatalytic activity, Nanoscale 7 (2015) 784–797.

[57] F.F. Li, Y.S. Jiang, M.S. Xia, M.M. Sun, B. Xue, D.R. Liu, et al., Effect of the P/Ti ratio on the visible-light photocatalytic activity of P-doped TiO_2, J. Phys. Chem. C 113 (2009) 18134–18141.

[58] R.Y. Zheng, L. Lin, J.L. Xie, Y.X. Zhu, Y.C. Me, State of doped phosphorus and its influence on the physicochemical and photocatalytic properties of P-doped titania, J. Phys. Chem. C 112 (2008) 15502–15509.

[59] M. Hamadanian, A. Reisi-Vanani, A. Majedi, Preparation and characterization of S-doped TiO_2 nanoparticles, effect of calcination temperature and evaluation of photocatalytic activity, Mater. Chem. Phys. 116 (2009) 376–382.

[60] R.G. Chaudhuri, S. Paria, Visible light induced photocatalytic activity of sulfur doped hollow TiO_2 nanoparticles, synthesized via a novel route, Dalton Trans. 43 (2014) 5526–5534.

[61] Z. Luo, A.S. Poyraz, C.H. Kuo, R. Miao, Y.T. Meng, S.Y. Chen, et al., Crystalline mixed phase (anatase/rutile) mesoporous titanium dioxides for visible light photocatalytic activity, Chem. Mater. 27 (2015) 6–17.

[62] P.S. Archana, A. Gupta, M.M. Yusoff, R. Jose, Tungsten doped titanium dioxide nanowires for high efficiency dye-sensitized solar cells, Phys. Chem. Chem. Phys. 16 (2014) 7448–7454.

[63] V.J. Babu, S. Vempati, T. Uyar, S. Ramakrishna, Review of one-dimensional and two-dimensional nanostructured materials for hydrogen generation, Phys. Chem. Chem. Phys. 17 (2015) 2960–2986.

[64] J.F. Zhu, W. Zheng, H.E. Bin, J.L. Zhang, M. Anpo, Characterization of Fe-TiO_2 photocatalysts synthesized by hydrothermal method and their photocatalytic reactivity for photodegradation of XRG dye diluted in water, J. Mol. Catal. A: Chem. 216 (2004) 35–43.

[65] K. Kalyanasundaram, M. Grätzel, Cyclic cleavage of water into H_2 and O_2 by visible-light with coupled redox catalysts, Angew. Chem. Int. Ed. 18 (1979) 701–702.

[66] K. Maeda, Z-Scheme water splitting using two different semiconductor photocatalysts, ACS Catal. 3 (2013) 1486–1503.

[67] K.L. Kelly, E. Coronado, L.L. Zhao, G.C. Schatz, The optical properties of metal nanoparticles: the influence of size, shape, and dielectric environment, J. Phys. Chem. B 107 (2003) 668–677.

[68] K. Yamada, K. Miyajima, F. Mafune, Thermionic emission of electrons from gold nanoparticles by nanosecond pulse-laser excitation of interband, J. Phys. Chem. C 111 (2007) 11246–11251.

[69] R.C. Jin, Y.W. Cao, C.A. Mirkin, K.L. Kelly, G.C. Schatz, J.G. Zheng, Photoinduced conversion of silver nanospheres to nanoprisms, Science 294 (2001) 1901–1903.

[70] M.M. Alvarez, J.T. Khoury, T.G. Schaaff, M.N. Shafigullin, I. Vezmar, R.L. Whetten, Optical absorption spectra of nanocrystal gold molecules, J. Phys. Chem. B 101 (1997) 3706–3712.

[71] S.K. Cushing, J. Li, F. Meng, T.R. Senty, S. Suri, M. Zhi, et al., Photocatalytic activity enhanced by plasmonic resonant energy transfer from metal to semiconductor, J. Am. Chem. Soc. 134 (2012) 15033–15041.

[72] S. Eustis, M.A. el-Sayed, Why gold nanoparticles are more precious than pretty gold: noble metal surface plasmon resonance and its enhancement of the radiative and nonradiative properties of nanocrystals of different shapes, Chem. Soc. Rev. 35 (2006) 209–217.

[73] S. Sarina, E.R. Waclawik, H. Zhu, Photocatalysis on supported gold and silver nanoparticles under ultraviolet and visible light irradiation, Green Chem. 15 (2013) 1814–1833.

[74] S. Linic, P. Christopher, D.B. Ingram, Plasmonic-metal nanostructures for efficient conversion of solar to chemical energy, Nat. Mater. 10 (2011) 911–921.

[75] P. Christopher, H. Xin, A. Marimuthu, S. Linic, Singular characteristics and unique chemical bond activation mechanisms of photocatalytic reactions on plasmonic nanostructures, Nat. Mater. 11 (2012) 1044–1050.

[76] P.K. Jain, X. Huang, I.H. El-Sayed, M.A. El-Sayed, Noble metals on the nanoscale: optical and photothermal properties and some applications in imaging, sensing, biology, and medicine, Acc. Chem. Res. 41 (2008) 1578–1586.

[77] L. Gunnarsson, T. Rindzevicius, J. Prikulis, B. Kasemo, M. Käll, S. Zou, et al., Confined plasmons in nanofabricated single silver particle pairs: experimental observations of strong interparticle interactions, J. Phys. Chem. B 109 (2005) 1079–1087.

[78] A. Grirrane, A. Corma, H. Garcia, Gold-catalyzed synthesis of aromatic Azo compounds from anilines and nitroaromatics, Science 322 (2008) 1661–1664.

[79] T.A. Westrich, K.A. Dahlberg, M. Kaviany, J.W. Schwank, High-temperature photocatalytic ethylene oxidation over TiO_2, J. Phys. Chem. C 115 (2011) 16537–16543.

[80] P. Pichat, M.N. Mozzanega, J. Disdier, J.M. Herrmann, Pt Content and temperature effects on the photocatalytic H_2 production from aliphatic-alcohols over Pt-TiO_2, New J. Chem. 6 (1982) 559–564.

[81] J. Disdier, J.M. Herrmann, P. Pichat, Platinum titanium-dioxide catalysts—a photoconductivity study of electron-transfer from the ultraviolet-illuminated support to the metal and of the influence of hydrogen, J. Chem. Soc. Faraday Trans. 79 (1983) 651–660.

[82] S. Sarina, H.Y. Zhu, Q. Xiao, E. Jaatinen, J.F. Jia, Y.M. Huang, et al., Viable photocatalysts under solar-spectrum irradiation: nonplasmonic metal nanoparticles, Angew. Chem. Int. Ed. Engl. 53 (2014) 2935–2940.

[83] S. Sarina, E. Jaatinen, Q. Xiao, Y.M. Huang, P. Christopher, J.C. Zhao, et al., Photon energy threshold in direct photocatalysis with metal nanoparticles: key evidence from the action spectrum of the reaction, J. Phys. Chem. Lett. 8 (2017) 2526–2534.

[84] X.J. Li, C.M. Zeng, G.Y. Fan, Ultrafast hydrogen generation from the hydrolysis of ammonia borane catalyzed by highly efficient bimetallic RuNi nanoparticles stabilized on $Ti_3C_2X_2$ (X = OH and/or F), Int. J. Hydrogen Energy 40 (2015) 3883–3891.

[85] G.C. Bond, The origins of particle size effects in heterogeneous catalysis, Surf. Sci. 156 (1985) 966–981.

[86] K. Mori, K. Miyawaki, H. Yamashita, Ru and Ru-Ni nanoparticles on TiO_2 support as extremely active catalysts for hydrogen production from ammonia-borane, ACS Catal. 6 (2016) 3128–3135.

[87] A. Islam, H. Sugihara, K. Hara, L.P. Singh, R. Katoh, M. Yanagida, et al., Sensitization of nanocrystalline TiO_2 film by ruthenium(II) diimine dithiolate complexes, J. Photochem. Photobiol. A Chem. 145 (2001) 135–141.

[88] M. Grätzel, Solar energy conversion by dye-sensitized photovoltaic cells, Inorg. Chem. 44 (2005) 6841–6851.

[89] H. Park, Y. Park, W. Kim, W. Choi, Surface modification of TiO_2 photocatalyst for environmental applications, J. Photochem. Photobiol. C Photochem. Rev. 15 (2013) 1–20.

[90] L.E. Oi, M.Y. Choo, H.V. Lee, H.C. Ong, S.B. Abd Hamid, J.C. Juan, Recent advances of titanium dioxide (TiO_2) for green organic synthesis, RSC Adv. 6 (2016) 108741–108754.

[91] F. Haber, Z. Elektrochem. Angew. Phys. Chem. 22 (1898) 506–514.

[92] A. Corma, P. Concepción, P. Serna, A different reaction pathway for the reduction of aromatic nitro compounds on gold catalysts, Angew. Chem. 119 (2007) 7404–7407.

[93] L.B. Wang, X.Q. Pan, Y. Zhao, Y. Chen, W. Zhang, Y.F. Tu, et al., A straightforward protocol for the highly efficient preparation of main-chain Azo polymers directly from bisnitroaromatic compounds by the photocatalytic process, Macromolecules 48 (2015) 1289–1295.

[94] H.Y. Zhu, X.B. Ke, X.Z. Yang, S. Sarina, H.W. Liu, Reduction of nitroaromatic compounds on supported gold nanoparticles by visible and ultraviolet light, Angew. Chem. Int. Ed. 49 (2010) 9657–9661.

[95] M. Boronat, P. Concepcion, A. Corma, S. Gonzalez, F. Illas, P. Serna, A molecular mechanism for the chemoselective hydrogenation of substituted nitroaromatics with nanoparticles of gold on TiO_2 catalysts: a cooperative effect between gold and the support, J. Am. Chem. Soc. 129 (2007) 16230–16237.

[96] U. Hartfelder, C. Kartusch, M. Makosch, M. Rovezzi, J. Sa, J.A. van Bokhoven, Particle size and support effects in hydrogenation over supported gold catalysts, Catal. Sci. Technol. 3 (2013) 454–461.

[97] B.X. Li, Y.G. Hao, X.K. Shao, H.D. Tang, T. Wang, J.B. Zhu, et al., Synthesis of hierarchically porous metal oxides and Au/TiO_2 nanohybrids for photodegradation of organic dye and catalytic reduction of 4-nitrophenol, J. Catal. 329 (2015) 368–378.

[98] K. Imamura, Y. Okubo, T. Ito, A. Tanaka, K. Hashimoto, H. Kominami, Photocatalytic hydrogenation of alkenes to alkanes in alcoholic suspensions of palladium-loaded titanium(IV) oxide without the use of hydrogen gas, RSC Adv. 4 (2014) 19883–19886.

[99] A. Elhage, A.E. Lanterna, J.C. Scaiano, Tunable photocatalytic activity of palladium-decorated TiO_2: non hydrogen-mediated hydrogenation or isomerization of benzylsubstituted alkenes, ACS Catal. 7 (2017) 250–255.

[100] T. Yui, A. Kan, C. Saitoh, K. Koike, T. Ibusuki, O. Ishitani, Photochemical reduction of CO_2 using TiO_2: effects of organic adsorbates on TiO_2 and deposition of Pd onto TiO_2, ACS Appl. Mater. Inter. 3 (2011) 2594–2600.

[101] O. Ola, M.M. Maroto-Valer, Review of material design and reactor engineering on TiO_2 photocatalysis for CO_2 reduction, J. Photochem. Photobiol. C Photochem. Rev. 24 (2015) 16–42.

[102] K. Wada, C.S.K. Ranasinghe, R. Kuriki, A. Yamakata, O. Ishitani, K. Maeda, Interfacial manipulation by rutile TiO_2 nanoparticles to boost CO_2 reduction into CO on a metal-complex/semiconductor hybrid photocatalyst, ACS Appl. Mater. Inter. 9 (2017) 23869–23877.

[103] V. Augugliaro, M. Bellardita, V. Loddo, G. Palmisano, L. Palmisano, S. Yurdakal, Overview on oxidation mechanisms of organic compounds by TiO_2 in heterogeneous photocatalysis, J. Photochem. Photobiol. C Photochem. Rev. 13 (2012) 224–245.

[104] M.A. Gondal, A. Hameed, Z.H. Yamani, A. Arfaj, Photocatalytic transformation of methane into methanol under UV laser irradiation over WO_3, TiO_2 and NiO catalysts, Chem. Phys. Lett. 392 (2004) 372–377.

[105] M.H. Ab Rahim, M.M. Forde, R.L. Jenkins, C. Hammond, Q. He, N. Dimitratos, et al., Oxidation of methane to methanol with hydrogen peroxide using supported gold-palladium alloy nanoparticles, Angew. Chem. Int. Ed. 52 (2013) 1280–1284.

[106] J. Yi, C. Bahrini, C. Schoemaecker, C. Fittschen, W. Choi, Photocatalytic decomposition of H_2O_2 on different TiO_2 surfaces along with the concurrent generation of HO_2 radicals monitored using cavity ring down spectroscopy, J. Phys. Chem. C. 116 (2012) 10090–10097.

[107] M. Buchalska, M. Kobielusz, A. Matuszek, M. Pacia, S. Wojtyla, W. Macyk, On oxygen activation at rutile- and anatase-TiO_2, ACS Catal. 5 (2015) 7424–7431.

[108] V. Augugliaro, V. Loddo, M.J. Lopez-Munoz, C. Marquez-Alvarez, G. Palmisano, L. Palmisano, et al., Home-prepared anatase, rutile, and brookite TiO_2 for selective photocatalytic oxidation of 4-methoxybenzyl alcohol in water: reactivity and ATR-FTIR study, Photoch. Photobio. Sci. 8 (2009) 663–669.

[109] C.J. Li, G.R. Xu, B.H. Zhang, J.R. Gong, High selectivity in visible-light-driven partial photocatalytic oxidation of benzyl alcohol into benzaldehyde over single-crystalline rutile TiO_2 nanorods, Appl. Catal. B. 115 (2012) 201–208.

[110] T. Reid, S.R. Chaganti, I.G. Droppo, C.G. Weisener, Novel insights into freshwater hydrocarbon-rich sediments using metatranscriptomics: opening the black box, Water Res. 136 (2018) 1–11.

[111] U.I. Gaya, A.H. Abdullah, Heterogeneous photocatalytic degradation of organic contaminants over titanium dioxide: a review of fundamentals, progress and problems, J. Photochem. Photobiol. C Photochem. Rev. 9 (2008) 1–12.

[112] J.F. Montoya, I. Ivanova, R. Dillert, D.W. Bahnemann, P. Salvador, J. Peral, Catalytic role of surface oxygens in TiO_2 photooxidation reactions: aqueous benzene photooxidation with $Ti^{18}O_2$ under anaerobic conditions, J. Phys. Chem. Lett. 4 (2013) 1415–1422.

[113] K. Selvam, H. Sakamoto, Y. Shiraishi, T. Hirai, Photocatalytic secondary amine synthesis from azobenzenes and alcohols on TiO_2 loaded with Pd nanoparticles, New J. Chem. 39 (2015) 2856–2860.

[114] M. Fagnoni, D. Dondi, D. Ravelli, A. Albini, Photocatalysis for the formation of the C-C bond, Chem. Rev. 107 (2007) 2725–2756.

[115] D. Ravelli, S. Protti, M. Fagnoni, Carbon-carbon bond forming reactions via photogenerated intermediates, Chem. Rev. 116 (2016) 9850–9913.

[116] C. Gambarotti, C. Punta, F. Recupero, T. Caronna, L. Palmisano, TiO_2 in organic photosynthesis: sunlight induced functionalization of heterocyclic bases, Curr. Org. Chem. 14 (2010) 1153–1169.

[117] L. Cermenati, C. Richter, A. Albini, Solar light induced carbon-carbon bond formation via TiO_2 photocatalysis, Chem. Commun. 0 (1998) 805–806.

[118] A.E. Lanterna, A. Elhage, J.C. Scaiano, Heterogeneous photocatalytic C-C coupling: mechanism of plasmon-mediated reductive dimerization of benzyl bromides by supported gold nanoparticles, Catal. Sci. Tech. 5 (2015) 4336–4340.

[119] A. Elhage, A.E. Lanterna, J.C. Scaiano, Light-induced sonogashira C–C coupling under mild conditions using supported palladium nanoparticles, ACS Sustain. Chem. Eng. 6 (2018) 1717–1722.

[120] Q. Yuan, Z.F. Wu, Y.K. Jin, L.S. Xu, F. Xiong, Y.S. Ma, et al., Photocatalytic cross-coupling of methanol and formaldehyde on a rutile $TiO_2(110)$ surface, J. Am. Chem. Soc. 135 (2013) 5212–5219.

[121] P.J. Yang, J.H. Zhao, B.Y. Cao, L. Li, Z.J. Wang, X.X. Tian, et al., Selective photocatalytic C-C coupling of bioethanol into 2,3-butanediol over Pt-decorated hydroxyl-group-tunable TiO_2 photocatalysts, ChemCatChem. 7 (2015) 2384–2390.

[122] C. Vila, M. Rueping, Visible-light mediated heterogeneous C-H functionalization: oxidative multi-component reactions using a recyclable titanium dioxide (TiO_2) catalyst, Green. Chem. 15 (2013) 2056–2059.
[123] Y.L. Liu, M.F. Zhang, C.H. Tung, Y.F. Wang, TiO_2 Photocatalytic cyclization reactions for the syntheses of aryltetralones, ACS Catal. 6 (2016) 8389–8394.
[124] J. Tang, G. Grampp, Y. Liu, B.X. Wang, F.F. Tao, L.J. Wang, et al., Visible light mediated cyclization of tertiary anilines with maleimides using Nickel(II) Oxide surface-modified titanium dioxide catalyst, J. Org. Chem. 80 (2015) 2724–2732.
[125] Z.J. Han, R. Eisenberg, Fuel from water: the photochemical generation of hydrogen from water, Acc. Chem. Res. 47 (2014) 2537–2544.
[126] K.C. Christoforidis, P. Fornasiero, Photocatalytic hydrogen production: a rift into the future energy supply, ChemCatChem. 9 (2017) 1523–1544.
[127] M.H. Zhou, Y.X. Li, S.Q. Peng, G.X. Lu, S.B. Li, Effect of epimerization of d-glucose on photocatalytic hydrogen generation over Pt/TiO_2, Catal. Commun. 18 (2012) 21–25.
[128] G. Zhang, C.S. Ni, X.B. Huang, A. Welgamage, L.A. Lawton, P.K.J. Robertson, et al., Simultaneous cellulose conversion and hydrogen production assisted by cellulose decomposition under UV-light photocatalysis, Chem. Commun. 52 (2016) 1673–1676.
[129] C.A. Schwengber, H.J. Alves, R.A. Schaffner, F.A. da Silva, R. Sequinel, V.R. Bach, et al., Overview of glycerol reforming for hydrogen production, Renew. Sust. Energ. Rev. 58 (2016) 259–266.
[130] K. Hashimoto, T. Kawai, T. Sakata, Photocatalytic reactions of hydrocarbons and fossil-fuels with water. Hydrogen production and oxidation, J. Phys. Chem. 88 (1984) 4083–4088.
[131] Y.X. Li, G.X. Lu, S.B. Li, Photocatalytic hydrogen generation and decomposition of oxalic acid over platinized TiO_2, Appl. Catal. A 214 (2001) 179–185.
[132] U.G. Akpan, B.H. Hameed, Parameters affecting the photocatalytic degradation of dyes using TiO_2-based photocatalysts: a review, J. Hazard. Mater. 170 (2009) 520–529.
[133] M. Rochkind, S. Pasternak, Y. Paz, Using dyes for evaluating photocatalytic properties: a critical review, Molecules 20 (2015) 88–110.
[134] J.F. Jacobs, I. van de Poel, P. Osseweijer, Sunscreens with titanium dioxide (TiO_2) nanoparticles: a societal experiment, Nanoethics 4 (2010) 103–113.
[135] A.P. Popov, A.V. Priezzhev, J. Lademann, R. Myllylä, TiO_2 nanoparticles as an effective UV-B radiation skin-protective compound in sunscreens, J. Phys. D. Appl. Phys. 38 (2005) 2564–2570.
[136] S. Tanemura, L. Miao, W. Wunderlich, M. Tanemura, Y. Mori, S. Toh, et al., Fabrication and characterization of anatase/rutile-TiO_2 thin films by magnetron sputtering: a review, Sci. Tech. Adv. Mater. 6 (2005) 11–17.
[137] X.W. Sun, H.S. Kwok, Optical properties of epitaxially grown zinc oxide films on sapphire by pulsed laser deposition, J. Appl. Phys. 86 (1999) 408–411.
[138] T.G. Smijs, S. Pavel, Titanium dioxide and zinc oxide nanoparticles in sunscreens: focus on their safety and effectiveness, Nanotechnol. Sci. App. 4 (2011) 95–112.
[139] M.J. Osmond-McLeod, Y. Oytam, A. Rowe, F. Sobhanmanesh, G. Greenoak, J. Kirby, et al., Long-term exposure to commercially available sunscreens containing nanoparticles of TiO_2 and ZnO revealed no biological impact in a hairless mouse model, Part. Fibre Toxicol. 13 (2016) 44–57.
[140] P.J. Lu, S.C. Huang, Y.P. Chen, L.C. Chiueh, D.Y. Shih, Analysis of titanium dioxide and zinc oxide nanoparticles in cosmetics, J. Food Drug Anal. 23 (2015) 587–594.

[141] J.H. Wu, W. Liu, C.B. Xue, S.C. Zhou, F.L. Lan, L. Bi, et al., Toxicity and penetration of TiO_2 nanoparticles in hairless mice and porcine skin after subchronic dermal exposure, Toxicol. Lett. 191 (2009) 1–8.

[142] V.A. Miyani, M.F. Hughes, Assessment of the in vitro dermal irritation potential of cerium, silver, and titanium nanoparticles in a human skin equivalent model, Cutan. Ocul. Toxicol. 36 (2017) 145–151.

[143] M. Morsella, N. d'Alessandro, A.E. Lanterna, J.C. Scaiano, Improving the sunscreen properties of TiO_2 through an understanding of its catalytic properties, ACS Omega 1 (2016) 464–469.

[144] M.D. Newman, M. Stotland, J.I. Ellis, The safety of nanosized particles in titanium dioxide- and zinc oxide-based sunscreens, J. Am. Acad. Dermatol. 61 (2009) 685–692.

[145] M. Rinnerthaler, J. Bischof, M.K. Streubel, A. Trost, K. Richter, Oxidative stress in aging human skin, Biomolecules 5 (2015) 545–589.

[146] D. Dondi, A. Albini, N. Serpone, Interactions between different solar UVB/UVA filters contained in commercial suncreams and consequent loss of UV protection, Photochem. Photobiol. Sci. 5 (2006) 835–843.

[147] X.H. Feng, S.K. Zhang, X. Lou, Controlling silica coating thickness on TiO_2 nanoparticles for effective photodynamic therapy, Colloid Surf. B. 107 (2013) 220–226.

[148] L. Rowenczyk, C. Duclairoir-Poc, M. Barreau, C. Picard, N. Hucher, N. Orange, et al., Impact of coated TiO_2-nanoparticles used in sunscreens on two representative strains of the human microbiota: effect of the particle surface nature and aging, Colloid Surf. B. 158 (2017) 339–348.

[149] X.B. Dong, Z.M. Sun, Y.Y. Liu, L. Jiang, S.L. Zheng, Insights into effects and mechanism of pre-dispersant on surface morphologies of silica or alumina coated rutile TiO_2 particles, Chem. Phys. Lett. 699 (2018) 55–63.

[150] H.Z. Guo, D. Klose, Y.H. Hou, G. Jeschke, I. Burgert, Highly efficient UV protection of the biomaterial wood by a transparent TiO_2/Ce xerogel, ACS Appl. Mater. Inter. 9 (2017) 39040–39047.

[151] A. Morlando, V. Sencadas, D. Cardillo, K. Konstantinov, Suppression of the photocatalytic activity of TiO_2 nanoparticles encapsulated by chitosan through a spray-drying method with potential for use in sun blocking applications, Powder Technol. 329 (2018) 252–259.

[152] P.K. Shetty, V. Venuvanka, H.V. Jagani, G.H. Chethan, V.S. Ligade, P.B. Musmade, et al., Development and evaluation of sunscreen creams containing morin-encapsulated nanoparticles for enhanced UV radiation protection and antioxidant activity, Int. J. Nanomed. 10 (2015) 6477–6491.

[153] H.J. Lee, S.E. Jeong, S. Lee, S. Kim, H. Han, C.O. Jeon, Effects of cosmetics on the skin microbiome of facial cheeks with different hydration levels, Microbiologyopen. 7 (2018). e00557.

[154] C. Wallen-Russell, S. Wallen-Russell, Meta analysis of skin microbiome: new link between skin microbiota diversity and skin health with proposal to use this as a future mechanism to determine whether cosmetic products damage the skin, Cosmetics 4 (2017) 14–33.

[155] K.D. Shah, H.W. Sarkas, P.G. Murray, Multifunctional coated powders and high solids dispersions, US9 139 (2015) 737B1.

[156] C.A. Downs, E. Kramarsky-Winter, R. Segal, J. Fauth, S. Knutson, O. Bronstein, et al., Toxicopathological effects of the sunscreen UV filter, oxybenzone (benzophenone-3), on coral planulae and cultured primary cells and its environmental contamination in Hawaii and the US Virgin Islands, Arch. Environ. Contam. Toxicol. 70 (2016) 265–288.

[157] M. Sendra, D. Sanchez-Quiles, J. Blasco, I. Moreno-Garrido, L.M. Lubian, S. Perez-Garcia, et al., Effects of TiO_2 nanoparticles and sunscreens on coastal marine microalgae: ultraviolet radiation is key variable for toxicity assessment, Environ. Int. 98 (2017) 62–68.

[158] Canada Natural and Non-prescription Health Products Directorate. <https://www.canada.ca/en/health-canada/services/drugs-health-products/public-involvement-consultations/natural-health-products/draft-guidance-document-sunscreen-monograph-consultation-document.html> (accessed 04.08).

Chapter 5

Nanoparticles within functional frameworks and their applications in photo(electro) catalysis

Weiyi Ouyang, Alain R. Puente Santiago, Kenneth Cerdán-Gómez and Rafael Luque*

Department of Organic Chemistry, University of Cordoba, Cordoba, Spain

*Corresponding author. e-mail: q62alsor@uco.es

Chapter Outline

5.1 Introduction

With the potential energy crisis and the rising awareness in environmental protection, photo(electro)catalysis, which utilizes the abundant light as energy source in the chemical transformations, has attracted intensive attention in the last few decades. Now it is considered as one of the most advanced, environmentally friendly, and prospective technologies because of its excellent advantages such as clean, efficient, cost-effective, and energy-saving properties. Various nanoparticles (NPs) [e.g., metal oxides,

Photoactive Inorganic Nanoparticles. DOI: https://doi.org/10.1016/B978-0-12-814531-9.00005-1
© 2019 Elsevier Inc. All rights reserved.

nanocomposites, metal organic frameworks (MOFs), g-C_3N_4, graphene oxides (GOs), etc.] were reported as heterogeneous photocatalysts with applications in water splitting, CO_2 reduction, organic transformations, water disinfection, and so on [1−9]. In this chapter, common methods for catalyst preparation will be briefly discussed, following by the introduction of nanostructured catalysts and their applications in photocatalysis and photoelectrocatalysis.

5.2 Synthesis methods

Various chemical and physical methods have been applied to synthesize nanomaterials. Physical methods, such as ball milling and electrodeposition, are advantageous in obtaining highly pure nanomaterials and applicable for large-scale production. Nanomaterials synthesized by chemical methods, such as sol−gel and solvothermal processes, are benefited from the controllable and uniform sizes and shapes. In the following sections, four common and representative methods are briefly discussed: sol−gel method, solvothermal method, mechanochemical method, and sonochemical method. However, the synthesis of nanomaterials with advanced functional frameworks requires not only one method but several methods might be combined in the synthetic procedure.

5.2.1 Sol−gel method

The sol−gel process is a widely used method for the fabrication of multicomponent ceramic and nanophase materials, especially metal oxides, while the most representative is the synthesis of titanium dioxide (TiO_2) and silica (SiO_2). With the rising application of semiconductor-based photocatalysts and photocatalytic processes using TiO_2 since the pioneering work of Fujishima and Honda in 1972 [10], the preparation of various forms of TiO_2 by sol−gel method and their application in photocatalysis were extensively reported in the recent decades.

Typically, the sol−gel process involves metal alkoxides precursors, M $(OR)_x$, where R is an alkyl group (R = Me, Et, Pr ...). In this chemical process, a gel-like diphasic system, containing polymer oxide networks is formed gradually from the solution by the hydrolysis and condensation of the precursors. Take the generation of TiO_2 by sol−gel method as an example. Its reaction pathway is described below:

$$\text{Hydrolysis: Ti-OR} + H_2O \rightarrow \text{Ti-OH} + \text{ROH} \tag{5.1}$$

$$\text{Condensation: Ti-OH} + \text{Ti-OR} \rightarrow \text{Ti-O-Ti} + \text{ROH} \tag{5.2}$$

Theoretically the hydrolysis process proceeds as follows:

$$Ti(OR)_4 + 4H_2O \rightarrow Ti(OH)_4 + 4ROH \quad (5.3)$$

$$Ti(OH)_4 \rightarrow TiO_2 + 2H_2O \quad (5.4)$$

However, the reaction does not proceed stoichiometrically as described in Eqs. (5.3) and (5.4), and it forms rapidly partially hydrolyzed species in the absence of peptizing agent, together with the proceeding of condensation reaction [11].

$$Ti(OR)_4 + xH_2O \rightarrow Ti(OH)_x(OR)_{4-x} + xROH \quad (5.5)$$

$$Ti(OH)_x(OR)_{4-x} + Ti(OR)_4 \rightarrow (OR)_{4-x}Ti\text{-}O_x\text{-}Ti(OR)_{4-x} + xROH \quad (5.6)$$

In the synthesis of TiO_2 with the hydrolytic sol–gel route, the hydrolysis ratio r (H_2O/TiO_2 ratio) plays a key factor in manipulating characteristics of the generated precipitates, such as size, morphology, and crystallinity. In detail, reactions with low hydrolysis ratios prefer partial hydrolysis route, while nearly complete hydrolysis or complete hydrolysis occurs with high ratios [12].

However, the application of the conventional hydrolytic sol–gel method has its limitations in the synthesis of nanocrystals in organic media: (1) incompatibility of the oxygen donor (H_2O) at high temperature, (2) difficulty in forming homogeneous reaction solution with organic solvents, and (3) difficulty in controlling the growth of nanocrystal in fast reaction between most of metal precursors and H_2O [13]. Apart from the hydrolytic pathway, sol–gel method can also proceed in a nonhydrolytic pathway, which starts from metal halides with oxygen donor, such as metal alkoxide, ether or alcohol instead of H_2O, while generating alkyl halides as byproducts [14,15].

$$\text{Using metal alkoxide: } MX_n + M(OR)_n \rightarrow X_{n-1}\text{-M-O-M-}(OR)_{n-1} + RX \quad (5.7)$$

$$\text{Using ether/alcohol: } MX_n + R\text{-O-}R' \rightarrow RO\text{-}MX_{n-1} + R'X \quad (5.8)$$

Condensation of the above intermediates will lead to the formation of metal oxide precipitates.

5.2.2 Solvothermal method

Solvothermal synthesis, which has been well developed in the last decade, is a versatile method for controlling the morphology of the well-defined nanostructure materials (e.g., metals, metal oxides, and MOFs). The synthesis by applying solvothermal method is performed in sealed containers where the temperature can exceed the boiling point of the solvents by the increase of the autogenous pressures [16], while the route using the water as solvent is also called hydrothermal synthesis. Increase of temperature and pressure in the solvothermal process promotes the solubility and the reactivity of the reagents, which might lead to unexpected reactions as compared with the classical routes [17].

Nuclei formation and growth are the two key steps in the solvothermal synthesis, which can be manipulated by varying some experimental factors, and therefore controlling the properties of the crystals:

1. Reaction temperatures have significant effect on the solvothermal synthesis. Properties of the used solvent can be greatly varied under different temperatures, such as viscosity and dielectric constant. Besides, the reagents can become more active under high temperature while the solubility in the solvent may also change along with the change in temperature.
2. The properties of solvents, including polarity, dielectric constants, and viscosity, have important effects on the solvothermal process. The reaction equilibriums, kinetics, and products could be significantly affected by the interactions of solvents with reactants, intermediates, and products. Water, dimethylformamide, polyols, oleylamine, and their mixture are commonly used as solvents due to the excellent compatibility with many reactants.
3. Facet-specific capping agents are usually applied in the synthesis of NPs exposing desired facets to promote the properties of the obtained products, which could be acids, solvents, surfactants, polymers, etc. For example, HF was used for the synthesis of TiO_2 nanosheets (NSs) with dominant {001} facets [18].

5.2.3 Mechanochemical method

As defined by the IUPAC, mechanochemistry is induced by the direct absorption of mechanical energy generated from shearing, stretching, and grinding [19]. Mechanochemistry has a long history, but it was mostly advanced in the recent two decades with its application in synthetic organic chemistry, and is developing toward the next platform with its application in the design of advanced nanomaterials [20–22]. As compared with traditional synthetic routes, mechanochemical method possesses several advantages in terms of cost, efficiency, sustainability, reproducibility, suitability in large-scale production, etc. In this aspect, mechanical ball milling is one of the most common used methods in the synthesis of various nanomaterials, such as metals, metal oxides, metal complexes, composite materials, and MOFs. Many of the reactions or syntheses can be achieved even under solvent-free condition due to the high-energy input and well mixing of the reactants in the ball mill process. Besides, ball milling process can also be applied in the exfoliation of bulk materials into NSs [e.g., g-C_3N_4, graphene (GR), WSe_2, and TaS_2] [23–25] because of the existence of strong shearing force between the balls and the solid compounds.

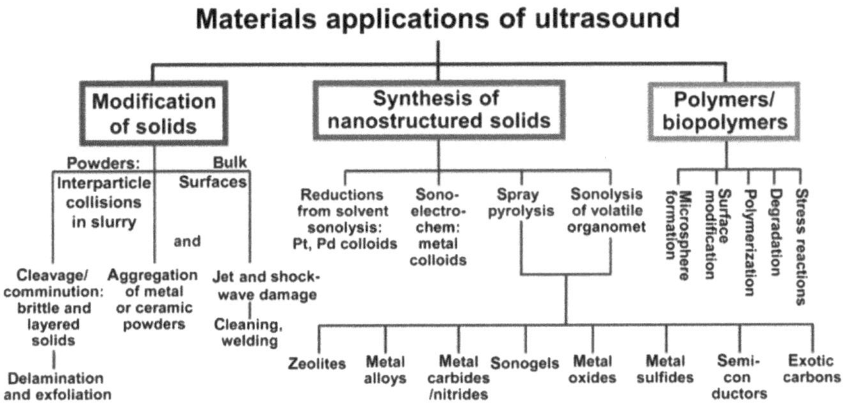

FIGURE 5.1 Applications of sonochemistry in material synthesis. *Reprinted with permission from Royal Society of Chemistry, H. Xu, B.W. Zeiger, K.S. Suslick, Sonochemical synthesis of nanomaterials, Chem. Soc. Rev. 42 (2013) 2555–2567. Available from: https://doi.org/10.1039/C2CS35282F.*

5.2.4 Sonochemical method

Sonochemical synthesis of nanostructured materials is initiated by ultrasonication in liquids, utilizing both the physical and chemical effects of ultrasound which arise from the implosive collapse of bubbles produced by acoustic cavitation [26,27]. Extremely high temperature and pressure can be generated in the cavitation bubbles, offering unique reaction conditions for synthesis in comparison to the other chemical processes. General applications of ultrasound in the synthesis of nanomaterials are summarized in Fig. 5.1.

5.3 Structures of the functional frameworks

The recombination of the photoexcited electrons and holes in the bulk materials is one of the key factors impacting the catalytic activity of photoactive materials. Tuning the structures and the morphologies of photoactive materials can be applied to promote the charge separation and increase the reaction sites to enhance their photocatalytic activity and photoefficiency. In this section, popular structures fulfilling the above purpose will be introduced, including encapsulation structure, core–shell and hollow structure, layer-by-layer (LbL) structure, and surface modified and composite materials.

5.3.1 Encapsulation of nanoentities in metal organic frameworks

MOFs, comprising bridging organic linkers (e.g., carboxylates, phosphonates, sulfonates, and heterocyclic compounds) combined with inorganic secondary building units, have been considered as highly promising porous materials because of their excellent properties including diverse chemical compositions, large surface areas, and porosity as compared with other porous materials [28–30]. Some MOFs are photoactive without any need of further modification such as MIL-125 (Ti) [31], MIL-101(Fe) [32], and NH_2-UiO-66 (Zr) [33], while the presence of functional groups in organic linkers can further improve the light absorption capability and photocatalytic activity of MOFs [33,34]. Nevertheless, the outstanding properties of MOFs, especially highly tunable pore properties (pore environment and size) and uniform pore structure with small windows, facilitate the incorporation of active phases for enhancing catalytic performance. Various nanoentities (NEs) can be encapsulated into MOFs, such as metal NPs (MNPs), quantum dots (QDs), metal oxides, and organic and metallorganic molecules (Fig. 5.2) [35].

There are mainly four strategies to encapsulate NEs into MOFs [35]:

1. Ship-in-bottle strategy (assembly of NEs inside MOFs). This strategy involves the assembly of NEs in an existing MOF via either postassembly or preassembly incorporation. In the preassembly incorporation method, NE precursors are anchored to the organic linker prior to the formation of the MOF, while NE precursors are introduced into the existing MOFs via postassembly incorporation methods.
2. Bottle-around-ship strategy (assembly of MOFs around NEs). NE cores with uniform sizes, shapes, and structures are first prepared in this strategy, followed by the growth of MOFs on NE surfaces.
3. Sandwich assembly strategy (embedding NEs between MOF layers). NEs are deposited on presynthesized MOF cores and a MOF shell is subsequently grown on the core.
4. In situ encapsulation (simultaneous synthesis of MOF and NEs). Both NE and MOF precursors are mixed together to carry out one-pot synthesis in which NE and MOFs are simultaneously generated.

Fang and coworkers recently reported significant enhancement in photocatalytic H_2 production after encapsulation of Pt into Al-TCPP using both ship-in-bottle strategy and bottle-around-ship strategies, namely, Al-TCPP-0.1Pt and Al-TCPP-PtNPs [36]. Al-TCPP-PtNPs showed H_2 production rate of 50 μmol/g//h over 33 times than pure MOF (Al-TCPP), while Al-TCPP-0.1Pt exhibited the best photocatalytic efficiency of 129 μmol/g//h, which resulted from the confined single Pt atom in the latter (also much smaller than the size of the former, ~3 nm). Fe_3O_4 encapsulated in MIL-100 (Fe) [Fe_3O_4@MIL-100 (Fe)] was reported to exhibit better photocatalytic activity than that of MIL-100 (Fe), while the magnetic property of Fe_3O_4 facilitates easy separation of Fe_3O_4@MIL-100 (Fe) by simply applying a magnetic field [37]. ZnO@ZIF-8 nanorod arrays synthesized with self-templating strategy was reported by Kong et al. which exhibited different photoelectrochemical responses to hole scavengers with various sizes from ZnO nanorod arrays owing to the molecular sieving capability of ZIF-8 shell [38]. Molecular Co catalyst encapsulated within NH_2-MIL-125 (Ti) using a

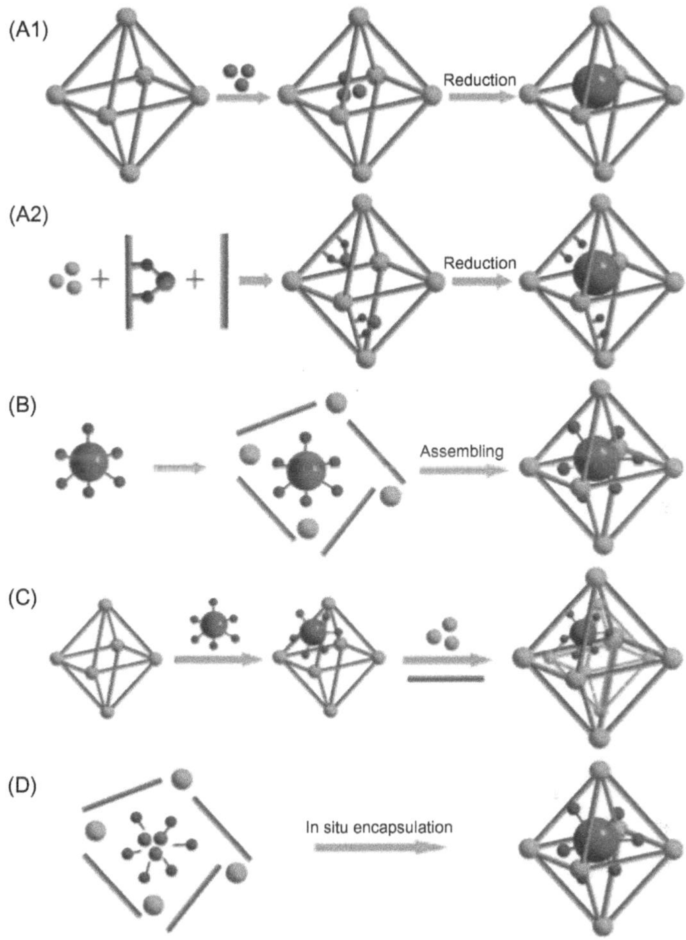

FIGURE 5.2 Strategies for encapsulation of NEs in MOFs. (A1) Post-assembly incorporation; (A2) pre-assembly incorporation; (B) bottle-around-ship, (C) sandwich assembly, and (D) in situ encapsulation. The octahedrons shown in the figure represent the unit structures of MOFs, while the balls within the structures represent different NEs. *Reproduced with permission from The Royal Society of Chemistry from L. Chen, R. Luque, Y. Li, Controllable design of tunable nanostructures inside metal–organic frameworks, Chem. Soc. Rev. 46 (2017) 4614–4630.*

ship-in-bottle strategy was reported by Gascon et al. [39]. As the dimensions of cobalt complex are larger than the size of the MOF windows, synthesis of Co@NH_2-MIL-125(Ti) involves two steps: (1) encapsulation of the organic ligand $C_{11}H_{20}N_4O_2$ into the pores of NH_2-MIL-125(Ti), and (2) addition of $CoBr_2$ to assemble the cobalt complex. Due to the different size of the complex dimensions and the pore windows, the Co complex can be confined within MOF pores. The combination of photoactive NH_2-MIL-125(Ti) and a

Co-based proton reduction catalyst provided up to 20-fold higher photocatalytic H_2 production rate compared to pristine NH_2-MIL-125(Ti). Hence the encapsulation of NEs in MOFs cannot only improve the stability of NEs and provide novel properties, but it also and more importantly improves the catalytic performance of MOFs.

5.3.2 Core–shell and hollow structure materials

Core–shell structure materials have attracted extensive attention since the 1980s, and have been vastly developed in the last decade [40]. Photoactive core–shell structure materials are usually combinations of different types of semiconductors or semiconductors with other materials (metals, metal oxides, metal sulfides, etc.), which can facilitate the separation and transportation of the photogenerated electron–hole pairs, while hollow structure materials can be categorized as derivatives of core–shell structure materials by eliminating the core materials. Some of the materials that compose the aforementioned structures can either act as cocatalysts in the photoreactions or functionalize the materials with unique properties, such as magnetism.

Magnetic materials (such as α-Fe_2O_3, γ-Fe_2O_3, Fe_3O_4, etc.) show great advantage in the synthesis of core–shell structure materials because they can promote the activities (charge separation and catalytic performance) of the as-synthesized photoactive materials as well as functionalize the resulted materials with magnetic properties which facilitate the recovery process by simply applying magnetic field. The magnetically separable $Fe_3O_4/SiO_2/TiO_2$ [41], α-Fe_2O_3/TiO_2 [42], and γ-Fe_2O_3/TiO_2 [43] core–shell structure materials were reported to have enhanced photocatalytic activity, even in visible light region, as compared with pure TiO_2. Meanwhile, Fe_2O_3/TiO_2 exhibits superior efficiency in photoelectrochemical water splitting to the single-metal oxides, which can be attributed to the formation of heterojunction between Fe_2O_3 and TiO_2, improving the separation efficiency of photogenerated electron–hole pairs with higher applied potential [44]. Apart from magnetic iron oxides, core@TiO_2 shell structure materials with various cores (SiO_2 spheres, GR oxide NSs, and carbon nanospheres) with tunable, uniform, and porous TiO_2 shell thickness were also synthesized successfully [45].

A SiO_2/Ta_3N_5 core–shell photocatalyst decorated with Pt NPs and IrO_2 or CoO_x on the inner and outer side of Ta_3N_5 was reported, in which Pt NPs act as electron collectors while IrO_2 or CoO_x act as hole collectors [46]. Furthermore, Pt@Ta_3N_5 hollow spheres were obtained by etching the SiO_2 cores with hot NaOH solution, which showed great enhancement in the H_2 evolution due to the increase of surface area and the easier access to PtNPs, the H_2 promoters. Similarly, Pt@TiO_2@MnO_x hollow spheres were also reported, in which MnO_x was deposited on the inner side of the TiO_2 shell as hole collector and Pt was deposited on the outer side as electron collector [47].

Remarkable spatial charge separation efficiency and photocatalytic activity in water oxidation were achieved with the aforementioned hollowed structure material. Therefore core–shell and hollow structure materials can greatly enhance the spatial separation of the photogenerated electron–hole pairs, and therefore improve the catalytic activity and photon efficiency.

5.3.3 Layer-by-layer structure materials

Significant advances in the fabrication of highly efficient photocatalytic multilayered nanomaterials have been achieved using bottom-up strategies such as the LbL assembly approach. LbL assembly is a versatile technique that has been widely used as a method for the construction of a number of nanoscale films such as advanced functional coatings [48], smart surfaces at the molecular level [49], and biocompatible coatings [50]. LbL process begins with the adsorption of a charged species onto a substrate of opposite charge, thereby reversing the substrate surface charge. Among the numerous advantages of the LbL technique appears the simplicity, the possibility to control the thickness at the molecular level, thus tuning the properties of the nanomaterials is obtained. The versatility of the LbL approach has allowed a broad range of materials (e.g., polymers, NPs, lipids, proteins, dye molecules) to be assembled on various substrates driven by electrostatic interactions, hydrogen bonding, hydrophobic interactions, covalent bonding, and complementary base pairing [51,52]. The large battery of inorganic shapes and composition of the components such as NPs, NSs, and nanowires (NWs), and the possibility to control the film thickness and architecture have considerably increased the fabrication ofLbL structure materials with significantly enhanced photoelectrochemical and photocatalytic activities under visible light irradiation. For example, Bo Liu et al. have assembled nanobuilding blocks of positively charged polymer-modified graphene nanosheets (GNs) and negatively charged CdS QDsto form a highly uniformed GNs–CdS QDs multilayered film that exhibits better photocatalytic properties in comparison with other GR-semiconductor composite materials [53]. They attributed the enhancement to the high interfacial contact reached between the nanobuilding blocks during the LbL assembly process (Fig. 5.3) [53].

Tae-Hyu-Kim et al. reported the synthesis of photocatalytic thin films composed by ionic polymers and positively charged TiO_2 NPs using LbL assembly method [54]. They characterized the film thickness and the film weight of each deposition using ellipsometry, UV–vis spectroscopy, and Quartz Crystal Microbalance. Also, the nanostructures constructed show a uniform distribution of TiO_2 NPs.

Likewise, Shi and coworkers [55] prepared LbL films from Bi_2WO_6 nanoplates, which showed a higher visible-light photoactivity ($\lambda > 420$ nm) than Bi_2WO_6 nanoplate powders. Thickness and photoactivity of the films

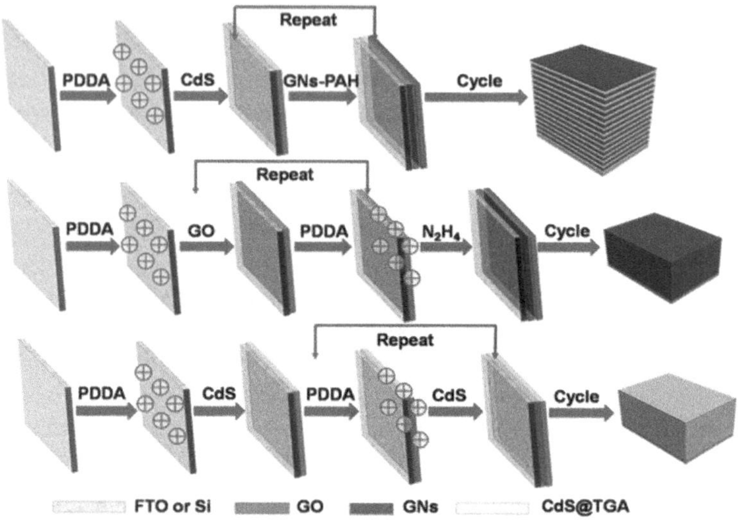

FIGURE 5.3 Schematic illustration of LbL self-assembly of GNs–CdS QDs, GNs, and CdS QDs multilayered films. *Reprinted with permission from American Chemical Society, F. Xiao, J. Miao, B. Liu, Layer-by-layer self-assembly of CdS quantum dots/graphene nanosheets hybrid films for photoelectrochemical and photocatalytic applications, J. Am. Chem. Soc. 136 (2014) 1559–1569 [53].*

were also modified by changing the deposition cycles. Bi_2WO_6 films showed spectral selectivity of the photocatalytic degradation of rhodamine B. Under a wavelength greater than 300 nm, the rhodamine B molecules tend to be transformed to rhodamine over Bi_2WO_6 films. In contrast, under shorter wavelength ($\lambda = 254$ nm) light irradiation, the rhodamine B molecules were completely photodegraded. The strategy of LbL assembly was also used for the synthesis of NWs that showed improved photocatalytic properties [56]. To prepare Au/TiO_2 NWs, ordered arrays were fabricated using LbL assembly of cationic polyelectrolytes and inorganic precursors on template-grown Au NWs.

5.3.4 Surface-modified materials

Among the most used photocatalysts, nanostructured TiO_2 has been considered one of the most promising materials owed to its high photocatalytic activity, low cost, and nontoxicity [57]. Nanosized TiO_2 is characterized by large surface areas, small particle sizes, and a number of surface defects, which provide large number of surface-active sites for the photocatalytic reaction. Despite the advantages mentioned, the high density of surface defects of these nanomaterials decreases their photocatalytic performances significantly due to the negative effects of the defects on the photogenerated charged separation processes. Also, the crystal phase of these nanostructured materials directly impinges on their photocatalytic behavior. At ambient conditions, rutile phase is the thermodynamically stable phase of TiO_2 whereas anatase and brookite are thermodynamically metastable. Besides, it is well known that anatase shows the highest photocatalytic activity among the three crystal phases of TiO_2. To solve these drawbacks, several surface-tuning strategies for TiO_2-based nanomaterials have been successfully developed [58–61]. Up to now, strategies to improve the photocatalytic properties of these nanosized semiconductor particles consist of choosing functional molecules to start surface binding or surface deposition, or surface modification agents to modify surface structures and surface properties of semiconductor nanocrystalline structures.

One of the most effective methods to maintain the stability of anatase TiO_2 phase, and in turn improve the photocatalytic performances of TiO_2 nanomaterials, is based on the surface coating of TiO_2 NPs with semiconductor oxide layers. Surface covering of TiO_2 anatase with La_2O_3 thin layers contributes to retain this phase and consequently its thermal stability. Zhang et al. have constructed TiO_2 anatase particles modified by highly disperse La_2O_3 layer and demonstrated that the phase transition from anatase to rutile in both the bulk and surface regions is significantly retarded, owing to avoiding direct contact of the anatase particles and occupying the surface defect sites of the anatase particles by La_2O_3 [62]. Likewise, surface modification with mesoporous SiO_2 synthesized by sol–hydrothermal processes and a posttreatment with appropriate amount of surfactant F127-modified silica sol, improved considerably the thermal stability range of anatase phase in TiO_2 NPs and in turn the photocatalytic activity (Fig. 5.4) of these nanomaterials [63]. In this direction, SiO_2-coated anatase NP photocatalysts have been synthesized from solutions of $TiOSO_4$ and tetraethyl orthosilicate by hydrolysis under hydrothermal conditions at 200°C [64]. The anatase to rutile phase transformation was retarded to maintain the anatase-type structure up to 1300°C. The –OTi groups in the region of tetrahedrally coordinated Si play a crucial role in the suppression of the phase transition.

Surface modification of TiO_2 NPs by thin carbon layers represents another successful approach to enhance the charge separation and prevent the recombination of photogenerated electron–hole pairs. The function of the thin carbon layers is mainly suppressed (the anatase–rutile phase transformation acting like a physical barrier). Zhao et al. created highly ordered anatase frameworks composed by mesoporous carbon–titania nanocomposites with crystalline-glass via an organic–inorganic–amphiphilic coassembly, followed by the in situ crystallization method [65]. The nanocomposites showed excellent photocatalytic performances toward the photodegradation of rhodamine B in aqueous solutions owed to the highly ordered frameworks and high adsorptive capacity arising from its high surface area (up to 465 m^2/g) and large uniform pore size (4.1 nm). Besides, selective GR coating supported on a mesoporous silica surface has been designed as a highly

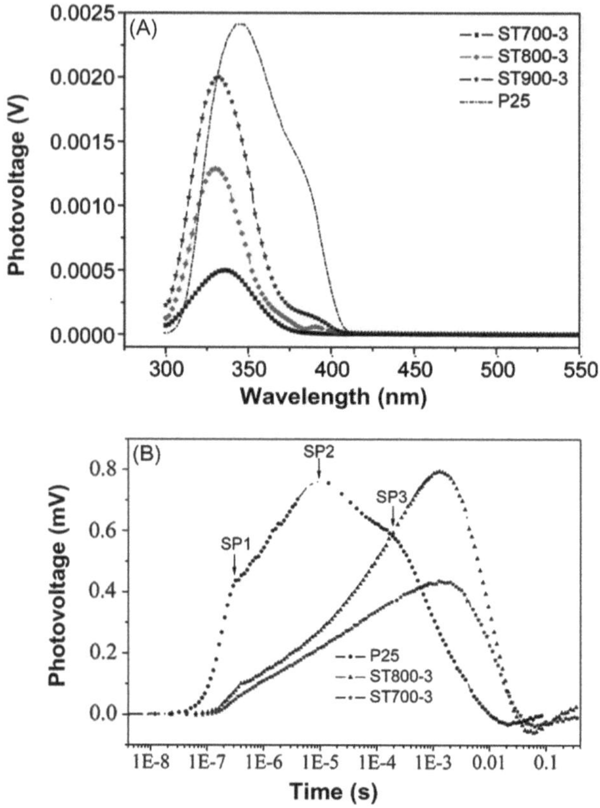

FIGURE 5.4 Surface photovoltage spectroscopy (SPS) responses (A) and transient state surface photovoltage spectroscopy (TS-SPS) response (B) of mesoporous SiO_2-modified TiO_2. *Reprinted with permission from the American Chemical Society, C. Kang, L. Jing, T. Guo, H. Cui, J. Zhou, H. Fu, Mesoporous SiO_2-modified nanocrystalline TiO_2 with high anatase thermal stability and large surface area as efficient photocatalyst, J. Phys. Chem. C. 113 (2009) 1006–1013.*

efficient photocatalytic framework [66]. After carbonization at 1073 K under a N_2 flow, TiO_2 NPs anchored on the mesoporous silica retained an anatase crystalline structure while the GR content and mesoporous material maintained the pore structure as well as the high surface area.

Finally the synthesis of plasmonic nanophotocatalysts using surface modification method, which is based on semiconductor nanomaterials modified by MNPs on their surface, has been widely explored in the last few years [7]. Among the most cited works, TiO_2 NPs modified with noble metals such as Pt, Pd, Ag, and Cu or even bimetallic Ag–Cu NPs or Ag–CuO nanoclusters, show higher photocatalytic yields [67–71]. This synthetic approach takes advantage of the surface plasmon resonance, which is a special and unique feature responsible for the activation of the photocatalyst under visible light [72].

5.3.5 Composite materials

The development of multicomponent hybrid nanocomposites with multifunctional and exciting properties has gained a lot of interest due to their outstanding photoactivity [73,74]. These multicomponent heterostructures usually integrate different kind of materials such as NPs, GR, conductive polymers, one-dimensional (1D) semiconductor materials among others, to achieve the better photocatalytic responses. For example, Kamat's group has developed an organic/inorganic ternary composite consisting of porphyrin, ZnO NPs, and GR through a bottom-up strategy [75]. This ternary nanocomposite shows a remarkably high photocurrent with an incident photon to current efficiency up to 70% because of the formation of building hierarchical electron transfer cascade channels between its components. Amal's group has synthesized a ternary nanocomposite (i.e., GR/$BiVO_4$−Ru/$SrTiO_3$: Rh) which can be used for Z-scheme photocatalytic water splitting under irradiation with visible light [76]. More recently, Bin Liu et al. have designed an Ag NP-modified and GR-wrapped TiO_2 nanobelt hierarchical heterostructure, in which the negatively charged GO was closely incorporated to the positively charged surface-coarsened TiO_2 nanobelt (SC-TNBs) framework driven by electrostatic attractive interactions followed by the deposition of Ag NPs via photochemical reduction [77]. The much higher photocurrent response of the Ag/GR/SC-TNBs ternary nanocomposite strongly confirmed that the hierarchical assembly of three components significantly improved both the efficiency in the separation and the lifetime of the photoexcited electron−hole pairs in comparison with its counterparts (Fig. 5.5A and B).

Among other interesting groups of photocatalytic active nanocomposites, the photoactive TiO_2 mesoporous materials modified with noble MNPs are

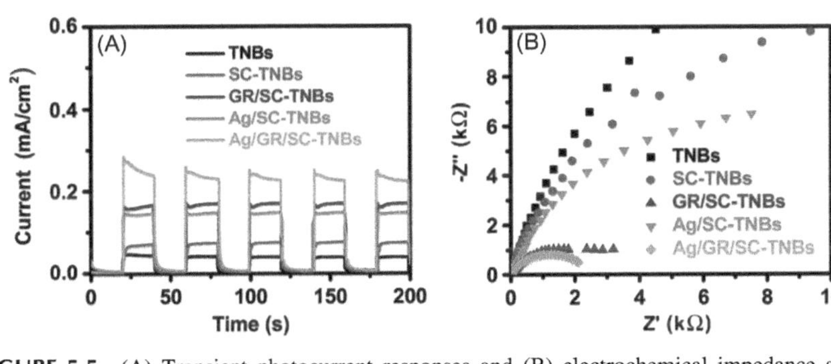

FIGURE 5.5 (A) Transient photocurrent responses and (B) electrochemical impedance spectroscopy Nyquist plots of the different samples in an aqueous solution of 0.1 M Na_2SO_4 (pH = 6.8). *Reprinted with permission from the Royal Society of Chemistry, J. Zhang, F.-X. Xiao, G. Xiao, B. Liu, Self-assembly of a Ag nanoparticle-modified and graphene-wrapped TiO_2 nanobelt ternary heterostructure: surface charge tuning toward efficient photocatalysis, Nanoscale. 6 (2014) 11293–11302.*

found. Combining the large surface area of the ordered mesoporous channels together with improving the efficiency of electron–hole separation facilitated by the noble MNPs, a large battery of promising photocatalysts and photovoltaics can be developed [78]. Ismail et al. have synthesized mesostructured Pt/TiO_2 nanocomposites at different Pt (0–2 wt.%) and anatase/rutile ratios by a simple one-step sol–gel method for methanol oxidation [79]. They demonstrated that the overall photocatalytic activities of 0.5 wt.% Pt/TiO_2 nanocomposites are three times higher than that of colloidal 0.5 wt. % Pt commercial photocatalysts. The large number of mesoporous channels interconnected and the bimetallic structure are the most important factors facilitating access to the reactive sites on the surface of the photocatalyst and in turn improving the photocatalytic activity.

A greener strategy to synthesize highly efficient photocatalysts using biomolecules is performed by a biomimetic synthetic approach. One remarkable advantage of this synthetic method is that proteins and other biomolecules can implement fine and accurate control over the production of inorganic materials, such as size, shape, and crystal structure [79]. Due to their versatile inducing/reducing properties, some proteins can be used to obtain multifunctional metal nanocomposites. In this direction, Chuang Liu et al. have synthesized a $TiO_2 - SiO_2 - Ag$ nanocomposite through a biomimetic method with enhanced visible-light photocatalytic activity by using a lysozyme as an inducing/reducing agent [80]. The resultant nanocomposites display a high photocatalytic activity for the degradation of Rhodamine B under visible-light irradiation. It can be attributed to both the plasmonic properties of Ag NPs and the elevated adsorption capacity for rhodamine B due to the high specific surface area of the framework by the SiO_2 incorporation.

5.4 Applications in photocatalysis

Photocatalysis is triggered mainly by the oxidation and reduction reactions on the surface of a photocatalyst material, mediated by the valence band (VB) such as holes (h^+) and conduction band (CB) such as electrons (e^-) generated by the absorption of UV–vis light. In the last decade, many photocatalytic nanosystems based on nanostructures within functional networks have been built for many potential applications such as the degradation of pollutants, direct solar energy conversion to storable fuels, and selective transformations for organic synthesis [81].

Photocatalytic processes can induce the formation of aggressive species such as hydroxyl (OH) or superoxide radicals from the humidity and atmospheric oxygen. The species formed are able to oxidize and decompose organic materials or smelling gas or eliminate bacteria, which have been widely employed as an efficient mechanism for the mineralization of toxic organic compounds, hazardous organic material, and microbial disinfection [82]. A number of high-efficient photocatalytic nanomaterials have been designed with the purpose of controlling the environmental toxic pollutants through degradation and mineralization of hazardous organic compounds to CO_2 and H_2O, reduction of toxic metal ions into nontoxic states, deactivation and destruction of all the water-borne microorganisms, decomposition of the air pollutants such as NO_2, CO, and NH_3 and degradation of the plastic waste, among other environmental applications [83].

Development of multicomponent 1D-based nanocomposites has shown a great potential in the photodegradation of organic compounds. Xiao has employed the LbL approach to synthesize a 1D noble metal/semiconductor [84,85]. The close interfacial contact between MNPs and semiconductor nanostructure is reached by the incorporation of polyelectrolyte layers. The nanocomposite heterostructures display remarkably enhanced photoactivity and outstanding photostability for the photodegradation of organic dyes (Fig. 5.6).

Zheng's group has constructed TiO_2/carbon nanotubes (CNTs) nanocomposites for the degradation of methyl orange (MO). The large surface area and porosity of the nanomaterials synthesized together with the suitable orientation in the attachment between TiO_2 nanocrystalline units, surface defects, and CNTs greatly suppressed the electron–hole recombination processes and in turn improved the photo-degradation of MO pollutant [86].

Among other significant photocatalytic applications of functional multicomponent nanostructures, the fabrication of nanosystems for photocatalytic production of H_2 has attracted a lot of attention due to the opportunity to generate hydrogen by a clean and low-cost approach [87]. Ju Zhang et al. have reported the assembly of a novel NiS NP-modified CdS nanorod $p–n$ junction photocatalysts for the hydrogen production from water [70]. Due to the differences between the Fermi levels of p-type NiS and n-type CdS, the

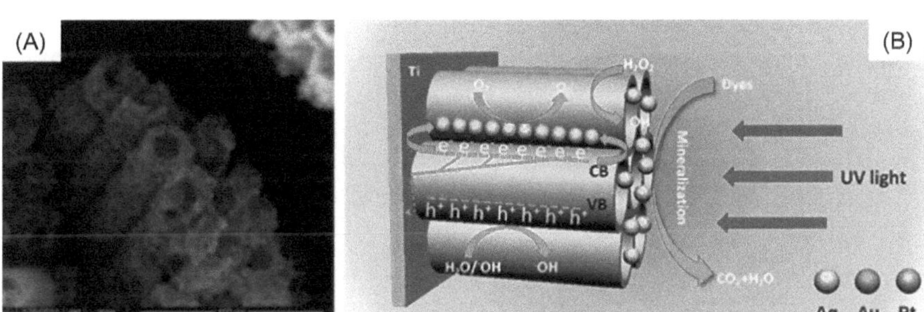

FIGURE 5.6 (A) Scanning electron microscopic image of the titania nanotube array substrate. (B) Schematic representation of the proposed mechanism for the photocatalytic degradation. *Reprinted with permission from the American Chemical Society, F. Xiao, Layer-by-layer self-assembly construction of highly ordered metal - TiO_2 nanotube arrays heterostructures (M/TNTs, M = Au, Ag, Pt) with tunable catalytic activities, J. Phys. Chem. C. 116 (2012) 16487–16498.*

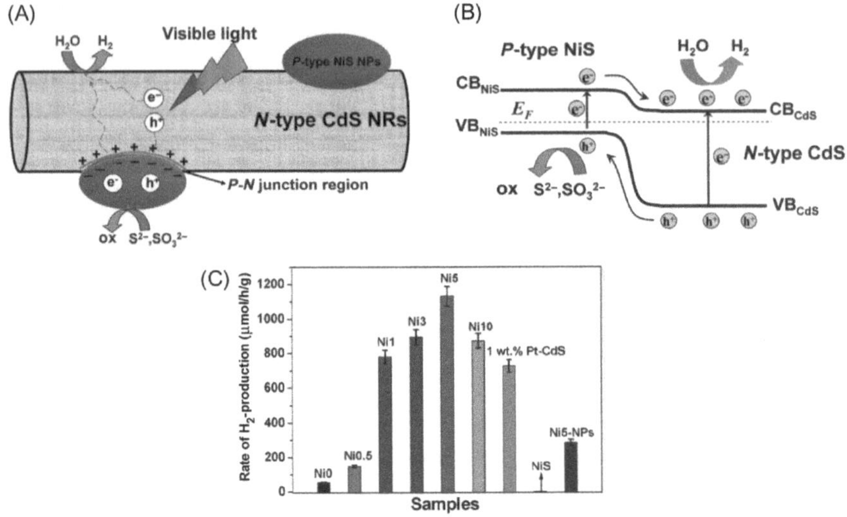

FIGURE 5.7 (A) Scheme of charge transfer and separation, (B) the *p–n* junction band structure, and schematic illustration of the electron–hole separation process. (C) Comparison of the visible light photocatalytic H_2-production activity of Ni0, Ni0.5, Ni1, Ni3, Ni5, Ni10, 1 wt.% Pt-CdS, NiS, and Ni5-NPs samples under visible light ($\lambda = 420$ nm). *Reprinted with permission from Royal Society of Chemistry, J. Zhang, S.Z. Qiao, L. Qi, J. Yu, Fabrication of NiS modified CdS nanorod p–n junction photocatalysts with enhanced visible-light photocatalytic H_2-production activity, Phys. Chem. Chem. Phys. 15 (2013) 12088.*

heterostructures undergo an inner electric field that increases the number of photoelectrons from CB of NiS into the CB of CdS (Fig. 5.7).

Photocatalytic nanosystems have also been employed in selective organic transformations such as selective oxidation and reduction processes, isomerization reactions, C–H bond activations, and C–C and C–N bond-forming reactions [88,89]. A promising 1D nanomaterial formed by CdS@TiO_2 core–shell nanocomposites for the selective oxidation of alcohols to aldehydes was developed by Yi-Jun Xu and coworkers (Fig. 5.8) [2]. They found that the TiO_2 shell of the 1D nanostructures plays a twofold role in the organic reaction blocking the pass of the photo-excited holes from the CdS core to the surface of TiO_2 shell and improving the motion of the photo-excited electrons.

5.5 Applications in photoelectrocatalysis

Hydrogen as a fuel is an interesting alternative to the conventional fossil fuels because it is storable, transportable, and can be transformed into electric energy efficiently. Photo(electro)chemical water splitting is one of the most investigated pathways nowadays to produce environmental favorable

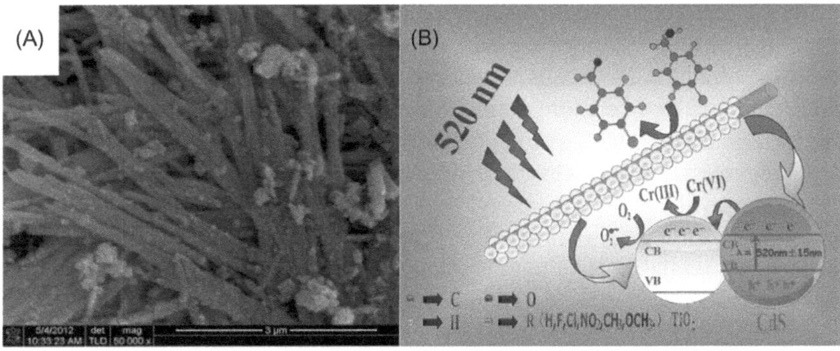

FIGURE 5.8 (A) Scanning electron microscopic image of the as-prepared 1D-CdS@TiO_2 CSNs; the inset is the corresponding schematic model, and (B) illustration of the proposed reaction mechanism. *Reprinted with permission from the American Chemical Society, S. Liu, N. Zhang, Z.R. Tang, Y.J. Xu, Synthesis of one-dimensional CdS@TiO_2 core-shell nanocomposites photocatalyst for selective redox: the dual role of TiO_2 shell, ACS Appl. Mater. Interfaces. 4 (2012) 6378–6385.*

hydrogen-based fuel. Nevertheless, it still is not a viable method because of its high cost and low yield compared with conventional fossil fuels.

Water photoelectrolysis was first reported in 1972 by Fujishima and Honda [10] using TiO_2 as photoanode. Since then, a wide research line has been developed in order to find a proper photoelectrode, which presents a band gap high enough for water splitting but small enough to absorb an important fraction of the visible spectrum, presents low electron–hole recombination and resists the corrosion [90]. Photoelectrochemical water splitting is based, mainly, on the conversion of light energy into electricity in a cell, based on two electrodes immersed in an aqueous electrolyte of which, at least one, must be composed of a semiconductor capable of absorbing the incoming radiative photons, creating electron–hole pairs responsible for the water-splitting reaction [91]. Conversion efficiency from sunlight to H_2 must be, at least, of 10% to make this kind of fuel practical and cost-effective. Water splitting is the chemical process which takes place at the photoanode/electrolyte interface, generating hydrogen and oxygen. H_2 generation from H_2O requires, at least, two catalytic steps:

- the oxidation of two H_2O molecules into O_2 and H^+ and
- the subsequent reduction of H^+ to H_2.

Holes energy at the photoanode surface is given by VB, whereas electrons energy at the metallic cathode surface is given by Fermi level of the metal. Thermodynamics dictates a minimum energy requirement of 1.23 eV for each incoming photon.

The most economical way employed for water splitting is via photoelectrochemical cells (PECs). In these devices, incoming photons in a semiconductor create electron–hole pairs, which create a potential electrical difference through the surface into the junctions of two different materials. The most effective PEC cells are called tandem cells or Z-scheme-type cells, which connect two or more semiconductors in series with different absorption ranges. In this way, semiconductors with lower gap bands can be used, so this device is able to absorb a higher solar spectrum fraction, obtaining a higher photovoltage [92]. Usually, in a tandem cell configuration, the photoanode is an *n*-type semiconductor, where the generated holes flow through the surface, oxidizing H_2O to O_2. Meanwhile, the generated electrons flow in the opposite direction to recombine with photocathode holes, which is a *p*-type semiconductor. On the contrary, the electrons flow through the surface of the photocathode, reducing H^+ to H_2, and the holes flow in the opposite direction for recombining with electrons from the photoanode.

In case of a tandem cell illustrated in Fig. 5.9, different photoelectrodes can be chosen, so that the photoanode presents more positive VB limit to obtain better results in the oxidation by the generated holes, whereas the photocathode needs to present more negative CB minimum to reduce H^+ to H_2 more efficiently [93]. In recent years, with the advance of nanoscience field, application of nanostructured materials in photoelectrochemical devices has been widely studied to achieve better efficiency for water splitting. The use of nanostructured materials makes the generated carriers reach earlier, from semiconductor surface, the electrolyte and at the back contact. This mechanism is usually carried out, in absence of an external electrical field, by a diffusive mechanism, so the mobility and lifetime of charge carriers must be

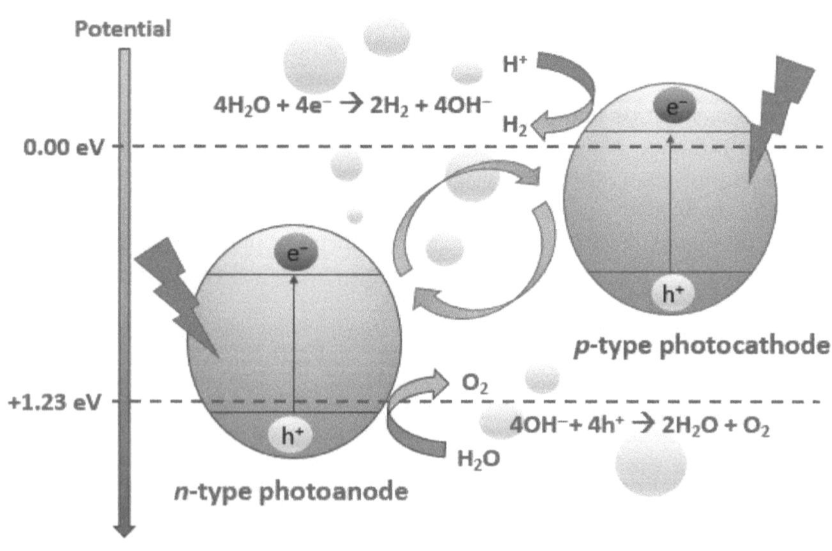

FIGURE 5.9 A dual absorber system consisting of a photoanode and a photocathode in a tandem configuration.

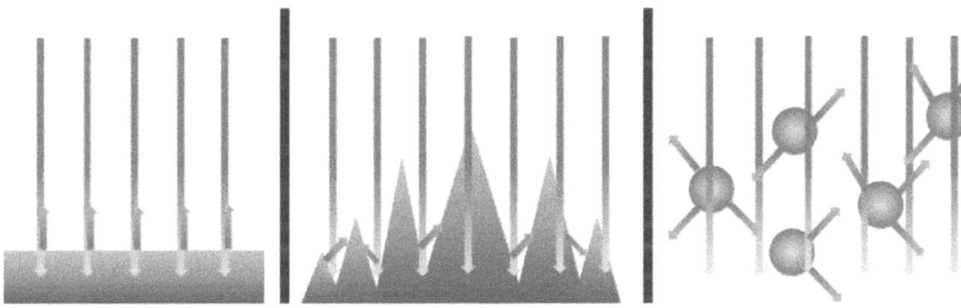

FIGURE 5.10 Light distribution in (A) flat surface, (B) rough surface, and (C) particle suspensions. Short arrows signify scattered or reflected light.

appropriate. This is useful for the first row transition metal oxides (i.e., MnO_2, Fe_2O_3), the holes of which present low mobility and lifetime, while presenting several roughness on their surface [94]. The roughness can "trap" incoming light horizontally by reflexive mechanisms (light scattering). In flat macrosurfaces, the light is totally reflected without absorption. In contrast, the light reflection is maximum in all possible directions in particle suspension that allows a greater absorption (Fig. 5.10).

Another advantage of nanosized particles is that particle sizes have a drastic effect on electronic states density because of quantum size confinement and therefore on the optic response. The space for electron movement in NPs is very small, around electron wavelength (De Broglie wavelength). The reduced region limits electron movement, thus electrons are confined. Electron confinement changes energy levels respect bulk material [95]. With increasing band gap, CB reaches more reductive potentials while VB reaches more oxidizing potentials, increasing the thermodynamic driving force of the interfacial charge transfer, thus achieving interesting results applicable to water splitting (i.e., CdS and TiO_2 nanoclusters [96]).

Potential determining ions (PDI), which are ions adsorbed on particle surface, have direct influence in energetic properties of PEC surfaces. The application of an external electric field can lead to energetic structure modification of the NP, controlling the interfacial charge transfer due to PDIs. The presence of sulfide ions shifts the band gap of Bi_2S_3 upward, enabling the injection of electrons from larger Bi_2S_3 NPs into TiO_2 [97]. Shift of the band edges of TiO_2 can be achieved by pH-induced protonation of the surface groups, thus the CB of TiO_2 shifts to more negative potentials with increasing pH [98]. The larger specific surface area of NPs promotes charge transfer through material interfaces of electrochemical devices, improving the kinetics of water splitting reactions (Fig. 5.11) [99].

Another advantage of the use of nanomaterials in solar cells is multiple exciton generation in the creation of multiple excitons (combination of an electron from CB with the respective hole from VB, enhanced by an electric

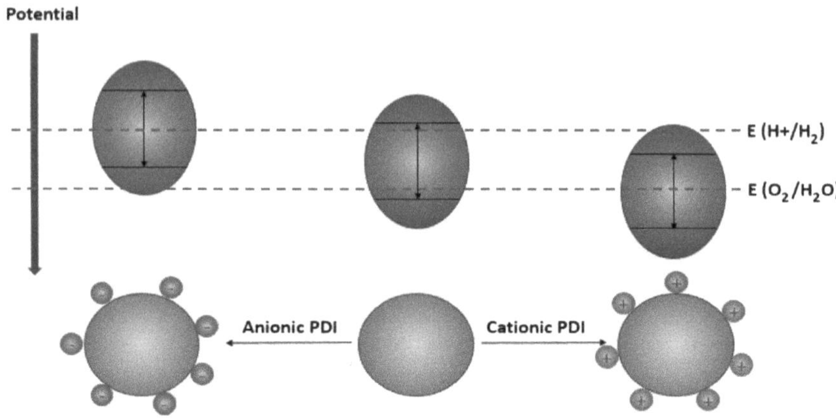

FIGURE 5.11 Effect of PDIs on nanocrystal energetics.

interaction) from only one incoming photon into the solar device. Conversion efficiency is increased because excess kinetic energy of the generated electron and holes can create additional electron–hole pairs. This phenomenon takes place in semiconductor nanocrystals (also called QDs), resulting in lower threshold photon energy, keeping energy by an electron–hole pair multiplication process [100].

Nanosized materials have been widely studied to increase the efficiency of PEC devices in water splitting. Water-splitting photocurrents over 3.0 mA/cm^2 were first time achieved at an applied potential of +1.23 V versus RHE under AM1.5G 100 mW cm^2 simulated solar, using nanostructure hematite deposited with IrO_2 as photoanode [101]. In addition, applications of other NPs (e.g., $BiVO_4$, ZnO, WO_3, GR, and g-C_3N_4) in PEC devices were also reported with excellent activities [102–106].

The applications of nanostructured materials in electrochemical devices entail numerous advantages for carrying out electrochemical transformations, but also show certain disadvantages. First, increase of semiconductor surface area also involves higher degree of surface electron–hole recombination, which can be resolved by reducing the amount of surface defects. Besides, the absorbed photon flux of the nanomaterials with higher surface area is lower than bulk materials, due to the surface defects. In detail, the photon flux per unit area received by the surface with defects is lower than that of the nondefect surface, reducing the thermodynamic driving force available for water electrolysis. Because of the small particle size in the nanocrystalline films, charge carriers move by diffusion instead of charge drift movement, which is slower than the latter. Additionally, drift movement delays the recombination of charge carriers [107].

In practice, H_2 production by direct water splitting still has not enough yield to make it as an efficient fuel source. Hence the presence of a

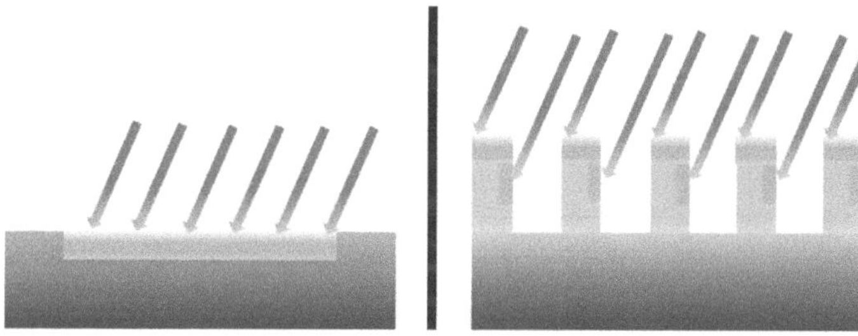

FIGURE 5.12 Increased surface roughness leads to reduced absorbed flux. This reduces the open circuit voltage, and the thermodynamic driving force for water electrolysis.

TABLE 5.1 Reduction potentials of energy conversion reactions [116].

Reaction	E^0 (V) vs RHE	$E^{0\prime}$ (V) vs NHF (pH = 7)
$CO_2 + e^- \rightarrow CO_2^{\cdot -}$	−1.48	−1.90
$CO_2 + 2H^+ + 2e^- \rightarrow CO + H_2O$	−0.11	−0.53
$CO_2 + 2H^+ + 2e^- \rightarrow HCOOH$	−0.19	−0.61
$CO_2 + 4H^+ + 4e^- \rightarrow HCHO + H_2O$	−0.06	−0.48
$CO_2 + 6H^+ + 6e^- \rightarrow CH_3OH + H_2O$	+0.04	−0.38
$CO_2 + 8H^+ + 8e^- \rightarrow CH_4 + 2H_2O$	+0.18	−0.24
$2H^+ + 2e^- \rightarrow H_2$	0.00	−0.42
$2H_2O \rightarrow O_2 + 4H^+ + 4e^-$	+1.23	+0.81

Source: J. Zhao, X. Wang, Z. Xu, J.S.C. Loo, Hybrid catalysts for photoelectrochemical reduction of carbon dioxide: a prospective review on semiconductor/metal complex co-catalyst systems, J. Mater. Chem. A. 2 (2014) 15228.

sacrificial agent is necessary because the sacrificial agent consumes irreversibly photogenerated holes for decreasing charge recombination, and therefore increases H_2 production efficiency (Fig. 5.12) [6,108].

Apart from H_2 evolution by water splitting, reduction of the abundant CO_2 into fuels and chemicals using solar energy in CO_2 is of great interest because the increase CO_2 emission resulting from the growing energy consumption has raised serious environmental issues, such as desertification, ocean acidification, and global warming. Photoelectric system for CO_2 reduction was first reported by Halmann in 1978, which transformed CO_2 into formic acid, formaldehyde, and methanol in a PEC with a GaP photocathode and a carbon rod as the counter electrode [109].

The mechanism for CO_2 reduction is similar to the mechanism of water splitting but the photogenerated electrons reduce CO_2 instead of H_2O, which is illustrated in Table 5.1. Therefore application of nanomaterials in CO_2

conversion has been widely investigated, although enough effective transformations have not been reached yet. Nanostructured semiconductor materials (GaP, TiO_2, ZnO, CdS, SiC, WO_3, etc.) such as nanorods, NWs, or nanotubes have shown promising results, achieving selectivity among 80% in their conversions, while incorporation on noble MNPs promotes the PEC activity [110–115]. It is very important for the CO_2 transformations that the redox potential of the CB electrons to be more negative than the respective redox pairs intending to reduce [110].

5.6 Conclusions

NPs within various functional frameworks together with their synthetic methods and their applications in photo(electro)catalysis are discussed in this chapter. The applications in chemical transformations and energy conversion using light irradiation show great potential in contributing to reduce the reliability on fossil fuels and integrating the society toward a more renewable and sustainable future. However, though the current results of photo(electro) catalysis are promising, more research efforts on the photo(electro)catalysts, especially photoactive nanostructure materials, are required to make them feasible. The best is yet to come.

Acknowledgments

Rafael Luque gratefully acknowledges funding from MINECO under project CTQ2016-78289-P. The European Union's Horizon 2020 research and innovation Program under the Marie Skłodowska-Curie is acknowledged for granting ITN Photo4Future, ITN-MSCA-2014-641861. Project website: http://www.photo4future.com/.

References

[1] J.C. Colmenares, W. Ouyang, M. Ojeda, E. Kuna, O. Chernyayeva, D. Lisovytskiy, et al., Mild ultrasound-assisted synthesis of TiO_2 supported on magnetic nanocomposites for selective photo-oxidation of benzyl alcohol, Appl. Catal. B Environ. 183 (2016) 107–112. Available from: https://doi.org/10.1016/j.apcatb.2015.10.034.

[2] S. Liu, N. Zhang, Z.R. Tang, Y.J. Xu, Synthesis of one-dimensional CdS@TiO_2 core-shell nanocomposites photocatalyst for selective redox: the dual role of TiO_2 shell, ACS Appl. Mater. Interfaces. 4 (2012) 6378–6385. Available from: https://doi.org/10.1021/am302074p.

[3] Y. Shen, T. Xiong, T. Li, K. Yang, Tungsten and nitrogen co-doped TiO_2 nano-powders with strong visible light response, Appl. Catal. B Environ. 83 (2008) 177–185. Available from: https://doi.org/10.1016/j.apcatb.2008.01.037.

[4] M. Li, L. Zhang, X. Fan, Y. Zhou, M. Wu, J. Shi, Highly selective CO_2 photoreduction to CO over g-C_3N_4/Bi_2WO_6 composites under visible light, J. Mater. Chem. A. 3 (2015) 5189–5196. Available from: https://doi.org/10.1039/C4TA06295G.

[5] S. Obregon, G. Colon, Improved O-2 evolution from a water splitting reaction over Er^{3+} and Y^{3+} co-doped tetragonal $BiVO_4$, Catal. Sci. Technol. 4 (2014) 2042–2050. Available from: https://doi.org/10.1039/C4cy00050a.

[6] V.M. Daskalaki, M. Antoniadou, G. Li Puma, D.I. Kondarides, P. Lianos, Solar light-responsive $Pt/CdS/TiO_2$ photocatalysts for hydrogen production and simultaneous degradation of inorganic or organic sacrificial agents in wastewater, Environ. Sci. Technol. 44 (2010) 7200–7205. Available from: https://doi.org/10.1021/es9038962.

[7] S. Linic, P. Christopher, D.B. Ingram, Plasmonic-metal nanostructures for efficient conversion of solar to chemical energy, Nat. Mater. 10 (2011) 911–921. Available from: https://doi.org/10.1038/nmat3151.

[8] K. Maeda, K. Teramura, D. Lu, T. Takata, N. Saito, Y. Inoue, et al., Photocatalyst releasing hydrogen from water, Nature. 440 (2006) 295. Available from: https://doi.org/10.1038/440295a.

[9] W. Ouyang, E. Kuna, A. Yepez, A. Balu, A. Romero, J. Colmenares, et al., Mechanochemical synthesis of TiO_2 nanocomposites as photocatalysts for benzyl alcohol photo-oxidation, Nanomaterials. 6 (2016) 93. Available from: https://doi.org/10.3390/nano6050093.

[10] A. Fujishima, K. Honda, TiO2 photoelectrochemistry and photocatalysis, Nature. 213 (1972) 8656. Available from: https://electrochem.org/dl/ma/203/pdfs/2729.pdf%5Cnpapers2://publication/uuid/F6B805BA-46E0-443A-B666-4F5841D3D337.

[11] D.A.H. Hanaor, I. Chironi, I. Karatchevtseva, G. Triani, C.C. Sorrell, Single and mixed phase TiO_2 powders prepared by excess hydrolysis of titanium alkoxide, Adv. Appl. Ceram. 111 (2012) 149–158. Available from: https://doi.org/10.1179/1743676111Y.0000000059.

[12] B.E. Yoldas, Hydrolysis of titanium alkoxide and effects of hydrolytic polycondensation parameters, J. Mater. Sci. 21 (1986) 1087–1092. Available from: https://doi.org/10.1007/BF01117399.

[13] J. Park, J. Joo, G.K. Soon, Y. Jang, T. Hyeon, Synthesis of monodisperse spherical nanocrystals, Angew. Chemie - Int. Ed. 46 (2007) 4630–4660. Available from: https://doi.org/10.1002/anie.200603148.

[14] A. Vioux, Nonhydrolytic sol – gel routes to oxides, Chem. Mater. 9 (1997) 2292–2299. Available from: https://doi.org/10.1021/cm970322a.

[15] K.H. Leong, P. Monash, S. Ibrahim, P. Saravanan, Solar photocatalytic activity of anatase TiO_2 nanocrystals synthesized by non-hydrolitic sol-gel method, Sol. Energy. 101 (2014) 321–332. Available from: https://doi.org/10.1016/j.solener.2014.01.006.

[16] J. Lai, W. Niu, R. Luque, G. Xu, Solvothermal synthesis of metal nanocrystals and their applications, Nano Today. 10 (2015) 240–267. Available from: https://doi.org/10.1016/j.nantod.2015.03.001.

[17] D.R. Modeshia, R.I. Walton, Solvothermal synthesis of perovskites and pyrochlores: crystallisation of functional oxides under mild conditions, Chem. Soc. Rev. 39 (2010) 4303. Available from: https://doi.org/10.1039/b904702f.

[18] H.G. Yang, G. Liu, S.Z. Qiao, C.H. Sun, Y.G. Jin, S.C. Smith, et al., Solvothermal synthesis and photoreactivity of anatase TiO_2 nanosheets with dominant {001} facets, J. Am. Chem. Soc. 131 (2009) 4078–4083. Available from: https://doi.org/10.1021/ja808790p.

[19] K. Horie, M. Barón, R.B. Fox, J. He, M. Hess, J. Kahovec, et al., Definitions of terms relating to reactions of polymers and to functional polymeric materials (IUPAC Recommendations 2003), Pure Appl. Chem. 76 (2004) 889–906. Available from: https://doi.org/10.1351/pac200476040889.

[20] V. Šepelák, A. Düvel, M. Wilkening, K.-D. Becker, P. Heitjans, Mechanochemical reactions and syntheses of oxides, Chem. Soc. Rev. 42 (2013) 7507. Available from: https://doi.org/10.1039/c2cs35462d.

[21] A.L. Garay, A. Pichon, S.L. James, Solvent-free synthesis of metal complexes, Chem. Soc. Rev. 36 (2007) 846. Available from: https://doi.org/10.1039/b600363j.

[22] S.L. James, C.J. Adams, C. Bolm, D. Braga, P. Collier, T. Friščić, et al., Mechanochemistry: opportunities for new and cleaner synthesis, Chem. Soc. Rev. 41 (2012) 413–447. Available from: https://doi.org/10.1039/C1CS15171A.

[23] H. Li, G. Lu, Y. Wang, Z. Yin, C. Cong, Q. He, et al., Mechanical exfoliation and characterization of single- and few-layer nanosheets of WSe_2, TaS_2, and $TaSe_2$, Small. 9 (2013) 1974–1981. Available from: https://doi.org/10.1002/smll.201202919.

[24] M. Yi, Z. Shen, A review on mechanical exfoliation for the scalable production of graphene, J. Mater. Chem. A. 3 (2015) 11700–11715. Available from: https://doi.org/10.1039/C5TA00252D.

[25] K. Zhu, W. Wang, A. Meng, M. Zhao, J. Wang, M. Zhao, et al., Mechanically exfoliated g-C_3N_4 thin nanosheets by ball milling as high performance photocatalysts, RSC Adv. 5 (2015) 56239–56243. Available from: https://doi.org/10.1039/C5RA09040G.

[26] H. Xu, B.W. Zeiger, K.S. Suslick, Sonochemical synthesis of nanomaterials, Chem. Soc. Rev. 42 (2013) 2555–2567. Available from: https://doi.org/10.1039/C2CS35282F.

[27] K.S. Suslick, The chemical effects of ultrasound, Sci. Am. 260 (1989) 80–86. Available from: https://doi.org/10.1038/scientificamerican0289-80.

[28] W. Xuan, C. Zhu, Y. Liu, Y. Cui, Mesoporous metal–organic framework materials, Chem. Soc. Rev. 41 (2012) 1677–1695. Available from: https://doi.org/10.1039/C1CS15196G.

[29] H. Deng, S. Grunder, K.E. Cordova, C. Valente, H. Furukawa, M. Hmadeh, et al., Large-pore apertures in a series of metal-organic frameworks, Science 336 (2012) 1018–1023. Available from: https://doi.org/10.1126/science.1220131.

[30] H. Furukawa, K.E. Cordova, M. O'Keeffe, O.M. Yaghi, The chemistry and applications of metal-organic frameworks, Science 341 (2013). Available from: https://doi.org/10.1126/science.1230444. 1230444–1230444.

[31] M. Dan-Hardi, C. Serre, T. Frot, L. Rozes, G. Maurin, C. Sanchez, et al., Crystalline highly porous titanium (IV) dicarboxylate, J. Am. Chem. Soc. 131 (2009) 10857–10859. Available from: https://doi.org/10.1021/ja903726m.

[32] Y. Horiuchi, T. Toyao, K. Miyahara, L. Zakary, D. Do Van, Y. Kamata, et al., Visible-light-driven photocatalytic water oxidation catalysed by iron-based metal–organic frameworks, Chem. Commun. 52 (2016) 5190–5193. Available from: https://doi.org/10.1039/C6CC00730A.

[33] C.G. Silva, I. Luz, F.X. Llabrés, I. Xamena, A. Corma, H. García, Water stable Zr-Benzenedicarboxylate metal-organic frameworks as photocatalysts for hydrogen generation, Chem. - A Eur. J. 16 (2010) 11133–11138. Available from: https://doi.org/10.1002/chem.200903526.

[34] C.H. Hendon, D. Tiana, M. Fontecave, C. Sanchez, L. D'arras, C. Sassoye, et al., Engineering the optical response of the titanium-MIL-125 metal–organic framework through ligand functionalization, J. Am. Chem. Soc. 135 (2013) 10942–10945. Available from: https://doi.org/10.1021/ja405350u.

[35] L. Chen, R. Luque, Y. Li, Controllable design of tunable nanostructures inside metal–organic frameworks, Chem. Soc. Rev. 46 (2017) 4614–4630. Available from: https://doi.org/10.1039/C6CS00537C.

[36] X. Fang, Q. Shang, Y. Wang, L. Jiao, T. Yao, Y. Li, et al., Single Pt Atoms confined into a metal-organic framework for efficient photocatalysis, Adv. Mater. 1705112 (2018) 1705112. Available from: https://doi.org/10.1002/adma.201705112.

[37] C. Zhang, L. Qiu, F. Ke, Y. Zhu, Y. Yuan, G. Xu, et al., A novel magnetic recyclable photocatalyst based on a core–shell metal–organic framework Fe_3O_4@MIL-100(Fe) for the decolorization of methylene blue dye, J. Mater. Chem. A. 1 (2013) 14329. Available from: https://doi.org/10.1039/c3ta13030d.

[38] W. Zhan, Q. Kuang, J. Zhou, X. Kong, Z. Xie, L. Zheng, Semiconductor@metal–organic framework core–shell heterostructures: a case of zno@zif-8 nanorods with selective photoelectrochemical response, J. Am. Chem. Soc. 135 (2013) 1926–1933. Available from: https://doi.org/10.1021/ja311085e.

[39] M.A. Nasalevich, R. Becker, E.V. Ramos-Fernandez, S. Castellanos, S.L. Veber, M.V. Fedin, et al., Co@NH_2-MIL-125(Ti): cobaloxime-derived metal–organic framework-based composite for light-driven H_2 production, Energy Environ. Sci. 8 (2015) 364–375. Available from: https://doi.org/10.1039/C4EE02853H.

[40] R. Ghosh Chaudhuri, S. Paria, Core/shell nanoparticles: classes, properties, synthesis mechanisms, characterization, and applications, Chem. Rev. 112 (2012) 2373–2433. Available from: https://doi.org/10.1021/cr100449n.

[41] M. Ye, Q. Zhang, Y. Hu, J. Ge, Z. Lu, L. He, et al., Magnetically recoverable core-shell nanocomposites with enhanced photocatalytic activity, Chem. - A Eur. J. 16 (2010) 6243–6250. Available from: https://doi.org/10.1002/chem.200903516.

[42] I. Jang, K.-E. You, Y.C. Kim, S.-G. Oh, Surfactant-assisted preparation of core-shell-type TiO_2 –Fe_2O_3 composites and their photocatalytic activities under room light irradiation, Appl. Surf. Sci. 316 (2014) 187–193. Available from: https://doi.org/10.1016/j.apsusc.2014.07.204.

[43] Y. Xia, L. Yin, Core–shell structured α-Fe_2O_3@TiO_2 nanocomposites with improved photocatalytic activity in the visible light region, Phys. Chem. Chem. Phys. 15 (2013) 18627. Available from: https://doi.org/10.1039/c3cp53178c.

[44] M. Wang, M. Pyeon, Y. Gönüllü, A. Kaouk, S. Shen, L. Guo, et al., Constructing Fe_2O_3/TiO_2 core–shell photoelectrodes for efficient photoelectrochemical water splitting, Nanoscale. 7 (2015) 10094–10100. Available from: https://doi.org/10.1039/C5NR01493J.

[45] W. Li, J. Yang, Z. Wu, J. Wang, B. Li, S. Feng, et al., Coating method to construct uniform porous TiO_2 shells for multifunctional core–shell structures, J. Am. Chem. Soc. 134 (2012) 11864–11867. Available from: https://doi.org/10.1021/ja3037146.

[46] D. Wang, T. Hisatomi, T. Takata, C. Pan, M. Katayama, J. Kubota, et al., Core/shell photocatalyst with spatially separated Co-catalysts for efficient reduction and oxidation of water, Angew. Chemie - Int. Ed. 52 (2013) 11252–11256. Available from: https://doi.org/10.1002/anie.201303693.

[47] A. Li, T. Wang, X. Chang, W. Cai, P. Zhang, J. Zhang, et al., Spatial separation of oxidation and reduction co-catalysts for efficient charge separation: Pt@TiO_2 @MnO_x hollow spheres for photocatalytic reactions, Chem. Sci. 7 (2016) 890–895. Available from: https://doi.org/10.1039/C5SC04163E.

[48] A.A. Mamedov, N.A. Kotov, Free-standing layer-by-layer assembled films of magnetite nanoparticles, Langmuir. 16 (2000) 5530–5533. Available from: https://doi.org/10.1021/la000560b.

[49] P.T. Hammond, Form and function in multilayer assembly: new applications at the nanoscale, Adv. Mater. 16 (2004) 1271–1293. Available from: https://doi.org/10.1002/adma.200400760.

[50] Z. Tang, Y. Wang, P. Podsiadlo, N.A. Kotov, Biomedical applications of layer-by-layer assembly: from biomimetics to tissue engineering, Adv. Mater. 18 (2006) 3203–3224. Available from: https://doi.org/10.1002/adma.200600113.

[51] W.B. Stockton, M.F. Rubner, Molecular-level processing of conjugated polymers. 4. Layer-by-layer manipulation of polyaniline via hydrogen-bonding interactions, Macromolecules. 30 (1997) 2717–2725. Available from: https://doi.org/10.1021/ma9700486.

[52] Z. Liang, Q. Wang, Multilayer assembly and patterning of poly(p-phenylenevinylene)s via covalent coupling reactions, Langmuir. 20 (2004) 9600–9606. Available from: https://doi.org/10.1021/la048828c.

[53] F. Xiao, J. Miao, B. Liu, Layer-by-layer self-assembly of cds quantum dots/graphene nanosheets hybrid films for photoelectrochemical and photocatalytic applications, J. Am. Chem. Soc. 136 (2014) 1559–1569. Available from: https://doi.org/10.1021/ja411651e.

[54] T.H. Kim, B.H. Sohn, Photocatalytic thin films containing TiO_2 nanoparticles by the layer-by-layer self-assembling method, Appl. Surf. Sci. 201 (2002) 109–114. Available from: https://doi.org/10.1016/S0169-4332(02)00612-8.

[55] S. Zhang, J. Shen, H. Fu, W. Dong, Z. Zheng, L. Shi, Bi_2WO_6 photocatalytic films fabricated by layer-by-layer technique from Bi_2WO_6 nanoplates and its spectral selectivity, J. Solid State Chem. 180 (2007) 1456–1463. Available from: https://doi.org/10.1016/j.jssc.2007.02.013.

[56] Y.G. Guo, L.J. Wan, C.L. Bai, Gold/titania core/sheath nanowires prepared by layer-by-layer assembly, J. Phys. Chem. B. 107 (2003) 5441–5444. Available from: https://doi.org/10.1021/jp034184q.

[57] N. Pinna, M. Niederberger, Surfactant-free nonaqueous synthesis of metal oxide nanostructures, Angew. Chem., Int. Ed. 47 (2008) 5292–5304. Available from: https://doi.org/10.1002/anie.200704541.

[58] J.B. Ma, B. Xu, J.H. Meng, X.N. Wu, X.L. Ding, X.N. Li, et al., Reactivity of atomic oxygen radical anions bound to titania and zirconia nanoparticles in the gas phase: Low-temperature oxidation of carbon monoxide, J. Am. Chem. Soc. 135 (2013) 2991–2998. Available from: https://doi.org/10.1021/ja311695t.

[59] H. Xu, S. Ouyang, P. Li, T. Kako, J. Ye, High-active anatase TiO_2 nanosheets exposed with 95% {100} facets toward efficient H_2 evolution and CO_2 photoreduction, ACS Appl. Mater. Interfaces. 5 (2013) 1348–1354. Available from: https://doi.org/10.1021/am302631b.

[60] G. Tian, Y. Chen, K. Pan, D. Wang, W. Zhou, Z. Ren, et al., Efficient visible light-induced degradation of phenol on N-doped anatase TiO_2 with large surface area and high crystallinity, Appl. Surf. Sci. 256 (2010) 3740–3745. Available from: https://doi.org/10.1016/j.apsusc.2010.01.016.

[61] L. Jing, W. Zhou, G. Tian, H. Fu, Surface tuning for oxide-based nanomaterials as efficient photocatalysts, Chem. Soc. Rev. 42 (2013) 9509. Available from: https://doi.org/10.1039/c3cs60176e.

[62] J. Zhang, M. Li, Z. Feng, J. Chen, C. Li, U.V. Raman Spectroscopic, Study on TiO_2. I. Phase transformation at the surface and in the bulk, J. Phys. Chem. B. 110 (2006) 927–935. Available from: https://doi.org/10.1021/jp0552473.

[63] C. Kang, L. Jing, T. Guo, H. Cui, J. Zhou, H. Fu, Mesoporous SiO_2-modified nanocrystalline tio_2 with high anatase thermal stability and large surface area as efficient photocatalyst, J. Phys. Chem. C. 113 (2009) 1006–1013. Available from: https://doi.org/10.1021/jp807552u.

[64] M. Hirano, K. Ota, H. Iwata, Direct formation of anatase (TiO_2)/silica (SiO_2) composite nanoparticles with high phase stability of 1300°C from acidic solution by hydrolysis under hydrothermal condition, Chem. Mater. 16 (2004) 3725–3732. Available from: https://doi.org/10.1021/cm040055q.

[65] R. Liu, Y. Ren, Y. Shi, F. Zhang, L. Zhang, B. Tu, et al., Controlled synthesis of ordered mesoporous C-TiO_2 nanocomposites with crystalline titania frameworks from organic-inorganic-amphiphilic coassembly, Chem. Mater. 20 (2008) 1140–1145. Available from: https://doi.org/10.1021/cm071470w.

[66] T. Kamegawa, D. Yamahana, H. Yamashita, Graphene coating of TiO_2 nanoparticles loaded on mesoporous silica for enhancement of photocatalytic activity, J. Phys. Chem. C. 114 (2010) 15049–15053. Available from: https://doi.org/10.1021/jp105526d.

[67] E. Kowalska, H. Remita, C. Colbeau-Justin, J. Hupka, J. Belloni, Modification of titanium dioxide with platinum ions and clusters: application in photocatalysis, J. Phys. Chem. C. 112 (2008) 1124–1131. Available from: https://doi.org/10.1021/jp077466p.

[68] Z. Hai, N. El Kolli, D.B. Uribe, P. Beaunier, M. José-Yacaman, J. Vigneron, et al., Modification of TiO_2 by bimetallic Au–Cu nanoparticles for wastewater treatment, J. Mater. Chem. A. 1 (2013) 10829. Available from: https://doi.org/10.1039/c3ta11684k.

[69] Z. Hai, N.E.L. Kolli, J. Chen, H. Remita, Radiolytic synthesis of Au–Cu bimetallic nanoparticles supported on TiO_2: application in photocatalysis, New J. Chem. 38 (2014) 5279–5286. Available from: https://doi.org/10.1039/C4NJ00883A.

[70] J. Zhang, S.Z. Qiao, L. Qi, J. Yu, Fabrication of NiS modified CdS nanorod p–n junction photocatalysts with enhanced visible-light photocatalytic H2-production activity, Phys. Chem. Chem. Phys. 15 (2013) 12088. Available from: https://doi.org/10.1039/c3cp50734c.

[71] M.L. Ovcharov, V.V. Shvalagin, V.M. Granchak, Photocatalytic reduction of CO_2 on mesoporous TiO_2 modified with Ag/Cu bimetallic nanostructures, Theor. Exp. Chem. 50 (2014) 175–180. Available from: https://doi.org/10.1007/s11237-014-9362-x.

[72] S. Banerjee, S.C. Pillai, P. Falaras, K.E. O'shea, J.A. Byrne, D.D. Dionysiou, New insights into the mechanism of visible light photocatalysis, J. Phys. Chem. Lett. 5 (2014) 2543–2554. Available from: https://doi.org/10.1021/jz501030x.

[73] W. Il Park, C.-H. Lee, J.M. Lee, N.-J. Kim, G.-C. Yi, Inorganic nanostructures grown on graphene layers, Nanoscale. 3 (2011) 3522. Available from: https://doi.org/10.1039/c1nr10370a.

[74] Q. Min, X. Zhang, H. Zhang, F. Zhou, J.-J. Zhu, Synthesis of Fe_3O_4–graphene–TiO_2 ternary composite networks for enhanced capture of phosphopeptides, Chem. Commun. 47 (2011) 11709. Available from: https://doi.org/10.1039/c1cc15151g.

[75] H. Hayashi, I.V. Lightcap, M. Tsujimoto, M. Takano, T. Umeyama, P.V. Kamat, et al., Electron transfer cascade by organic/inorganic ternary composites of porphyrin, zinc oxide nanoparticles, and reduced graphene oxide on a tin oxide electrode that exhibits efficient photocurrent generation, J. Am. Chem. Soc. 133 (2011) 7684–7687. Available from: https://doi.org/10.1021/ja201813n.

[76] A. Iwase, Y.H. Ng, Y. Ishiguro, A. Kudo, R. Amal, Reduced graphene oxide as a solid-state electron mediator in Z-scheme photocatalytic water splitting under visible light, J. Am. Chem. Soc. 133 (2011) 11054–11057. Available from: https://doi.org/10.1021/ja203296z.

[77] J. Zhang, F.-X. Xiao, G. Xiao, B. Liu, Self-assembly of a Ag nanoparticle-modified and graphene-wrapped TiO_2 nanobelt ternary heterostructure: surface charge tuning toward

efficient photocatalysis, Nanoscale. 6 (2014) 11293–11302. Available from: https://doi.org/10.1039/C4NR03115F.

[78] J.C. Yu, X. Wang, L. Wu, W. Ho, L. Zhang, G. Zhou, Sono- and photochemical routes for the formation of highly dispersed gold nanoclusters in mesoporous titania films, Adv. Funct. Mater. 14 (2004) 1178–1183. Available from: https://doi.org/10.1002/adfm.200305145.

[79] A.A. Ismail, D.W. Bahnemann, One-step synthesis of mesoporous platinum/titania nanocomposites as photocatalyst with enhanced photocatalytic activity for methanol oxidation, Green Chem. 13 (2011) 428–435.

[80] C. Liu, D. Yang, Y. Jiao, Y. Tian, Y. Wang, Z. Jiang, Biomimetic synthesis of TiO_2-SiO_2-Ag nanocomposites with enhanced visible-light photocatalytic activity, ACS Appl. Mater. Interfaces. 5 (2013) 3824–3832. Available from: https://doi.org/10.1021/am4004733.

[81] B. Weng, S. Liu, Z.-R. Tang, Y.-J. Xu, One-dimensional nanostructure based materials for versatile photocatalytic applications, RSC Adv. 4 (2014) 12685. Available from: https://doi.org/10.1039/c3ra47910b.

[82] M.M. Khin, A.S. Nair, V.J. Babu, R. Murugan, S. Ramakrishna, A review on nanomaterials for environmental remediation, Energy Environ. Sci. 5 (2012) 8075. Available from: https://doi.org/10.1039/c2ee21818f.

[83] M. Pelaez, N.T. Nolan, S.C. Pillai, M.K. Seery, P. Falaras, A.G. Kontos, et al., A review on the visible light active titanium dioxide photocatalysts for environmental applications, Appl. Catal. B Environ. 125 (2012) 331–349. Available from: https://doi.org/10.1016/j.apcatb.2012.05.036.

[84] F. Xiao, Layer-by-layer self-assembly construction of highly ordered metal- TiO_2 nanotube arrays heterostructures (M/TNTs, M = Au, Ag, Pt) with tunable catalytic activities, J. Phys. Chem. C. 116 (2012) 16487–16498. Available from: https://doi.org/10.1021/am302462d.

[85] F. Xiao, Construction of highly ordered ZnO – TiO_2 nanotube arrays (ZnO / TNTs) heterostructure for photocatalytic application, ACS Appl. Mater. Interfaces. 4 (2012) 7055–7063. Available from: https://doi.org/10.1021/am302462d.

[86] B. Liu, H.C. Zeng, Carbon nanotubes supported mesoporous mesocrystals of anatase TiO_2, Chem. Mater. 20 (2008) 2711–2718. Available from: https://doi.org/10.1021/cm800040k.

[87] H.G. Kim, D.W. Hwang, J.S. Lee, An undoped, single-phase oxide photocatalyst working under visible light, J. Am. Chem. Soc. 126 (2004) 8912–8913. Available from: https://doi.org/10.1021/ja049676a.

[88] M.J. Kim, I.Y. Jeon, J.M. Seo, L. Dai, J.B. Baek, Graphene phosphonic acid as an efficient flame retardant, ACS Nano. 8 (2014) 2820–2825. Available from: https://doi.org/10.1021/nn4066395.

[89] R. Liu, G. Yuan, C.L. Joe, T.E. Lightburn, K.L. Tan, D. Wang, Silicon nanowires as photoelectrodes for carbon dioxide fixation, Angew. Chemie - Int. Ed. 51 (2012) 6709–6712. Available from: https://doi.org/10.1002/anie.201202569.

[90] G.K. Mor, H.E. Prakasam, O.K. Varghese, K. Shankar, C.A. Grimes, Vertically oriented Ti – Fe – O nanotube array films: toward a useful material architecture for solar spectrum water photoelectrolysis, Nano Lett. 7 (2007) 2356–2364. Available from: https://doi.org/10.1021/nl0710046.

[91] N.S. Lewis, G. Crabtree, A.J. Nozik, M.R. Wasielewski, P. Alivisatos, H. Kung, et al., Basic Research Needs for Solar Energy Utilization. Report of the Basic Energy Sciences Workshop on Solar Energy Utilization, April 18–21, 2005, 2005. https://doi.org/10.2172/899136.

[92] F.E. Osterloh, Inorganic nanostructures for photoelectrochemical and photocatalytic water splitting, Chem. Soc. Rev. 42 (2013) 2294–2320. Available from: https://doi.org/10.1039/C2CS35266D.

[93] W.A. Smith, I.D. Sharp, N.C. Strandwitz, J. Bisquert, Interfacial band-edge energetics for solar fuels production, Energy Environ. Sci. 8 (2015) 2851–2862. Available from: https://doi.org/10.1039/C5EE01822F.

[94] M.N. Huda, M.M. Al-Jassim, J.A. Turner, Mott insulators: an early selection criterion for materials for photoelectrochemical H 2 production, J. Renew. Sustain. Energy. 3 (2011) 53101. Available from: https://doi.org/10.1063/1.3637367.

[95] P. Salvador, Semiconductors' photoelectrochemistry: a kinetic and thermodynamic analysis in the light of equilibrium and nonequilibrium models, J. Phys. Chem. B. 105 (2001) 6128–6141. Available from: https://doi.org/10.1021/jp0033869.

[96] P.A. Sant, P.V. Kamat, Interparticle electron transfer between size-quantized CdS and TiO_2 semiconductor nanoclusters dedicated to Professor Frank Wilkinson on the occasion of his retirement, Phys. Chem. Chem. Phys. 4 (2002) 198–203. Available from: https://doi.org/10.1039/b107544f.

[97] L.M. Peter, K.G.U. Wijayantha, D.J. Riley, J.P. Waggett, Band-edge tuning in self-assembled layers of Bi_2S_3 nanoparticles used to photosensitize nanocrystalline TiO_2, J. Phys. Chem. B. 107 (2003) 8378–8381. Available from: https://doi.org/10.1021/jp0303341.

[98] V. Chakrapani, K. Tvrdy, P.V. Kamat, Modulation of electron injection in CdSe − TiO_2 system through medium alkalinity, J. Am. Chem. Soc. 132 (2010) 1228–1229. Available from: https://doi.org/10.1021/ja909663r.

[99] A.J. Cowan, J. Tang, W. Leng, J.R. Durrant, D.R. Klug, Water splitting by nanocrystalline TiO_2 in a complete photoelectrochemical cell exhibits efficiencies limited by charge recombination, J. Phys. Chem. C. 114 (2010) 4208–4214. Available from: https://doi.org/10.1021/jp909993w.

[100] M.C. Beard, K.P. Knutsen, P. Yu, J.M. Luther, Q. Song, W.K. Metzger, et al., Multiple exciton generation in colloidal silicon nanocrystals, Nano Lett. 7 (2007) 2506–2512. Available from: https://doi.org/10.1021/nl071486l.

[101] S.D. Tilley, M. Cornuz, K. Sivula, M. Grätzel, Light-induced water splitting with hematite: improved nanostructure and iridium oxide catalysis, Angew. Chemie Int. Ed. 49 (2010) 6405–6408. Available from: https://doi.org/10.1002/anie.201003110.

[102] M.R. Waller, T.K. Townsend, J. Zhao, E.M. Sabio, R.L. Chamousis, N.D. Browning, et al., Single-crystal Tungsten oxide nanosheets: photochemical water oxidation in the quantum confinement regime, Chem. Mater. 24 (2012) 698–704. Available from: https://doi.org/10.1021/cm203293j.

[103] Y.H. Ng, A. Iwase, A. Kudo, R. Amal, Reducing graphene oxide on a visible-light $BiVO_4$ photocatalyst for an enhanced photoelectrochemical water splitting, J. Phys. Chem. Lett. 1 (2010) 2607–2612. Available from: https://doi.org/10.1021/jz100978u.

[104] J. Zhang, M. Grzelczak, Y. Hou, K. Maeda, K. Domen, X. Fu, et al., Photocatalytic oxidation of water by polymeric carbon nitride nanohybrids made of sustainable elements, Chem. Sci. 3 (2012) 443–446. Available from: https://doi.org/10.1039/C1SC00644D.

[105] G. Xie, K. Zhang, B. Guo, Q. Liu, L. Fang, J.R. Gong, Graphene-based materials for hydrogen generation from light-driven water splitting, Adv. Mater. 25 (2013) 3820–3839. Available from: https://doi.org/10.1002/adma.201301207.

[106] C.X. Guo, Y. Dong, H. Bin Yang, C.M. Li, Graphene quantum dots as a green sensitizer to functionalize ZnO nanowire arrays on F-doped SnO_2 glass for enhanced photoelectrochemical water splitting, Adv. Energy Mater. 3 (2013) 997–1003. Available from: https://doi.org/10.1002/aenm.201300171.

[107] A. Hagfeldt, H. Lindström, S. Södergren, S.-E. Lindquist, Photoelectrochemical studies of colloidal TiO_2 films: the effect of oxygen studied by photocurrent transients, J. Electroanal. Chem. 381 (1995) 39–46. Available from: https://doi.org/10.1016/0022-0728(94)03622-A.

[108] D. Ke, T. Peng, L. Ma, P. Cai, P. Jiang, Photocatalytic water splitting for O_2 production under visible-light irradiation on $BiVO_4$ nanoparticles in different sacrificial reagent solutions, Appl. Catal. A Gen. 350 (2008) 111–117. Available from: https://doi.org/10.1016/j.apcata.2008.08.003.

[109] M. Halmann, Photoelectrochemical reduction of aqueous carbon dioxide on p-type gallium phosphide in liquid junction solar cells, Nature. 275 (1978) 115–116. Available from: https://doi.org/10.1038/275115a0.

[110] K. Li, D. Martin, J. Tang, Conversion of solar energy to fuels by inorganic heterogeneous systems, Chinese J. Catal. 32 (2011) 879–890. Available from: https://doi.org/10.1016/S1872-2067(10)60209-4.

[111] A. Primo, T. Marino, A. Corma, R. Molinari, H. García, Efficient visible-light photocatalytic water splitting by minute amounts of gold supported on nanoparticulate CeO_2 obtained by a biopolymer templating method, J. Am. Chem. Soc. 133 (2011) 6930–6933. Available from: https://doi.org/10.1021/ja2011498.

[112] D. Gao, H. Zhou, J. Wang, S. Miao, F. Yang, G. Wang, et al., Size-dependent electrocatalytic reduction of CO_2 over Pd nanoparticles, J. Am. Chem. Soc. 137 (2015) 4288–4291. Available from: https://doi.org/10.1021/jacs.5b00046.

[113] Q. Kong, D. Kim, C. Liu, Y. Yu, Y. Su, Y. Li, et al., Directed assembly of nanoparticle catalysts on nanowire photoelectrodes for photoelectrochemical CO_2 reduction, Nano Lett. 16 (2016) 5675–5680. Available from: https://doi.org/10.1021/acs.nanolett.6b02321.

[114] H. Park, H.-H. Ou, A.J. Colussi, M.R. Hoffmann, Artificial photosynthesis of C1–C3 hydrocarbons from water and CO_2 on titanate nanotubes decorated with nanoparticle elemental copper and CdS quantum dots, J. Phys. Chem. A. 119 (2015) 4658–4666. Available from: https://doi.org/10.1021/jp511329d.

[115] Y. Wei, L. Ke, E.S.P. Leong, H. Liu, L.-L. Liew, J.H. Teng, et al., Enhanced photoelectrochemical performance of bridged ZnO nanorod arrays grown on V-grooved structure, Nanotechnology. 23 (2012) 365704. Available from: https://doi.org/10.1088/0957-4484/23/36/365704.

[116] J. Zhao, X. Wang, Z. Xu, J.S.C. Loo, Hybrid catalysts for photoelectrochemical reduction of carbon dioxide: a prospective review on semiconductor/metal complex co-catalyst systems, J. Mater. Chem. A. 2 (2014) 15228. Available from: https://doi.org/10.1039/C4TA02250E.

Chapter 6

Photoactive nanoparticles capped with macrocycles as platforms and hosts

Zheng Li[a], Xi Li[a] and Ying-Wei Yang[a,*]

[a]*International Joint Research Laboratory of Nano-Micro Architecture Chemistry (NMAC), College of Chemistry, Jilin University, Changchun, P.R. China, *Corresponding author.email address: ywyang@jlu.edu.cn*

Chapter Outline

6.1 Pillararene-capped nanoparticles

6.1.1 Quantum dots

Quantum dots (QDs), as high-efficiency fluorescent nanomaterials, have shown a wide application prospect in materials science and nanotechnology for their fascinating engineered optical and physicochemical properties including high fluorescence quantum yields, broad excitation, size-tunable emission wavelength, precise tunability of narrow and symmetric emissions, long fluorescence lifetimes, negligible photobleaching, and excellent

Photoactive Inorganic Nanoparticles. DOI: https://doi.org/10.1016/B978-0-12-814531-9.00006-3
© 2019 Elsevier Inc. All rights reserved.

photochemical stability [1–9]. Semiconductor QDs have always been regarded as proper phosphors for photoluminescence, electroluminescence, and cathodoluminescence. The sensitivity of QDs toward multiple signals has stimulated a tremendous upsurge of interest in fluorescent sensors, biological imaging, and labeling for long-term and simultaneous detection. The great potential of surface engineering endows the QDs with desirable surface functionalities and stability as a result of introduction of various organic ligands. In recent years, more functional capping agents on the surface of QDs have been proposed or studied in depth, allowing for further cooperative actions and synergetic effect of two components [10].

Pillararenes (or pillarenes) are a relatively new generation of macrocyclic host compounds with rigid cylinder architectures, which consist of hydroquinone units linked by methylene ($-CH_2-$) bridges at their *para*-positions. They can be readily modified with a wide variety of functional groups and their attractive properties in host–guest chemistry pave an avenue for the construction of supramolecular architectures (including catenanes, rotaxanes, pseudorotaxanes, and polyrotaxanes), self-assembly materials, hybrid materials, supramolecular organic frameworks and metal organic frameworks, and their applications in nanotechnology, molecular recognition, ion transport and heterogeneous asymmetric catalysis [11]. As an example of host–guest interactions based on pillarenes, water-soluble pillar[5]arene (WP5) modified with carboxylic acid groups can form host–guest complex with biomolecules such as amino acids, taking the advantages of their electron-rich cavity and intermolecular hydrogen bonds between carboxyl group and amino acids, which has addressed potential for amino acids detection [12]. The surface modification of QDs with synthetic macrocyclic compounds has received considerable attention due to the enhanced photochemical properties of QDs as well as retaining the molecular recognizing capacity of the macrocyclic capping agents.

Liquid QDs, which are solvent-free, have been developed to improve the performance of solid-state QDs such as sensitivity toward fluorescence quenching agents. Therefore enormous effort has been devoted to the preparation of liquid QDs and enhancement of their fluorescent quantum yield. Li and coworkers developed a fluorescent film sensor for lysine (Lys) based on the liquid QDs in 2016 [12]. WP5 synthesized by a "thiol-ene" click reaction was successfully introduced to the surface of polyenthylene glycol (PEG)-1810-modified CdTe-QDs, denoted as WP5-1810-QDs, via forming a host–guest complex between the alkyl chain of PEG-1810 and the hydrophobic cavity of WP5. WP5-1810-QDs exhibited a clear enhancement of the fluorescence intensity in comparison with the 1810-QDs, as a result of the suppression of quenching path or a nonradiative process through a core protection mechanism. Lys as the competing agent was added into the system of WP5-1810-QDs, thereafter, the alkyl chain of Lys could enter into the cavity of WP5 through hydrophobic interactions. The ^1H NMR spectra verified that the interactions between WP5 and Lys were stronger than those with PEG-1810. With the competitive effect of Lys, an obvious fluorescence quenching of WP5-1810-QDs was observed as a result of the release of WP5 and the destruction of core protection. A smart sensor of WP5-capped QDs was constructed. Lys is a kind of ubiquitous amino acid and plays an important role in biological processes such as promoting the growth, strengthening the immunologic function, and improving the organizational function of the central nervous system. This macroscopic Lys-responsive system based on liquid QDs has provided a promising alternative as biological sensor, which may contribute to biochemistry and medical science.

Similar to gold nanoparticles (AuNPs) stabilized with multithiolated compounds, QDs can also be stabilized by multicarboxylated compounds through their modification on the surface of QDs. Our research group first utilized carboxylatopillar[5]arene (CP[5]A) to stabilize CdTe QDs via a ligand-exchange process in 2013 as shown in Fig. 6.1. This organic–inorganic composite nanomaterial exhibited high stability in aqueous media at room temperature and the CP[5]A rings effectively fine-tuned the surface environment of QDs and affected the photophysical properties of this hybrid material. The decrease in the fluorescence intensity of CP[5]A@QDs system upon the addition of methyl viologen (MV) or bridged *bis*-MV was at a modest pace, which is in a sharp contrast with the dramatic decrease in the case

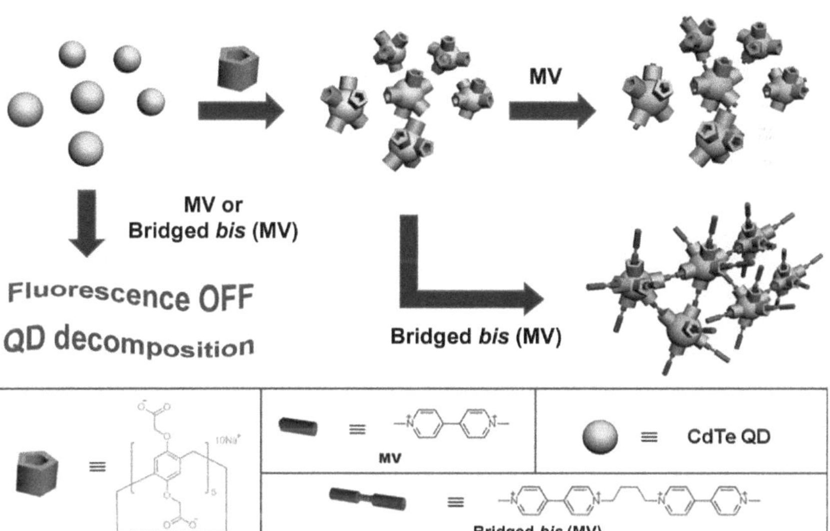

FIGURE 6.1 Schematic description of CP[5]A@QDs, the photophysical properties, and the process of self-assembly mediated by viologens. *Reproduced with permission from D.X. Chen, Y.L. Sun, Y. Zhang, J.Y. Cui, F.Z. Shen, Y.-W. Yang, Supramolecular self-assembly and photophysical properties of pillar[5]arene-stabilized CdTe quantum dots mediated by viologens, RSC Adv. 3 (2013) 5765–5768. Copyright 2013 The Royal Society of Chemistry.*

of CdTe QDs system stabilized with 3-mercaptopropionic acid (MPA@QDs). Furthermore, the addition of bridged MV guest could induce an aggregated supramolecular network via supramolecular host–guest interactions. This nanocomposite displayed great potential in detecting herbicide as an efficient chemical sensor [13].

6.1.2 Gold and silver nanoparticles

Under the domain of photoactive inorganic NPs, metal NPs, such as gold and silver NPs, possess a wide array of potential applications in sensing, biomedicine, optoelectronic devices, catalysis, and surface-enhanced Raman scattering (SERS) spectroscopy owing to their unique properties in optics, electronics, and biology [14]. Of great interest is the assembly of metal NPs due to the resultant improvement of the abovementioned characteristics. In general, gold NPs (AuNPs) can be assembled via host–guest interactions, hydrogen bonding, ion pairing, and metal–ligand interactions, among which host–guest interactions have made a significant impact on the reversible assembly of AuNPs [15]. Tremendous effort has been made in the construction of NP–macrocycle hybrids, which could be achieved through proper surface modification of synthesized NPs. These macrocycles not only allow the self-assembly of nanocomposite at the molecular level but also improve the stability and biocompatibility of the inorganic NPs, whereas the introduction of macrocyclic host molecules has provided special selectivity toward various guest molecules for detection, involving SERS for plasmonic-sensing applications.

Inspired by the stabilizing ligands in the process of synthesis and stabilization of AuNPs, such as amine ($-NH_2$), carboxyl ($-COOH$), or sulfhydryl ($-SH$) groups, water-soluble carboxylatopillar[5]arene sodium salts (CP[5]A) with five carboxyl groups on each rim have already been used as stabilizers for AuNPs [16]. The resultant AuNPs with narrow size distributions (3.1 ± 0.5 nm) could be obtained through *in situ* preparation, that is, the reduction of $HAuCl_4$ with $NaBH_4$ in the presence of CP[5]A. Herein, CP[5]A acted as a macrocyclic capping agent for the stabilization of AuNPs which could not be achieved in the absence of organic ligands. It is noteworthy to mention that the carboxylate anions on both rims of CP[5]A endow the AuNPs with negatively charged surface for the sake of binding cationic guests such as 1,1′-dimethyl-4,4′-bipyridinium salt (viologen I), namely paraquat, used as herbicide. With the mediation of viologen I and II (dimeric viologen I), CP[5]A modified AuNPs underwent supramolecular self-assembly to the aggregated state, which could be proved by solution color changes or the change of surface plasmon resonance (SPR) absorption, according to the work of our group in 2013 (Fig. 6.2). This nanosystem can be used as optical probe for herbicide detection.

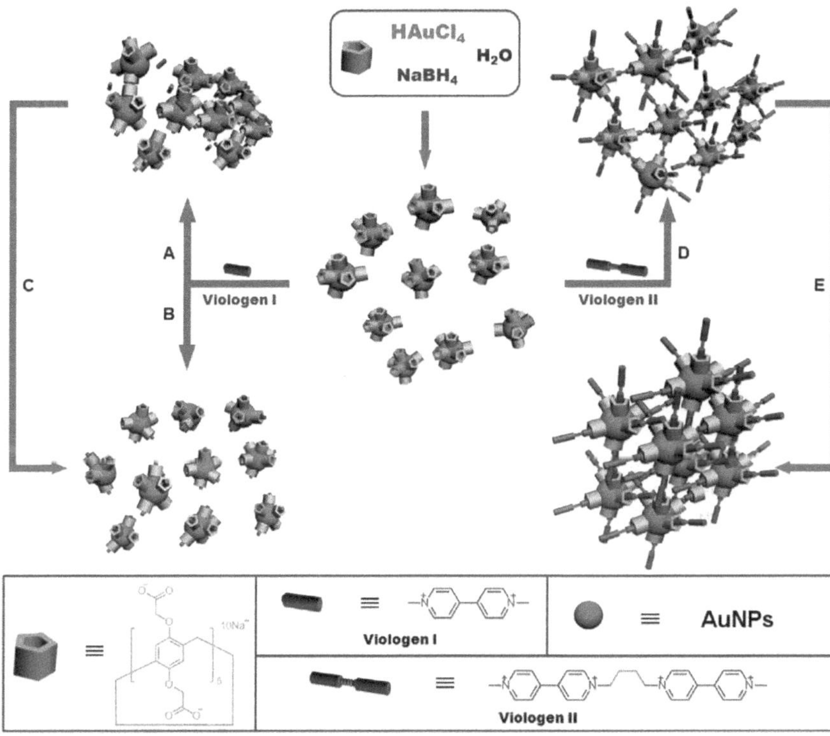

FIGURE 6.2 Schematic description of CP[5]A@AuNPs and the process of self-assembly mediated by viologen molecules (I and II). *Reproduced with permission from H. Li, D.X. Chen, Y.L. Sun, Y.B. Zheng, L.L. Tan, P.S. Weiss, et al., Viologen-mediated assembly of and sensing with carboxylatopillar[5]arene-modified gold NPs, J. Am. Chem. Soc. 135 (2013) 1570–1576 [16]. Copyright 2012 American Chemical Society.*

One of the most important synergetic effects between macrocycles and metal NPs relies on the precise control over their size and shape via mediation of supramolecular chemistry. The cooperative action of relevant nanosystems based on the combination of supramolecular chemistry and metal NPs can be explained by SERS detection. According to the work by Pastoriza-Santos and coworkers in 2014, ammonium pillar[5]arene (AP[5]A) has successfully been used as capping agents in a seeded-growth preparation of AuNPs in a wide size range up to 120 nm [17]. The size-controllable mechanism was achieved through the complexation of the Au salt with the ammonium groups, resulting in the inhibition of Au nucleation. Taking into account the hydrophobic and electrostatic interactions between carboxylates and cationic ammonium groups of AP[5]A, AP[5]A on the surface of AuNPs was used to bind 2-naphthoic acid (2NA). The host–guest interactions have been investigated by ^{1}H NMR spectroscopic titrations. SERS spectroscopy exhibited desirable spectrum signals after adding 2NA to the

AP[5]A-stabilized AuNPs. The several characteristic Raman signals of 2NA could not be detected by SERS spectroscopy under the system of citrate-stabilized AuNPs, but on the contrary, clear SERS signals could be observed in the presence of AP[5]A-stabilized AuNPs. A polycyclic aromatic hydrocarbon, pyrene, was also utilized to evaluate SERS signals on the basis of this inorganic–organic nanosystem. The experimental results demonstrated that the pillarenes macrocycles could serve as promising platforms for AuNPs in SERS detection.

AuNPs capped with pillarenes can absorb near-infrared (NIR) light and, as a consequence, produce photothermal effects that can be used for tumor photothermo-chemotherapy. Thus, CP[5]A and phosphonated pillar[5]arenes (PPA[5]) have been constructed as supramolecular nanovalves on mesoporous silica nanoparticles (MSNs) containing AuNPs or nanorods (GNRs) [18–22]. Mechanized MSNs have gained substantial attention in the field of biomedicine due to excellent biocompatibility and robust storage capacity as well as tunable pore size [23–25]. In the example of PPA[5]-capped GNR-embedded MSNs (GNR@MSNs), MSNs were modified with choline and

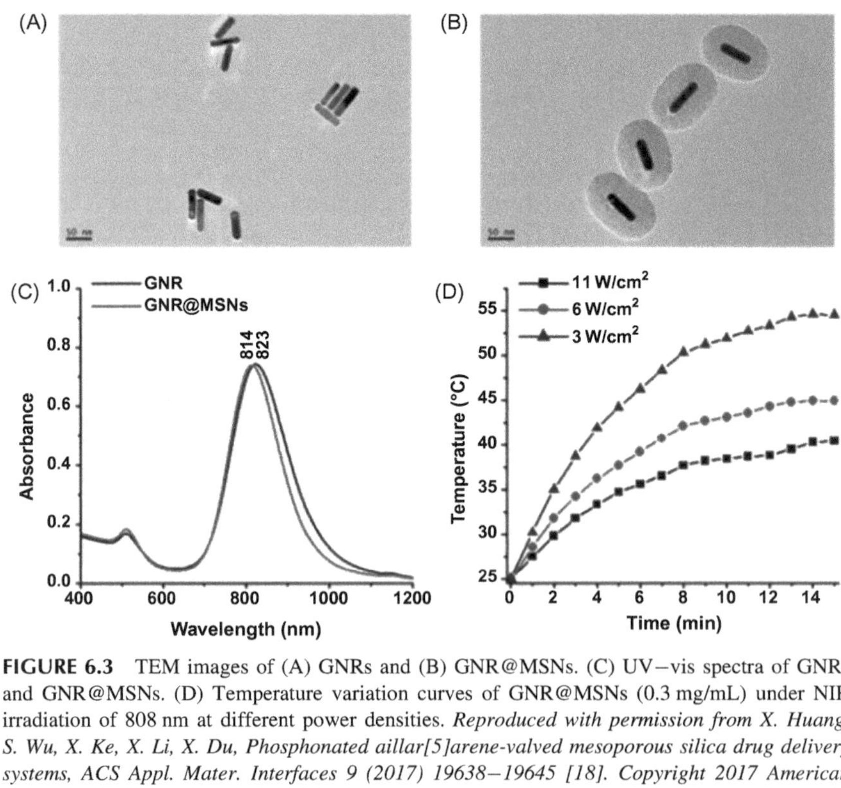

FIGURE 6.3 TEM images of (A) GNRs and (B) GNR@MSNs. (C) UV–vis spectra of GNRs and GNR@MSNs. (D) Temperature variation curves of GNR@MSNs (0.3 mg/mL) under NIR irradiation of 808 nm at different power densities. *Reproduced with permission from X. Huang, S. Wu, X. Ke, X. Li, X. Du, Phosphonated aillar[5]arene-valved mesoporous silica drug delivery systems, ACS Appl. Mater. Interfaces 9 (2017) 19638–19645 [18]. Copyright 2017 American Chemical Society.*

pyridinium as stalks and macrocyclic host conjugated with MSNs via ion pairing between the phosphonate and quaternary ammonium moieties (Fig. 6.3). This multiresponsive nanosystem has proved to be sensitive toward pH, metal ion coordination, competitive agents, and light irradiation. Upon 808 nm NIR irradiation, GNR@MSNs exhibited a photothermal effect at different laser power densities, which could weaken host–guest interactions and result in the removal of macrocyclic nanovalves for tumor therapy such as drug delivery.

Silver NPs modified with WP5 could be induced from nonaggregated state to aggregated state by the addition of guest molecules such as spermine analogues as can be seen in Fig. 6.4 [26]. The electron-rich cyclic cavity of WP5 could form complexes with amino groups of spermine analogues through electrostatic, C–H$\cdots\pi$ interactions and hydrogen bonds at a ratio of 2:1, which could cause the aggregation of silver NPs. And the closer distance of silver NPs would lead to the coupling plasmon absorbance with a distinct color change from yellow to black, providing a clue for visual detection of spermine and its analogs.

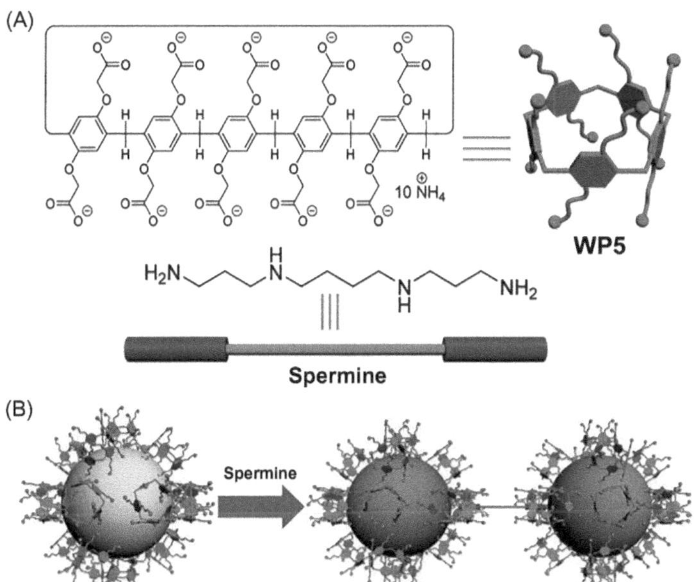

FIGURE 6.4 Schematic description of (A) chemical structures of WP5 and spermine. (B) Schematic representation of visual detection of spermine based on WP5-modified silver NPs. *Reproduced with permission from Y. Yao, Y. Zhou, J. Dai, S. Yue, M. Xue, Host–guest recognition-induced color change of water-soluble pillar[5]arene modified silver NPs for visual detection of spermine analogues, Chem. Commun. 50 (2014) 869–871. Copyright 2014 The Royal Society of Chemistry.*

6.2 Calixarene-capped nanoparticles

6.2.1 Gold nanoparticles

Calixarenes, as the third generation of macrocyclic hosts after crown ethers and cyclodextrins, are constructed with phenolic structural blocks linked by methylene bridges in the arrangement of cyclic arrays and have garnered substantial attention in the molecular recognition, self-assembly and sensing because of their ability to be easily modified on either the wide upper or narrow lower rim of the supramolecular skeleton and other advantages involving hydrophobic cavity for forming inclusion complexes [27,28]. Their potential complexation ability toward biomolecules, such as neurotransmitter acetylcholine, endowed them great possibility in biochemical sensing and the design of rational platforms for biomolecular inhibitor screening [29,30]. The great potential of AuNPs in surface chemistry has expanded their applications' scope, for example, as colorimetric assays, benefitting from their optical properties and size-dependent SPR properties [31,32]. Colorimetric biosensors derived from calixarenes-capped AuNPs have also been investigated, especially the mechanistic basis of the AuNPs aggregation and their ability to detect biological or environmental molecules such as creatinine, diamine oxidase (DAO) and herbicide [29,30,33]. In such system, the aggregation of calixarenes-capped AuNPs can be induced upon the addition of targeted molecules via electrostatic interaction, hydrogen bonding or $\pi-\pi$ stacking, yielding a drastic red-shift in absorption spectra and an obvious color change from red to blue, which can be distinguished by naked eyes. In 2017, Chen and coworkers [30] reported *p*-sulfonatecalix[6]arene (pSC_6) modified AuNPs as a sensitive colorimetric biosensor for DAO assay and inhibitor screening. 1,6-hexanediamine (HMD) could aggregate pSC_6−AuNPs and cause color change from wine red to blue through host−guest interactions including electrostatic forces, cation−π and $\pi-\pi$ interactions between the protonated amino residue of HMD and the electron-rich cavity or the SO_3^- groups of pSC_6. However, this process could be inhibited in the presence of DAO due to its oxidation effect toward amino groups and then DAO successfully prevented pSC6−AuNPs from aggregation, which could not be achieved in the presence of guanidine, an inhibitor of DAO.

According to the work by Kailasa and coworkers in 2016 [33], metsulfuron−methyl herbicide could aggregate dithiocarbamate-*p*-tertbutylcalix[4] arene (DTC−PTBCA)-capped AuNPs via noncovalent interactions between the cavity of the macrocycle and benzene or triazine units of metsulfuron−methyl herbicide at a ratio of 2:1. The aggregated state could not only cause an obvious color change from red to blue but also induce a red-shift of absorption spectrum from 524 to 572 nm. In short, DTC-PTBCA functionalized AuNPs could be used as a sensitive plasmonic probe for colorimetric sensing toward metsulfuron−methyl herbicide; this system exhibited a good linearity with the concentration of metsulfuron−methyl in the range from 1.0 to 50 μM and a detection limit of 1.9×10^{-7} M ($S/N = 3$).

6.2.2 Quantum dots

In the first two decades of the 20th century, calixarene−QD systems have elicited considerable concerns regarding their synergetic superior properties than individual QDs or calixarene, especially the enhancement and preservation of optical properties and the improvement of stability [34,35]. There are many chemosensors and biosensors constructed for pesticides, amino acids, neurotransmitters, organic molecules, and metals in the light of fluorescence changes of calixarene coated QDs. In general, the fluorescence efficiency of semiconductor QDs relies on the nature of their surface adsorbates. Sulfonatocalix[4]arene (SC[4]A) was selected as suitable coating of ZnS to prepare a highly stable optical probe for the discrimination of menadione (VK_3), and sulfonatocalix[4,6]arene was selected to modify CdS QDs to fine-tune the resultant Au electrodes with enhanced photocurrents [36,37]. SC[4]A afforded effective trapping effect toward surface defects of QDs, resulting in a stable FL suspension. Upon the addition of VK_3, the FL intensity was clearly quenched by almost 95% on the basis of selectively supramolecular host−guest interactions between the hydrophobic cavity of SC[4]A and VK_3 in comparison with other vitamins. The Stern−Volmer model was used to investigate the quenching behaviors of VK_3, a static mechanism was observed till the concentration of 1×10^{-6} M of VK_3 and followed by a dynamic mechanism at higher concentrations quenching process. Additionally, core−shell CdSe/ZnS QDs could be capped by *p*-tertbutylcalix[8]arene to obtain the highly fluorescent and stable nanosensor for determination of C_{60} via a selective luminescence quenching mechanism [38]. Indeed, CdTe QDs in silica spheres capped with calixarenes were also constructed as fluorescent probes for pesticides [39].

6.3 Cyclodextrin-capped nanoparticles

6.3.1 Rare-earth-doped upconversion nanoparticles

Cyclodextrins (CDs), regarded as a class of water-soluble cyclic oligosaccharides with truncated cone-like structures that possess hydrophilic exteriors and hydrophobic cavities for internalizing guest molecules in water, have been extensively investigated in the construction of versatile nonviral vectors or biomedical diagnoses and treatment because of their excellent water solubility, low toxicity, and penetration capability toward biological membranes. Their selective association with targeted guests based on host−guest chemistry and capability to be used as surface capping agents for controllable self-assembly of inorganic NPs have been widely exploited. For example, thiolated CDs have been applied to modify AuNPs and regulate their self-assembly upon the introduction of proper guest molecules. CDs have also broadened the scope of optical devices with the assistance of semiconductor QDs, such as fluorescent carbon quantum dots (CQDs), for chiroselective sensing, molecular recognition, ions detection, enzyme activity sensing, and light energy conversion devices [40]. Upconversion nanoparticles (UCNPs), which are codoped with rare-earth ions, have been capped with CDs and widely investigated for various fluorescence properties of these nanocomposites.

Differing from single photon excitation of high energy, a two-photon absorption process consists of the absorption of two photons simultaneously and results in one photon of higher energy. These processes have triggered great interest owing to the desirable penetration depth ($>500\ \mu m$) and a lower tissue autofluorescence [41]. Compared with two-photon fluorescence, photon upconversion is derived from the absorption of two or multiphoton sequentially and regarded as an intense anti-Stokes photoluminescence [42]. It is noteworthy to mention that upconversion luminescence (UCL) shows promising application potential in biomedicine as a result of narrow emission spectra, low photodamage, and long lifetimes. UCNPs, which emerge as a kind of fluorophores consisting of a matrix codoped with rare-earth ions, have been paid great efforts due to their fascinating properties such as sharp emission bands, strong photoluminescence, good photostability, and low organisms autofluorescence upon the irradiation of low-energy NIR photons, and they can convert light from low energy to high energy via multiphoton mechanism [43–45]. UCNPs are considered as ideal donors in various förster resonance energy transfer (FRET)-based processes and hold great potential as optical sensors for detecting Hg^{2+}, Cu^{2+}, CN^{-}, or in biological sciences including labeling tumor tissue slices and bioimaging [46]. Generally, among various lanthanide-doped nanocrystals capable of emitting strong NIR-to-vis UCL, $NaYF_4$ codoped with Yb and Er has been regarded as the most efficient UCNP, whereas the Yb^{3+} ions can absorb multiple NIR photons, the Er^{3+} ions can emit visible light [47]. However, the as-prepared NPs exhibit low solubility because of the hydrophobic organic ligands used for their synthesis, and this is the main limitation for their further biomedical application. In order to solve this restriction, many hydrophilic biocompatible molecules functionalized with $-COOH$, $-NH_2$, and $-OH$ groups have been introduced to the surface of UCNPs through ligand exchange and ligand oxidation or after removal of the ligand to lead to naked UCNPs. To meet the need of developing new surface ligands for UCNPs, macrocyclic hosts such as CDs have been tested. This strategy not only facilitates the improvement of water-solubility and the biocompatibility in biosystems but also imparts the capability for further conjugation with guest molecules [48].

Phosphonated CD, for example, 6-phosphate-6-deoxy-β-cyclodextrin (βPCD) served as surface ligand in the fabrication of an upconversion luminescent nanoplatform via a two-step process (Fig. 6.5) [42]. Ligand exchange was conducted upon the addition of 2-azidoethylphosphonic acid ligands for displacing oleic acid ligands on $NaYF_4$ UCNPs, driven by the strong interactions between negatively charged phosphates and surface lanthanide ions. Then, water-dispersible βPCD/UCNPs were obtained through a cyclo-addition "click" reaction between azidoethylphosphonate ligands and alkyne-functionalizedβ-cyclodextrin (β-CD) catalyzed by copper.

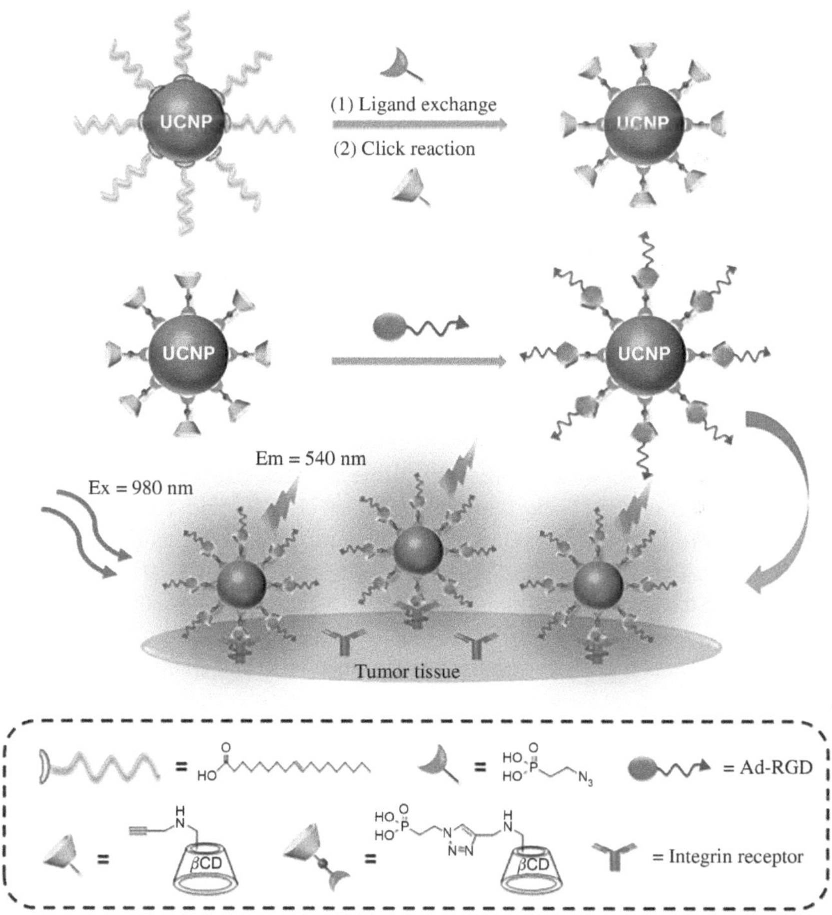

FIGURE 6.5 Schematic representation of the construction of CD-modified UCNP for biomedical imaging. *Reproduced with permission from C. Ma, T. Bian, S. Yang, C. Liu, T. Zhang, J. Yang, et al., Fabrication of versatile cyclodextrin-functionalized upconversion luminescence nanoplatform for biomedical imaging, Anal. Chem. 86 (2014) 6508–6515. Copyright 2014 American Chemical Society.*

Benefiting from the peculiar hydrophobic cavities of macrocycles and suitable spectral overlap between the absorption spectrum of rhodamine B (RB)-conjugated adamantane (Ad-RB) and the green emission band of βPCD/UCNPs, Ad-RB was selected as desired functionalities for further conjugation and construction of FRET-based upconversion nanosensors. Furthermore, c(RGDyK) peptide conjugated with adamantane (Ad-RGD) was also selected as functional models for the special recognition toward $\alpha_v\beta_3$ and $\alpha_v\beta_5$ integrin receptors overexpressed by specific cells or tumor tissues, indicative of their potential in bioimaging. Upon 980 nm laser irradiation, green luminescence of βPCD/UCNPs colloids was observed due to the green emissions of Er^{3+} ion of UCNPs. The upconversion emissions were located at 520, 540, and 650 nm, respectively. After the introduction of Ad-RB, host–guest interactions based on βPCD accounted for the closer proximity between the βPCD/UCNPs (energy donor) and the Ad-RB (energy acceptor), and resulted in energy transfer from UCNPs to Ad-RB. This process was verified by the decrease of green emission signals at 520 and 540 nm of UCNPs and an emerging broad emission band at 590 nm of Ad-RB. Based on this system, confocal upconversion luminescence images were used to evaluate the applicability of βPCD/UCNPs in imaging diagnosis of cervical cancer; impressively, cervical tumor tissue slice treated with Ad-RGD/*β*PCD/UCNPs exhibited an intense luminescence signal compared with normal cervical tissue slice or cervical tumor tissue slice treated with βPCD/UCNPs. The experimental results have addressed a promising nanoplatform for biomedical science and provided more possibilities in the diagnosis of diseases [49].

Another upconversion fluorescent probe for detecting cholesterol (Cho) was also based on β-CD derivative modified UCNPs and RB via a FRET process, where UCNPs served as the donor and RB served as the quencher. According to the work of Zhu and coworkers in 2014, β-CD (CD-Cit) modified with citric acid coordinated to the rare-earth ions on the surface of $NaYF_4$ UCNPs (Fig. 6.6) [46]. The macrocyclic hosts facilitated RB guests to enter and caused a closer contact between donor and quencher to induce a FRET process. Cho held stronger binding capacity with macrocyclic hosts than RB via competitive host–guest interactions. The introduction of Cho could release RB from the hydrophobic inner cavities and switch off FRET. The recovered fluorescence of UCNPs was followed by UCL intensity.

In 2017, Li and coworkers used β-CD to improve the stability and biocompatibility of UCNPs, which were encapsulated in the shell of mesoporous SiO_2, for detecting nitric oxide (NO) in physiological fluids, where β-CD could not only prevent biomacromolecules from entering but also allowed NO to penetrate [47]. RB-derived molecules were encapsulated in the pores of SiO_2 shell and the *O*-phenylenediamine moiety could react with NO, resulting in the opening of spiro-ring accompanied with RB absorption band and

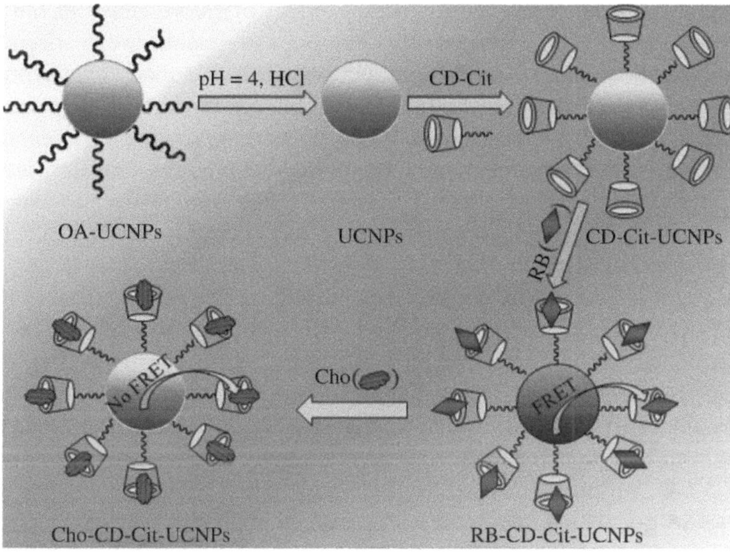

FIGURE 6.6 Schematic representation of the preparation of RB-CD-Cit-UCNPs and the UCL response toward Cho. *Reproduced with permission from Y.J. Ding, H. Zhu, X.X. Zhang, J.G. Gao, E.S. Abdel-Halim, L.P. Jiang, et al., An upconversion nanocomposite for fluorescence resonance energy transfer based cholesterol-sensing in human serum, Nanoscale 6 (2014) 14792–14798. Copyright 2014 The Royal Society of Chemistry.*

spectral overlap with emission spectrum of UCNPs. NO could be quantitatively measured by the changes of luminescence intensity ratio (I_{656}/I_{540}).

6.3.2 Quantum dots

CDs are considered as one of the most desirable macrocyclic hosts in terms of covering QDs for further applications [50–69]. In the second decade of the 20th century, QDs modified by CDs have been studied in the field of molecular sensing, ions detection, and enantioselective fluorescent analysis. The resultant nanocomposites not only enhance the photochemical properties of QDs as energy donor, but they also preserve the macrocyclic capacity of forming inclusion complexes with suitable guests that could serve as energy acceptors. Derived from molecular recognition on the basis of fluorescent variation, many optical sensors have been developed for the discrimination of *p*-nitrophenol (*p*-NP), β-galactosidase activity, adenosine-5′-triphosphate (ATP), acid or alkaline phosphatase, vanillin, phenolic compounds, megestrol acetate, C_{60} fullerenes, and separation of amino acids enantiomers, mainly via FRET or ET process. β-CD-CdSe/ZnS QDs hold promising feasibility in selecting penicillamine (PA) enantiomers, L-PA and D-PA, relying on the absolutely different fluorescence changes [70]. Experimental results exhibited an enhancement and quenching effect of fluorescence in the presence of D-PA and L-PA, respectively. The opposite results were arisen from the selective host–guest interactions between PA and macrocyclic pockets which were governed by host-size of hydrophobic pockets, adsorption processes, different host–guest interactions depending on the different functional groups of enantiomers, and the followed electron transfer process. Taking the different physiological effect of L-PA and D-PA into consideration, this study has provided an alternative for determination of medical and pharmaceutical analytes.

Host–guest complexes could form between CDs and *p*-NP due to the specific hydrophobic interactions and size-matching effect. β-CD-CQDs were fabricated for the detection of β-galactosidase activity depending on the catalytic activity toward glycoconjugates [71,72]. In this work, glycoconjugates could be hydrolyzed to obtain glycose and *p*-NP under the catalysis of glycosidase. And the resultant *p*-NP could induce the fluorescent quenching of β-CD-CQDs through the host–guest interaction between CDs and *p*-NP via a static quenching mechanism. Compared with the weakly fluorescent quenching effect of bare CQDs, β-CD-CQD nanoprobe exhibited high fluorescent quenching efficiency of 75% with a much smaller amount of *p*-NP (90.0 μM); however, 52.0 mM was necessary for the same quenching degree of bare CQDs. Another example for clinical enzymes detection was based on the boronic acid-functionalized CQDs which were further modified with hydropropyl β-CD via B–O bonds. Alkaline phosphatase could catalyze the hydrolytic process of *p*-NP phosphate disodium salt to produce *p*-NP [73]. The inclusion complex formed by β-CD-CQD and *p*-NP has been confirmed as good electron donor/acceptor pair, a photo-induced electron transfer (PET) process could be induced to quench the fluorescence of β-CD-CQD, which has been used to detect alkaline phosphatase.

6.3.3 Gold nanoparticles

Arising from the excitation of localized surface plasmon resonance, AuNPs display unique optical properties such as photothermal effect and photocatalytic activity [74–76]. There are three possible mechanisms accounting for the photocatalytic activity of plasmonic AuNPs: (1) an obvious increase of the excitation light intensity induced by near-field enhancement; (2) the photothermal effect can facilitate the chemical processes which are triggered by thermal environment; and (3) the hot charge carriers generated around the particles could be injected into molecules nearby to induce the chemical reactions [77]. AuNPs covered with macrocyclic ligands, which could tether the reactants via supramolecular host–guest interactions on the surface of AuNPs, can restrict separation and preserve the plasmon-based photocatalytic activities. β-CD derivative, heptathiolated β-CD, was used as coating on the surface of AuNPs via multiple sulfur bonds through a ligand exchange process. The photocatalytic activity has been proven by the catalytic process of Z→E isomerization of stilbenes.

A fluorescent nanosensor was developed for dopamine (DA) based on the gold nanoclusters (AuNCs) capped with β-CD [78]. PET was induced through supramolecular interactions between DA and β-CD and a significant fluorescence quenching of β-CD@AuNCs demonstrated the feasibility of this fluorescent probe for DA. The fluorescence spectra of β-CD@AuNCs exhibited a green emission at 450 nm upon the excitation of 370 nm. With the increase of the DA concentrations, fluorescence intensity decreased gradually with a good linear relationship in the range from 80.0 to 100.0 nM. Polydopamine and *o*-dopaquinone, which were obtained by polymerization and oxidation of DA, respectively, held the potential in interfering with the electron shuttling process and triggering the PET, thus resulting in the fluorescence quenching.

6.4 Crown ether-capped nanoparticles

With the rapid advances in the region of nanotechnology and supramolecular chemistry, some intriguing formulations for the construction of smart nanosensors or optical devices arising from crown ethers and QDs have been proposed. Among the macrocyclic hosts, crown ethers are well known for their coordinating ability with metal ions and these macrocycles can form sandwich complex with alkali metal ions. Upon the combination of metal ions and fluorophore, these nanosensors are designed for recognizing different metal ions via FRET and PET, which can be characterized by the variation of photophysical properties such as spectral shifts and the changes of fluorescence intensity or lifetime. Owing to two pivotal factors that influence the FRET process, that is, the distance between energy donor and energy acceptor and the spectral overlapping between the emission spectrum of energy donor and the absorption spectrum of energy acceptor, 15-crown-5-ether-capped CdSe/ZnS QDs (QDCE) and 15-crown-5-ether attached RB (RBCE) have already been prepared for use as donor and acceptor, respectively. QDCE and RBCE can be connected through electrostatic interactions. This water-soluble FRET probe performs good sensitivity toward K^+ ions. Benefiting from the charge−electron binding force, metal ions can form sandwich complex with CE moieties from QDCE and RBCE. For this reason, the introduction of metal ions facilitates the energy transfer process from QDCE to RBCE as a result of the closer proximity between these two parts. Fluorescence titration was utilized to monitor the fluorescence behaviors of QDCE−RBCE conjugate after addition of K^+ ions. QDCE−RBCE conjugate showed dual fluorescence emission peaks at 530 and 575 nm in the buffer solution of pH 8.3, in accordance with the fluorescence spectra of QDCE and RBCE, respectively. Fluorescence intensity exhibited a decreasing and an increasing variation of QDCE and RBCE, respectively, due to the energy transfer process between QDCE and RBCE. FRET was further confirmed by the time-resolved fluorescence measurements. The presence of K^+ ions induced a reduction of QDCE lifetime and an increase of RBCE lifetime, which verified the nonradiative ET process. K^+ ions are one of the most important bio-related metal ions in physiological systems, which play essential roles in the function of enzymes and living cells as well as the transmission of nerve signals. This NP–macrocycle hybrid system provides an effective method for metal ions sensing in the field of materials science and biomedicine.

In 2016, Lin and coworkers reported another fluorescence sensor based on silicon QDs (SiQDs) which were capped with three different aza-crown ethers via covalent interactions, namely, SiQDs/aza15C5, SiQDs/aza18C6, and SiQDs/diaza18C6, as can be seen in Fig. 6.7 [79]. The fluorescence quenching of SiQDs was attributed to the PET from lone pair electrons on

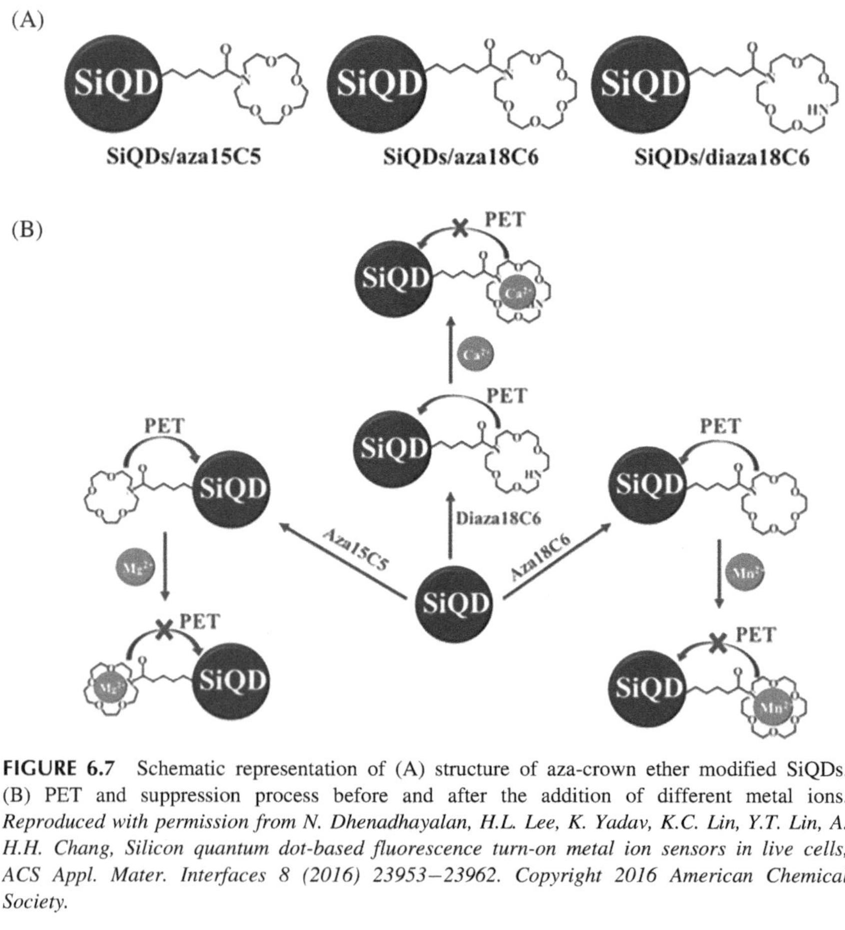

FIGURE 6.7 Schematic representation of (A) structure of aza-crown ether modified SiQDs. (B) PET and suppression process before and after the addition of different metal ions. *Reproduced with permission from N. Dhenadhayalan, H.L. Lee, K. Yadav, K.C. Lin, Y.T. Lin, A. H.H. Chang, Silicon quantum dot-based fluorescence turn-on metal ion sensors in live cells, ACS Appl. Mater. Interfaces 8 (2016) 23953–23962. Copyright 2016 American Chemical Society.*

the nitrogen atom of crown ether to the valence band of SiQDs. Subsequent addition of metal ions could regain the probability of electron-hole recombination of QDs core to induce PET suppression and result in a fluorescence enhancement. Fluorescence titration revealed that the fluorescence intensity of SiQDs/aza15C5 exhibited an increasing enhancement upon the increasing of Mg^{2+} concentration as compared with other metal ions. The same phenomena were observed for the special selectivity and sensitivity of SiQDs/aza18C6 and SiQDs/diaza18C6 toward Mn^{2+} and Ca^{2+}, respectively. Based on the variation of fluorescence intensity, the association of SiQDs/CE and metal ions was evaluated by Benesi–Hildebrand plot of $1/(F - F_0)$ versus $1/[M^{n+}]$, where F_0 and F represented the fluorescence intensity of SiQDs/CE before and after the introduction of metal ions. The electrostatic interactions and size-matching effect are the main driving forces for the trapping of metal ions. Afterward, confocal fluorescence images were used to monitor the practicability of this sensor for detecting Mg^{2+}, Mn^{2+}, and Ca^{2+} ions within live cells individually. The resultant hybrid system addressed a potential metal-ions-sensing platform.

6.5 Cucurbituril-capped nanoparticles

6.5.1 Gold nanoparticles

Cucurbit[n]urils (CB[*n*]s, $n = 5-8$, 10) are a family of rigid, hollow, and pumpkin-shaped macrocycles which are composed of 5, 6, 7, 8, or 10 glycoluril repeating units and synthesized through a condensation process in the presence of formaldehyde and acidic conditions [80–83]. Their hydrophobic cavities combined with two symmetric carbonyl-rich portals render CB[*n*]s as barrel vessels that are capable of accommodating polar and nonpolar organic guests through hydrophobic and ion-dipole interactions [84]. On account of their larger host–guest binding constants, CB[*n*]s have emerged as promising alternatives for CDs to be a new class of macrocyclic hosts. The remarkable behavior of host–guest interactions and unique structures with excellent solubility in water or salt water make CB[*n*]s a prime candidate in the field of supramolecular chemistry and polymeric smart materials including dynamic supramolecular self-assemblies, supramolecular polymer, surface engineering, colloidal clusters, and supramolecular hydrogel networks [85–88]. With the cutting-edge development of NPs in diversified applications such as photothermal therapy, drug delivery and magnetic resonance imaging (MRI), a wide range of macrocycles such as CB[*n*]s have been investigated as surface modifications to improve the properties and enrich the functions of NPs. As exemplified by magnetic iron oxide NPs, microwave heating has been utilized to synthesize CB[7]-iron oxide nanocomposites with dual-function of magnetic property for imaging and carrier capability for drug delivery [89–93].

Derived from the assemblies of discrete AuNPs, NP separation and interstitial nanojunctions play key roles in the localized electric field enhancement for SERS detection [94–99]. CB[*n*]s have attracted great attention in controlling the nanojunction spacing between AuNPs, which could be precisely fixed to 0.9 nm (Fig. 6.8). Taking advantages of the selectively interior cavities of CB[*n*]s and uniform distribution of AuNPs, the resulting self-assembled AuNPs linked by CB[*n*]s could serve as reliable in situ SERS substrate for qualitative and quantitative monitoring of target molecules that were selectively trapped in the cavities of CB[*n*]s through exposure to the plasmonic filed. CB[5] is not only Raman active which could create near-field strength within the center of junctions via their own SERS, but also could induce aggregation and define the rigid nanojunction separations of AuNPs. As shown in Fig. 6.8, CB[5] rings were bound to the surface of AuNPs through carbonyl groups, and

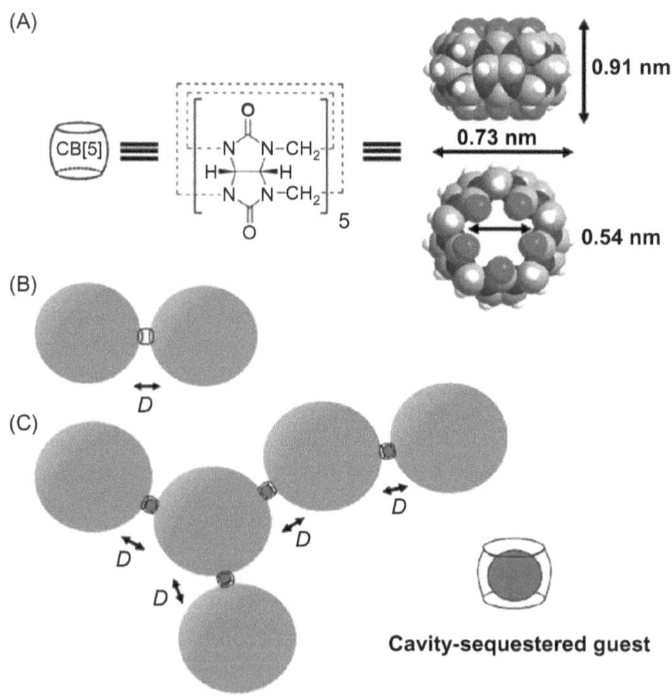

FIGURE 6.8 (A) Structural information of cucurbit[5]uril. (B) AuNPs dimer linked by CB[*n*] with fixed nanoparticle distance at 0.9 nm. (C) SERS substrate constructed by the AuNPs assemblies glued by CB[*n*] as plasmonic junctions to sequestrate guests. *Reproduced with permission from R.W. Taylor, T.C. Lee, O.A. Scherman, R. Esteban, J. Aizpurua, F.M. Huang, et al., Precise subnanometer plasmonic junctions for SERS within gold nanoparticle assemblies using cucurbit[n]uril "glue", ACS Nano (5) 2011 3878–3887. Copyright 2011 American Chemical Society.*

meanwhile, the hydrophobic cavity with appropriate size could hold the guests in the center of nanogaps between the coupled AuNPs [100]. As a result, the guest molecules located in the confined electric field (hot spot) could be optimally sensed by SERS.

On the other hand, nanocomposites assembled by AuNPs and CB[*n*]s also could function as potential SERS-nanoreactors [101]. CB[*n*]s acted as linkages between AuNPs as well as nanoscale reaction vessels for sequestering and templating the photoreaction of stilbene. Diaminostilbene (DAS) has been selected as photochemical reactant as a result of its UV-irradiated transformations, which could be templated by the size selectivity of the cavity as shown in Fig. 6.9. As a result of the size matching effect, CB[8] and CB[7] formed 2:1 and 1:1 inclusion complexes with DAS, respectively. The enhanced Raman scattering provided a powerful monitoring approach toward the selective photoisomerism and photodimerization process in the cavities of vessels in real time. Upon photoirradiation, the SERS fingerprint demonstrated different pathways of photodimerization and *trans–cis* isomerization

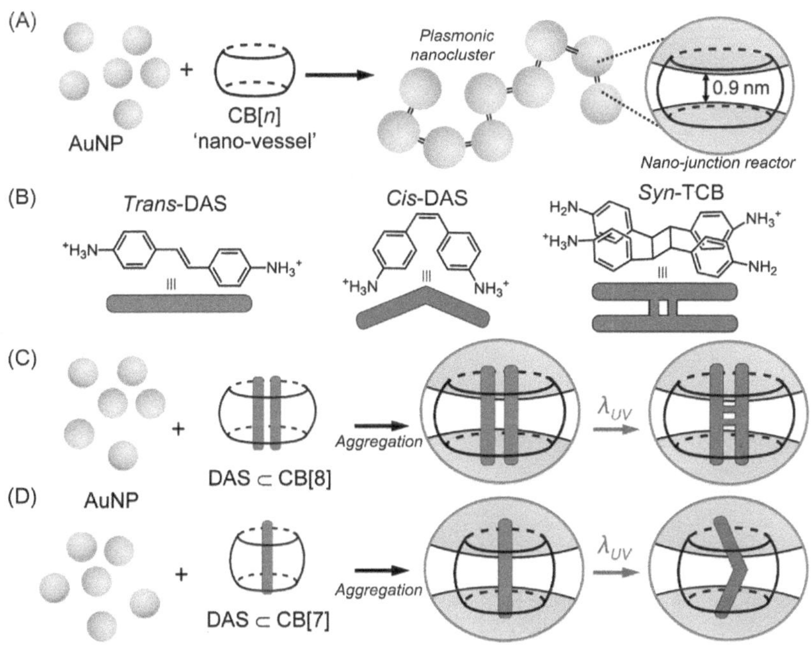

FIGURE 6.9 (A) Schematic description of the self-assembly of AuNPs linked by CB[*n*]. (B) Structural representation of photoactive *trans*-DAS, *cis*-DAS, and *syn*-TCB obtained through photodimerization. (C) CB[8] and (D) CB[7] in combination with double and single DAS followed by templated photodimerization and photoisomerization reaction, respectively. *Reproduced with permission from R.W. Taylor, R.J. Coulston, F. Biedermann, S. Mahajan, J.J. Baumberg, O.A. Scherman, In situ SERS monitoring of photochemistry within a nanojunction reactor, Nano Lett 13 (2013) 5985–5990. Copyright 2013 American Chemical Society.*

within the nanogaps of AuNPs-CB[8] and AuNPs-CB[7] aggregates, respectively. The presence of CBs has lowered the detection limit to ppb concentrations and strengthened the binding effect between DAS and AuNPs surface.

6.5.2 Quantum dots

Great effort has been paid on the host–guest complex interactions between CBs and quenchers of fluorescence QDs, which has promoted the development of turn on/off fluorescence of QDs systems. In these systems, there are two main mechanisms of fluorescence transformation of CBs as fluorescence switching. On the one hand, CBs function as linkers between QDs and quenchers and result in the fluorescence quenching through interfacial electron transfer from the excited QDs to the quenchers. As exemplified by the work of Majima and coworkers in 2011, thiol-modified CB[6]s were utilized to form, relying on the functions, a complex with the aromatic quencher, *N*-(6-aminohexyl)-4-nitrobenzenamine (NBzA), through specific host–guest interactions with the hexane diamine group of NBzA, and then bound on the surface of water-soluble CdTe QDs via two thiol groups [102]. The successful modification of CBs not only improved the stability of QDs but also provided host cavities to the guest NBzA for the further process of PET (Fig. 6.10). The steady-state and time-resolved fluorescence and absorption spectroscopies have been used to determine the quenching effect. Upon the addition of NBzA/CB [6] (10 μM), the fluorescence quenching extent could reach up to 60%.

On the other hand, CBs could also act as sequestrations to remove the quenchers from the surface of QDs, which prevent the photoinduced electron

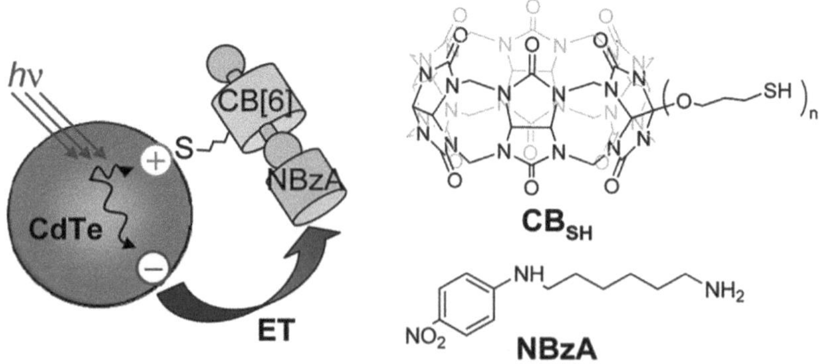

FIGURE 6.10 Schematic illustration of PET process between CdTe QDs and NBzA/CB and structural descriptions of thiol-modified CB[6] (CB_{SH}) and NBzA. *Reproduced with permission from S.C. Cui, T. Tachikawa, M. Fujitsuka, T. Majima, Photoinduced electron transfer in a quantum dot – cucurbituril supramolecular complex, J. Phys. Chem. C 115 (2011) 115 1824–1830. Copyright 2011 American Chemical Society.*

or energy transfer and lead to the recovery of luminescence. CdSe–ZnS core–shell quantum dots modified with hydrophilic mercaptoacetate groups could bind to bipyridinium dications through electrostatic interactions, resulting in the luminescence quenching through electron transfer process [103]. CBs function as receptors to bind with and remove bipyridinium dications from the surface of QDs. The prevention of ET process led to the luminescence recovery of QDs. According to the work by Patra and coworkers in 2013, a kind of negatively charged alloy QDs ($Cd_{1-x}Zn_xS$) capped with glutathione were synthesized to attach with positively charged 5-(4-aminophenyl)-10,15,20-triphenyl-21H,23H-porphyrin (APTPP), followed with a resonance energy transfer process between QDs and APTPP [104]. A decreasing fluorescence intensity of QDs was observed with the increase of APTPP. As shown in Fig. 6.11, the photoluminescence quenching was restored with the addition of CB[7]s as a result of their encapsulation and

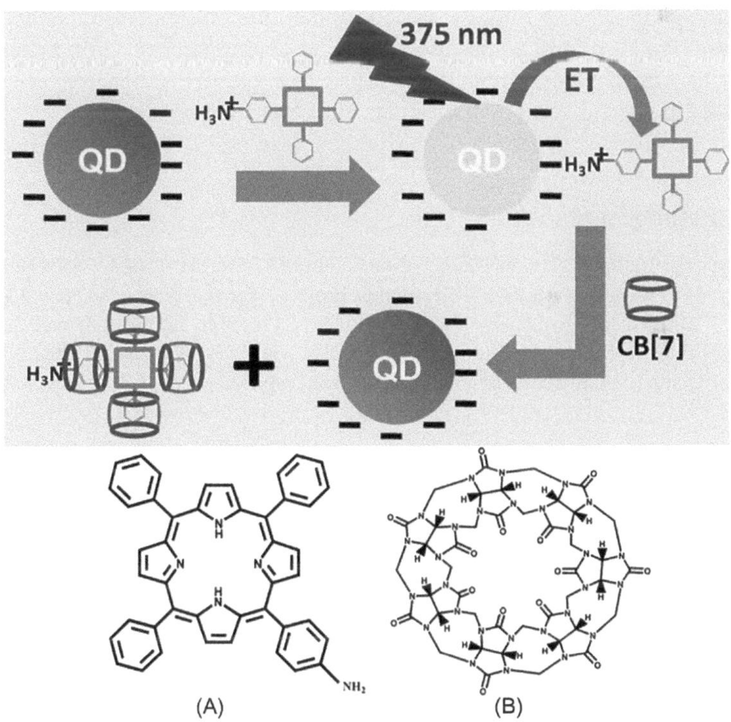

FIGURE 6.11 Schematic illustration of energy transfer from QDs to porphyrin and the luminescence recovery with the addition of CB[7]s and the structural description of (A) APTPP and (B) CB[7]. *Reproduced with permission from S. Mandal, M. Rahaman, S. Sadhu, S.K. Nayak, A. Patra, Fluorescence switching of quantum dot in quantum dot–porphyrin–cucurbit [7] Uril Assemblies, J. Phys. Chem. C 117 (2013), 3069–3077. Copyright 2013 American Chemical Society.*

removal ability toward porphyrin. This property makes it possible to be used as optical sensors for porphyrin or other bioapplications.

6.6 Concluding remarks

We have described various nanosystems based on a series of photoactive inorganic NPs and macrocyclic hosts as capping agents and platforms, including pillararenes, calixarenes, CDs, crown ethers and CEs. These nanocomposites not only expand the scope of inorganic–organic hybrid nanomaterials but also occupy an important position in the development of host–guest chemistry, and simultaneously hold great promise for the design and application of optical devices in the field of fluorescent sensing, bioimaging, chiral separation, stimuli–responsive drug-delivery systems, noninvasive optical diagnosis, and photothermal therapy. Besides these traditional macrocycles, there are also some novel macrocyclic molecules stimulating a tremendous upsurge of interest, including the blue box reported by the group of Fraser Stoddart, and some others such as resorcinarene, coronarenes, and cyanostar macrocycles [105–108]. These new classes of macrocycles have addressed powerful potency toward the future of materials science.

Acknowledgments

We thank the National Natural Science Foundation of China (51673084, 21871108), Jilin Province-University Cooperative Construction Project—Special Funds for New Materials (SXGJSF2017-3), Jilin University Talents Cultivation Program, the Fundamental Research Funds for the Central Universities, and the JLU Cultivation Fund for the National Science Fund for Distinguished Young Scholars for financial support.

References

[1] M.K. Choi, J. Yang, D.C. Kim, Z. Dai, J. Kim, H. Seung, et al., Extremely vivid, highly transparent, and ultrathin quantum dot light-emitting diodes, Adv. Mater. 30 (2018) 1703279–1703285.

[2] Z.-W. Ren, J.-K. Sun, H. Li, P. Mao, Y.-Z. Wei, X.-H. Zhong, et al., Bilayer PbS quantum dots for high-performance photodetectors, Adv. Mater. 29 (2017) 1702055–1702061.

[3] Z.-F. Wang, F.-L. Yuan, X.-H. Li, Y.-C. Li, H.-Z. Zhong, L.-Z. Fan, et al., 53% Efficient red emissive carbon quantum dots for high color rendering and stable warm white-light-emitting diodes, Adv. Mater. 29 (2017) 1702910–1702916.

[4] Z.-Y. Huang, Z.-H. Xu, M. Mahboub, X. Li, J.W. Taylor, W.H. Harman, et al., PbS/CdS core-shell quantum dots suppress charge transfer and enhance triplet transfer, Angew. Chem. Int. Ed. 56 (2017) 16583–16587.

[5] B.C.M. Martindale, E. Joliat, C. Bachmann, R. Alberto, E. Reisner, Clean donor oxidation enhances the H2 evolution activity of a carbon quantum dot-molecular catalyst photosystem, Angew. Chem. Int. Ed. 55 (2016) 9402–9406.

[6] J.-L. Shen, Q. Tang, L. Li, J. Li, X.-L. Zuo, X.-M. Qu, et al., Valence-engineering of quantum dots using programmable DNA scaffolds, Angew. Chem. Int. Ed. 56 (2017) 16077–16081.

[7] H.-C. Wang, W.-G. Wang, A.-C. Tang, H.-Y. Tsai, Z. Bao, T. Ihara, et al., High-performance $CsPb_{1-x}$ Sn_x Br_3 perovskite quantum dots for light-emitting diodes, Angew. Chem. Int. Ed. 56 (2017) 13650–13654.

[8] M.-Y. Ye, Z.-H. Zhao, Z.-F. Hu, L.-Q. Liu, H.-M. Ji, Z.-R. Shen, et al., 0D/2D Heterojunctions of vanadate quantum dots/graphitic carbon nitride nanosheets for enhanced visible-light-driven photocatalysis, Angew. Chem. Int. Ed. 56 (2017) 8407–8411.

[9] M.-Y. Zhao, D.-X. Wang, M.-X. Wang, Synthesis, structure, and properties of corona[6]arenes and their assembly with anions in the crystalline state, J. Org. Chem. 83 (2018) 1502–1509.

[10] R.A. Sperling, W.J. Parak, Surface modification, functionalization and bioconjugation of colloidal inorganic nanoparticles, Phil. Trans. R. Soc. A 368 (2010) 1333–1383.

[11] N. Song, Y.-L. Sun, Y.-W. Yang, Switchable host-guest systems on surfaces, Acc. Chem. Res. 47 (2014) 1950–1960.

[12] J.-K. Ma, F.-D. Shi, D.-M. Tian, H.-B. Li, Macroscopic responsive liquid quantum dots constructed via pillar[5]arene-based host-guest interactions, Chem. Eur. J. 22 (2016) 13805–13809.

[13] D.-X. Chen, Y.-L. Sun, Y. Zhang, J.-Y. Cui, F.-Z. Shen, Y.-W. Yang, Supramolecular self-assembly and photophysical properties of pillar[5]arene-stabilized CdTe quantum dots mediated by viologens, RSC Adv. 3 (2013) 5765–5768.

[14] G.V. Hartland, Optical studies of dynamics in noble metal nanostructures, Chem. Rev. 111 (2011) 3858–3887.

[15] Q. Zhou, B. Zhang, D. Han, R. Chen, F. Qiu, J. Wu, et al., Photo-responsive reversible assembly of gold nanoparticles coated with pillar[5]arenes, Chem. Commun. 51 (2015) 3124–3126.

[16] H. Li, D.-X. Chen, Y.-L. Sun, Y.B. Zheng, L.-L. Tan, P.S. Weiss, et al., Viologen-mediated assembly of and sensing with carboxylatopillar[5]arene-modified gold nanoparticles, J. Am. Chem. Soc. 135 (2013) 1570–1576.

[17] V. Montes-Garcia, C. Fernandez-Lopez, B. Gomez, I. Perez-Juste, L. Garcia-Rio, L.M. Liz-Marzan, et al., Pillar[5]arene-mediated synthesis of gold nanoparticles: size control and sensing capabilities, Chem. Eur. J. 20 (2014) 8404–8409.

[18] X. Huang, S. Wu, X. Ke, X. Li, X. Du, Phosphonated pillar[5]arene-valved mesoporous silica drug delivery systems, ACS Appl. Mater. Interfaces 9 (2017) 19638–19645.

[19] X. Wang, L.-L. Tan, X. Li, N. Song, Z. Li, J.-N. Hu, et al., Smart mesoporous silica nanoparticles gated by pillararene-modified gold nanoparticles for on-demand cargo release, Chem. Commun. 52 (2016) 13775–13778.

[20] X. Huang, X. Du, Pillar[6]arene-valved mesoporous silica nanovehicles for multiresponsive controlled release, ACS Appl. Mater. Interfaces 6 (2014) 20430–20436.

[21] N. Song, Y.-W. Yang, Molecular and supramolecular switches on mesoporous silica nanoparticles, Chem. Soc. Rev. 44 (2015) 3474–3504.

[22] H. Li, L.-L. Tan, P. Jia, Q.-L. Li, Y.-L. Sun, J. Zhang, et al., Near-infrared light-responsive supramolecular nanovalve based on mesoporous silica-coated gold nanorods, Chem. Sci. 5 (2014) 2804–2808.

[23] Y.-J. Chen, H. Zhang, X.-Q. Cai, J.-B. Ji, S.-W. He, G.-X. Zhai, Multifunctional mesoporous silica nanocarriers for stimuli-responsive target delivery of anticancer drugs, RSC Adv. 6 (2016) 92073–92091.

[24] S. Kim, H.-K. Na, C. Wonb, D.-H. Min, In-depth study on the gene silencing capability of silica nanoparticles with different pore sizes: degree and duration of RNA interference, RSC Adv. 6 (2016) 27143–27150.

[25] J.-X. Shao, M.-J. Xuan, H.-Y. Zhang, X.-K. Lin, Z.-G. Wu, Q. He, Chemotaxis-guided hybrid neutrophil micromotors for targeted drug transport, Angew. Chem. Int. Ed. 56 (2017) 12935–12939.

[26] Y. Yao, Y. Zhou, J. Dai, S. Yue, M. Xue, Host-guest recognition-induced color change of water-soluble pillar[5]arene modified silver nanoparticles for visual detection of spermine analogues, Chem. Commun. 50 (2014) 869–871.

[27] A.N. Sokolov, M.E. Roberts, O.B. Johnson, Y. Cao, Z. Bao, Induced sensitivity and selectivity in thin-film transistor sensors via calixarene layers, Adv. Mater. 22 (2010) 2349–2353.

[28] M. Yamada, M.R. Gandhi, U.M.R. Kunda, F. Hamada, Thiacalixarenes: emergent supramolecules in crystal engineering and molecular recognition, J. Incl. Phenom. Macrocycl. Chem. 85 (2016) 1–18.

[29] P.G. Sutariya, A. Pandya, A. Lodha, S.K. Menon, A simple and rapid creatinine sensing via DLS selectivity, using calix[4]arene thiol functionalized gold nanoparticles, Talanta 147 (2016) 590–597.

[30] B.-P. Wu, F. Zou, X.-X. Wang, K. Koh, K.-M. Wang, H.-X. Chen, The colorimetric assay of diamine oxidase activity with high sensitivity based on calixarene derivative-capped gold nanoparticles, Anal. Methods 9 (2017) 2153–2158.

[31] H. Valkenier, V. Malytskyi, P. Blond, M. Retout, A. Mattiuzzi, J. Goole, et al., Controlled functionalization of gold nanoparticles with mixtures of calix[4]arenes revealed by infrared spectroscopy, Langmuir 33 (2017) 8253–8259.

[32] F. Zou, B. Wu, X. Wang, Y. Chen, K. Koh, K. Wang, et al., Signal amplification and dual recognition strategy for small-molecule detection by surface plasmon resonance based on calix[4]arene crown ether-modified gold nanoparticles, Sens. Actuators B Chem. 241 (2017) 160–167.

[33] J.V. Rohit, R.K. Singhal, S.K. Kailasa, Dithiocarbamate-calix[4]arene functionalized gold nanoparticles as a selective and sensitive colorimetric probe for assay of metsulfuron-methyl herbicide via non-covalent interactions, Sens. Actuators B Chem. 237 (2016) 1044–1055.

[34] H.J. Kim, M.H. Lee, L. Mutihac, J. Vicens, J.S. Kim, Host-guest sensing by calixarenes on the surfaces, Chem. Soc. Rev. 41 (2012) 1173–1190.

[35] M. Oszajca, C. Lincheneau, M. Amelia, M. Baroncini, S. Silvi, K. Szaciłowski, et al., Supramolecular assemblies of semiconductor quantum dots and a bis(bipyridinium) derivative: luminescence quenching and aggregation phenomena, RSC Adv. 4 (2014) 29847–29854.

[36] S. Sayin, H. Azak, H.B. Yildiz, P. Camurlu, G.U. Akkus, L. Toppare, et al., Calixarene assembly with enhanced photocurrents using P(SNS-NH_2)/CdS nanoparticle structure modified Au electrode systems, Phys. Chem. Chem. Phys. 17 (2015) 19911–19918.

[37] K.V. Joshi, B.K. Joshi, A. Pandya, P.G. Sutariya, S.K. Menon, Calixarene capped ZnS quantum dots as an optical nanoprobe for detection and determination of menadione, Analyst 137 (2012) 4647–4650.

[38] C. Carrillo-Carrion, B. Lendl, B.M. Simonet, M. Valcarcel, Calix[8]arene coated CdSe/ZnS quantum dots as C60-nanosensor, Anal. Chem. 83 (2011) 8093–8100.

[39] K. Wu, Y.-Y. Zhou, T. Li, J.-Y. Sun, X.-Q. Xing, Q. Song, Cadmium tellurium quantum dots in sol-gel-derived silica spheres coated with calix[6]arene as an optical probe for Dalapon, Anal. Lett. 45 (2012) 2749–2762.

[40] S. Mondal, P. Purkayastha, α-Cyclodextrin functionalized carbon dots: pronounced photoinduced electron transfer by aggregated nanostructures, J. Phys. Chem. C 120 (2016) 14365–14371.

[41] P. Chen, S.-D. Hou, Y. Yang, Z.-R. Chen, L.-Y. Yang, J.-Y. Li, et al., ITO nanoparticles enhanced upconversion luminescence in Er^{3+}/Yb^{3+}-codoped silica glasses, Nanoscale 10 (2018) 3299–3306.

[42] C. Ma, T. Bian, S. Yang, C. Liu, T. Zhang, J. Yang, et al., Fabrication of versatile cyclodextrin-functionalized upconversion luminescence nanoplatform for biomedical imaging, Anal. Chem. 86 (2014) 6508–6515.

[43] S. Chen, A.Z. Weitemier, X. Zeng, L. He, X. Wang, Y. Tao, et al., Near-infrared deep brain stimulation via upconversion nanoparticle-mediated optogenetics, Science 359 (2018) 679–684.

[44] H.T.T. Duong, Y. Chen, S.A. Tawfik, S. Wen, M. Parviz, O. Shimoniab, et al., Systematic investigation of functional ligands for colloidal stable upconversion nanoparticles, RSC Adv. 8 (2018) 4842–4849.

[45] M. Chan, Y. Pan, Y. Chan, M. Hsiao, C. Chen, L. Sun, et al., Nanobubble-embedded inorganic 808 nm excited-upconversion nanocomposites for tumorous multiple imaging and treatment, Chem. Sci. 9 (2018) 3141–3151.

[46] Y.-J. Ding, H. Zhu, X.-X. Zhang, J.-G. Gao, E.S. Abdel-Halim, L.-P. Jiang, et al., An upconversion nanocomposite for fluorescence resonance energy transfer based cholesterol-sensing in human serum, Nanoscale 6 (2014) 14792–14798.

[47] N.-N. Wang, X.-Y. Yu, K. Zhang, C.A. Mirkin, J.-S. Li, Upconversion nanoprobes for the ratiometric luminescent sensing of nitric oxide, J. Am. Chem. Soc. 139 (2017) 12354–12357.

[48] A. Wang, W. Jin, E. Chen, J. Zhou, L. Zhou, S. Wei, Drug delivery function of carboxymethyl-β-cyclodextrin modified upconversion nanoparticles for adamantine phthalocyanine and their NIR-triggered cancer treatment, Dalton Trans. 45 (2016) 3853–3862.

[49] D. Kim, N. Lee, Y.I. Park, T. Hyeon, Recent advances in inorganic nanoparticle-based NIR luminescence imaging: semiconductor nanoparticles and lanthanide nanoparticles, Bioconjugate Chem. 28 (2017) 115–123.

[50] S. Geng, S.M. Lin, S.G. Liu, N.B. Li, H.Q. Luo, A new fluorescent sensor for detecting p-nitrophenol based on β-cyclodextrin-capped ZnO quantum dots, RSC Adv. 6 (2016) 86061–86067.

[51] W. Na, Q. Liu, B. Sui, T. Hu, X. Su, Highly sensitive detection of acid phosphatase by using a graphene quantum dots-based forster resonance energy transfer, Talanta 161 (2016) 469–475.

[52] Y. Wei, H. Li, H. Hao, Y. Chen, C. Dong, G. Wang, β-Cyclodextrin functionalized Mn-doped ZnS quantum dots for the chiral sensing of tryptophan enantiomers, Polym. Chem. 6 (2015) 591–598.

[53] K. Dwiecki, A. Piasecka, G. Neunert, M. Nogala-Kałucka, K. Polewski, Mechanism study of selected phenolic compounds determination using β-cyclodextrin-coated CdSe/ZnS quantum dots, J. Lumin. 192 (2017) 1119–1126.

[54] L.-Y. Dong, L.-Y. Wang, X.-F. Wang, Y. Liu, H.-L. Liu, M.-X. Xie, Development of fluorescent FRET probe for determination of glucose based on β-cyclodextrin modified ZnS-quantum dots and natural pigment 3-hydroxyflavone, Dyes Pigments 128 (2016) 170–178.

[55] L.-Y. Wang, L.-Y. Dong, L. Chen, Y.-B. Fan, J. Wu, X.-F. Wang, et al., A novel water-soluble quantum dot–neutral red fluorescence resonance energy transfer probe for the selective detection of megestrol acetate, New J. Chem. 39 (2015) 555–565.

[56] A. Cayuela, M.L. Soriano, M. Valcárcel, β-Cyclodextrin functionalized carbon quantum dots as sensors for determination of water-soluble C_{60} fullerenes in water, Analyst 141 (2016) 2682–2687.

[57] J. Zhou, Y. Liu, Z. Zhang, S. Yang, J. Tang, W. Liu, et al., Cyclodextrin-clicked silica/CdTe fluorescent nanoparticles for enantioselective recognition of amino acids, Nanoscale 8 (2016) 5621–5626.

[58] C. Yang, R.P. Thomsen, R. Ogaki, J. Kjems, B.M. Teo, Ultrastable green fluorescence carbon dots with a high quantum yield for bioimaging and use as theranostic carriers, J. Mater. Chem. B 3 (2015) 4577–4584.

[59] T. Hu, W. Na, X. Yan, X. Su, Sensitive fluorescence detection of ATP based on host-guest recognition between near-infrared β-cyclodextrin-$CuInS_2$ QDs and aptamer, Talanta 165 (2017) 194–200.

[60] J. Li, W.Y. Lee, T. Wu, J. Xu, K. Zhang, G. Li, et al., Multifunctional quantum dot nanoparticles for effective differentiation and long-term tracking of human mesenchymal stem cells in vitro and in vivo, Adv. Healthc Mater. 5 (2016) 1049–1057.

[61] H. Li, C. Han, Sonochemical synthesis of cyclodextrin-coated quantum dots for optical detection of pollutant phenols in water, Chem. Mater. 20 (2008) 6053–6059.

[62] R. Freeman, T. Finder, L. Bahshi, I. Willner, β-Cyclodextrin-modified CdSe/ZnS quantum dots for sensing and chiroselective analysis, Nano Lett. 9 (2009) 2073–2076.

[63] H. Bavireddi, R. Kikkeri, Glyco-β-cyclodextrin capped quantum dots: synthesis, cytotoxicity and optical detection of carbohydrate-protein interactions, Analyst 137 (2012) 5123–5127.

[64] Z.B. Shang, S. Hu, Y. Wang, W.J. Jin, Interaction of β-cyclodextrin-capped CdSe quantum dots with inorganic anions and cations, Luminescence 26 (2011) 585–591.

[65] M. Hu, Y. Yang, X. Gu, Y. Hu, J. Huang, C. Wang, One-pot synthesis of photoluminescent carbon nanodots by carbonization of cyclodextrin and their application in Ag^+ detection, RSC Adv. 4 (2014) 62446–62452.

[66] L. Jia, J.P. Xu, D. Li, S.-P. Pang, Y. Fang, Z.-G. Song, et al., Fluorescence detection of alkaline phosphatase activity with β-cyclodextrin-modified quantum dots, Chem. Commun. 46 (2010) 7166–7168.

[67] S.-Y. Liu, H. Wang, T. He, L. Qi, Z.-Q. Zhang, Sensitive fluorimetric assays for alpha-glucosidase activity and inhibitor screening based on β-cyclodextrin-coated quantum dots, Luminescence 31 (2016) 96–101.

[68] M.-X. Zhao, Q. Xia, X.-D. Feng, X.-H. Zhu, Z.-W. Mao, L.-N. Ji, et al., Synthesis, biocompatibility and cell labeling of L-arginine-functional β-cyclodextrin-modified quantum dot probes, Biomaterials 31 (2010) 4401–4408.

[69] X. Ai, L. Niu, Y. Li, F. Yang, X. Su, A novel β-cyclodextrin-QDs optical biosensor for the determination of amantadine and its application in cell imaging, Talanta 99 (2012) 409–414.

[70] G.M. Durán, C. Abellán, A.M. Contento, Á. Ríos, Discrimination of penicillamine enantiomers using β-cyclodextrin modified CdSe/ZnS quantum dots, Microchim. Acta 184 (2017) 815–824.

[71] C. Tang, J. Zhou, Z. Qian, Y. Ma, Y. Huang, H. Feng, A universal fluorometric assay strategy for glycosidases based on functional carbon quantum dots: β-galactosidase activity detection in vitro and in living cells, J. Mater. Chem. B 5 (2017) 1971–1979.

[72] Z. Zhang, J. Zhou, Y. Liu, J. Tang, W. Tang, Cyclodextrin capped CdTe quantum dots as versatile fluorescence sensors for nitrophenol isomers, Nanoscale 7 (2015) 19540–19546.

[73] C. Tang, Z. Qian, Y. Huang, J. Xu, H. Ao, M. Zhao, et al., A fluorometric assay for alkaline phosphatase activity based on β-cyclodextrin-modified carbon quantum dots through host-guest recognition, Biosens. Bioelectron. 83 (2016) 274–280.

[74] M. Padilla, F. Peccati, J.L. Bourdelande, X. Solans-Monfort, G. Guirado, M. Sodupe, et al., Enhanced photocatalytic activity of gold nanoparticles driven by supramolecular host-guest chemistry, Chem. Commun. 53 (2017) 2126–2129.

[75] Z. Yu, M. Wang, W. Pan, H. Wang, N. Li, B. Tang, Tumor microenvironment-triggered fabrication of gold nanomachines for tumor-specific photoacoustic imaging and photothermal therapy, Chem. Sci. 8 (2017) 4896–4903.

[76] T. Huang, F. Meng, L. Qi, Facile synthesis and one-dimensional assembly of cyclodextrin-capped gold nanoparticles and their applications in catalysis and surface-enhanced Raman scattering, J. Phys. Chem. C 113 (2009) 13636–13642.

[77] P. Herves, M. Perez-Lorenzo, L.M. Liz-Marzan, J. Dzubiella, Y. Lu, M. Ballauff, Catalysis by metallic nanoparticles in aqueous solution: model reactions, Chem. Soc. Rev. 41 (2012) 5577–5587.

[78] M.I. Halawa, F. Wu, T.H. Fereja, B. Lou, G. Xu, One-pot green synthesis of supramolecular β-cyclodextrin functionalized gold nanoclusters and their application for highly selective and sensitive fluorescent detection of dopamine, Sens. Actuators B Chem. 254 (2018) 1017–1024.

[79] N. Dhenadhayalan, H.-L. Lee, K. Yadav, K.-C. Lin, Y.-T. Lin, A.H.H. Chang, Silicon quantum dot-based fluorescence turn-on metal ion sensors in live cells, ACS Appl. Mater. Interfaces 8 (2016) 23953–23962.

[80] Y. Lan, Y.-C. Wu, A. Karas, O.A. Scherman, Photoresponsive hybrid raspberry-like colloids based on cucurbit[8]uril host-guest interactions, Angew. Chem. Int. Ed. 53 (2014) 2166–2169.

[81] Y.-L. Sun, B.-J. Yang, S.X.-A. Zhang, Y.-W. Yang, Cucurbit[7]uril pseudorotaxane-based photoresponsive supramolecular nanovalve, Chem. Eur. J. 18 (2012) 9212–9216.

[82] Q.-L. Li, Y.-F. Sun, Y.-L. Sun, J.-J. Wen, Y. Zhou, Q.-M. Bing, et al., Mesoporous silica nanoparticles coated by layer-by-layer self-assembly using cucurbit[7]uril for in vitro and in vivo anticancer drug release, Chem. Mater. 26 (2014) 6418–6431.

[83] X.-L. Qiu, Y. Zhou, X.-Y. Jin, A.-D. Qi, Y.-W. Yang, One-pot solvothermal synthesis of biocompatible magnetic nanoparticles mediated by cucurbit[n]urils, J. Mater. Chem. C 3 (2015) 3517–3521.

[84] X.-H. Ren, Y.-C. Wu, D.E. Clarke, J. Liu, G.-L. Wu, O.A. Scherman, Surface-bound cucurbit[8]uril catenanes on magnetic nanoparticles exhibiting molecular recognition, Chem. Asian J. 11 (2016) 2382–2386.

[85] C.S.Y. Tan, J. Liu, A.S. Groombridge, S.J. Barrow, C.A. Dreiss, O.A. Scherman, Controlling spatiotemporal mechanics of supramolecular hydrogel networks with highly branched cucurbit[8]uril polyrotaxanes, Adv. Funct. Mater. 28 (2018) 1702994–1702999.

[86] C. Hu, Y. Lan, F. Tian, K.R. West, O.A. Scherman, Facile method for preparing surface-mounted cucurbit[8]uril-based rotaxanes, Langmuir 30 (2014) 10926–10932.

[87] D. Das, O.A. Scherman, Cucurbituril: at the interface of small molecule host-guest chemistry and dynamic aggregates, Isr. J. Chem. 51 (2011) 537–550.

[88] J. Liu, C.S.Y. Tan, Y. Lan, O.A. Scherman, Aqueous polymer self-assembly based on cucurbit[n]uril-mediated host-guest interactions, Macromol. Chem. Phys. 217 (2016) 319–332.

[89] F. Benyettou, K. Nchimi-Nono, M. Jouiad, Y. Lalatonne, I. Milosevic, L. Motte, et al., Viologen-templated arrays of cucurbit[7]uril-modified iron-oxide nanoparticles, Chem. Eur. J. 21 (2015) 4607–4613.

[90] F. Benyettou, H. Fahs, R. Elkharrag, R.A. Bilbeisi, B. Asma, R. Rezgui, et al., Selective growth inhibition of cancer cells with doxorubicin-loaded CB[7]-modified iron-oxide nanoparticles, RSC Adv. 7 (2017) 23827–23834.

[91] F. Benyettou, I. Milosevic, Y. Lalatonne, F. Warmont, R. Assah, J.-C. Olsen, et al., Toward theranostic nanoparticles: CB[7]-functionalized iron oxide for drug delivery and MRI, J. Mater. Chem. B 1 (2013) 5076–5082.

[92] F. Benyettou, L. Motte, H. Traboulsi, J. Mazher, R. Pasricha, J.-C. Olsen, et al., Palladium-loaded cucurbit[7]uril-modified iron oxide nanoparticles for C-C cross-coupling reactions, Chem. Eur. J 24 (2018) 2349–2353.

[93] F. Benyettou, M. Alhashimi, M. O'Connor, R. Pasricha, J. Brandel, H. Traboulsi, et al., Sequential delivery of doxorubicin and zoledronic acid to breast cancer cells by cb[7]-modified iron oxide nanoparticles, ACS Appl. Mater. Interfaces 9 (2017) 40006–40016.

[94] N. Hüsken, R.W. Taylor, D. Zigah, J.-C. Taveau, O. Lambert, O.A. Scherman, et al., Electrokinetic assembly of one-dimensional nanoparticle chains with cucurbit[7]uril controlled subnanometer junctions, Nano Lett. 13 (2013) 6016–6022.

[95] T.-C. Lee, O.A. Scherman, Formation of dynamic aggregates in water by cucurbit[5]uril capped with gold nanoparticles, Chem. Commun. 46 (2010) 2438–2440.

[96] S. Mahajan, T.-C. Lee, F. Biedermann, J.T. Hugall, J.J. Baumberga, O.A. Scherman, Raman and SERS spectroscopy of cucurbit[n]urils, Phys. Chem. Chem. Phys. 12 (2010) 10429–10433.

[97] T.-C. Lee, O.A. Scherman, A facile synthesis of dynamic supramolecular aggregates of cucurbit[n]uril (n = 5-8) capped with gold nanoparticles in aqueous media, Chem. Eur. J 18 (2012) 1628–1633.

[98] S.T. Jones, R.W. Taylor, R. Esteban, E.K. Abo-Hamed, P.H.H. Bomans, N.A.J.M. Sommerdijk, et al., Gold nanorods with sub-nanometer separation using cucurbit[n]uril for SERS applications, Small 10 (2014) 4298–4303.

[99] S.T. Jones, J.M. Zayed, O.A. Scherman, Supramolecular alignment of gold nanorods via cucurbit[8]uril ternary complex formation, Nanoscale 5 (2013) 5299–5302.

[100] R.W. Taylor, T.-C. Lee, O.A. Scherman, R. Esteban, J. Aizpurua, F.M. Huang, et al., Precise subnanometer plasmonic junctions for SERS within gold nanoparticle assemblies using cucurbit[n]uril "glue", ACS Nano 5 (2011) 3878–3887.

[101] R.W. Taylor, R.J. Coulston, F. Biedermann, S. Mahajan, J.J. Baumberg, O.A. Scherman, In situ SERS monitoring of photochemistry within a nanojunction reactor, Nano Lett. 13 (2013) 5985–5990.

[102] S.-C. Cui, T. Tachikawa, M. Fujitsuka, T. Majima, Photoinduced electron transfer in a quantum dot – cucurbituril supramolecular complex, J. Phys. Chem. C 115 (2011) 1824–1830.

[103] I. Yildiz, M. Tomasulo, F.M. Raymo, A mechanism to signal receptor-substrate interactions with luminescent quantum dots, PNAS 103 (2006) 11457–11460.

[104] S. Mandal, M. Rahaman, S. Sadhu, S.K. Nayak, A. Patra, Fluorescence switching of quantum dot in quantum dot–porphyrin–cucurbit [7] uril assemblies, J. Phys. Chem. C 117 (2013) 3069–3077.

[105] C.-H. Sue, S. Basu, A.C. Fahrenbach, A.K. Shveyd, S.K. Dey, Y.Y. Botrosacd, et al., Enabling tetracationic cyclophane production by trading templates, Chem. Sci. 1 (2010) 119–125.

[106] M.T. Nguyen, M.D. Krzyaniak, M. Owczarek, D.P. Ferris, M.R. Wasielewski, J.F. Stoddart, A boat-shaped tetracationic macrocycle with a semiconducting organic framework, Angew. Chem. Int. Ed. 56 (2017) 5795–5800.

[107] E.M. Fatila, M. Pink, E.B. Twum, J.A. Karty, A.H. Flood, Phosphate–phosphate oligomerization drives higher order co-assemblies with stacks of cyanostar macrocycles, Chem. Sci. 9 (2018) 2863–2872.
[108] C. Gropp, B.L. Quigley, F. Diederich, Molecular recognition with resorcin[4]arene cavitands: switching, halogen-bonded capsules, and enantioselective complexation, J. Am. Chem. Soc. 140 (2018) 2705–2717.

Chapter 7

Photoactive nanoparticles capped with functional polymers

Olga García[a] and Marta Liras[b,*]
[a]*Instituto de Ciencia y Tecnología de Polímeros, Consejo Superior de Investigaciones Científicas ICTP-CSIC, Madrid, Spain,* [b]*IMDEA Energy Institute, Móstoles, Spain,*
**Corresponding author.email address: Marta Liras marta.liras@imdea.org*

Chapter Outline

7.1 Introduction

At present, there is a great interest in the study and knowledge of new nanostructured hybrid materials that respond to the development of advanced applications in fields such as biomedicine, biotechnology, energy storage and conversion, optoelectronics, catalysis, and water purification. In particular, inorganic nanoparticles (NPs) offer a wide and exclusive range of physical–chemical, optical, electronic, magnetic, and photothermal conversion properties, revealing themselves as valuable tools at the forefront of new applications [1]. Despite the great potential of inorganic NPs, the industrial development of applications based on them is limited as they are colloidal NPs. Therefore, many efforts are being devoted to the development of hybrid materials that combine the properties of inorganic materials and the versatility of polymeric materials to obtain hybrids with better and tuneable properties at the level of solubility, reduction of toxicity, improvement of biocompatibility and functionalization capacity, transport or controlled administration of bioactive and therapeutic agents and drugs, and/or response

Photoactive Inorganic Nanoparticles. DOI: https://doi.org/10.1016/B978-0-12-814531-9.00007-5
© 2019 Elsevier Inc. All rights reserved.

to external stimuli (temperature, pH, or redox condition). Interestingly, the unique properties of a wide range of new materials arise from the synergy of the individual components [2,3]. In this context, the development of new processes for the functionalization of NPs by applying surface engineering is essential for the design of new hybrid nanomaterials [4–6].

The design of new hybrid nanostructured systems that integrate components of different nature, such as metallic NPs, nanoclusters (Au, Ag, Cu, etc.), and photoactive NPs [quantum dots (QDs), UCNPs, etc.] with organic polymers, offers new challenges for the Materials Science due to its high interest in technology and with the view of their application in different fields [1,7–13].

To achieve this goal, it is necessary to act simultaneously on several aspects, such as on the design of new monomers and polymers with suitable functionalities by applying adequate polymerization techniques. It is unavoidable to develop novel polymer functionalization strategies that facilitate and enhance polymer–NP interactions, using new, simpler, scalable, economically profitable and, without any doubt, more versatile protocols [14].

Nowadays, the synthesis, characterization, and subsequent application of hybrid NPs composed of an inorganic NP and a polymer capping is a growing research field [15]. The nature of the polymer capping is crucial for the final properties and use of the nanohybrid. Although the main features of the nanohybrid, such as size, shape, photophysical, thermal, or magnetic properties, are derived from the inorganic core, it is undoubtable that the shell is directly in contact with the environment. Therefore, the shell will be responsible of some of the final features, such as dispersibility, biocompatibility, and response to changes of the media and so on. If this shell has a polymeric nature, it can be tailor-made to meet specific application, thanks to the synthetic organic tools available up to now. The polymeric shell could carry additional functional groups that can be used to link biomolecules, fluorophores, or other inorganic NPs with smaller size than the core, etc. [4,5]. In this sense, the versatility of the polymer coating is higher than that of the inorganic core itself and it can confer the desired properties to the final nanohybrid (Fig. 7.1).

It is crucial to have into account the interface between both materials, the inorganic core, and the polymeric shell. The interface is going to be dependent on the synthetic route used for the synthesis of the hybrid NP (Fig. 7.2). In general, the inorganic NPs are prepared from colloidal synthetic approaches using high temperatures and are surrounded by organic ligands of relative small molecular weight such as citrate, oleic acid, dodecylamine, trioctylphosphine (TOP) and trioctylphosphine oxide (TOPO). There are a lot of approaches for NP–polymer integration [5,6,17–23], for instance, (1) promoting covalent and non-covalent interactions with mono-/multidentate polymers; (2) favoring the encapsulation of NPs using polymers of amphiphilic nature [24] or new block copolymers [25], and (3) favoring the NPs

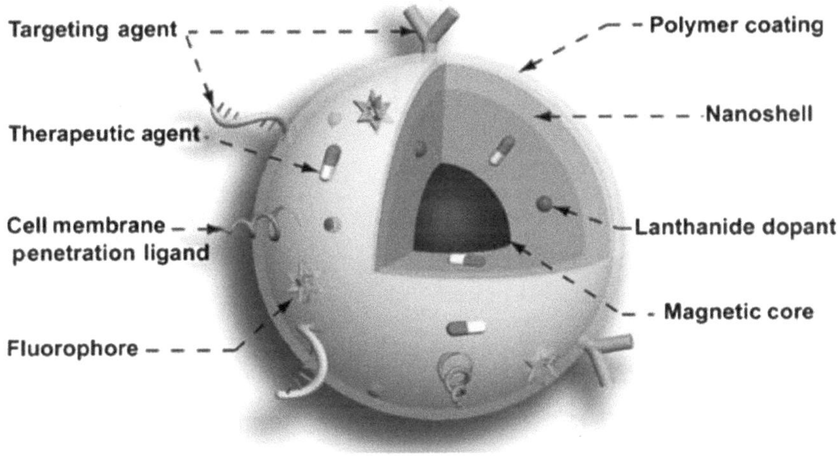

FIGURE 7.1 Complete hybrid nanomaterial where the polymeric shell includes the desired functionalities such as targeting agents, therapeutic agent, and fluorophores. *Reprint with permission from X. Xue, F. Wang, X. Liu, Emerging functional nanomaterials for therapeutics, J. Mater. Chem. 21 (2011) 13107 [16]. Copyright (2011) by Royal Society.*

incorporation into hydrogels or polymeric nanogels that protect the NPs from the surrounding [26,27], thus providing excellent colloidal stability and serving as dielectric spacers to control the interparticle distances ranging from several tens to hundreds of nanometers [27,28]. More specifically, the different synthetic approaches can be grouped into four types attending to the methodologies employed, as indicated below:

- The *first synthetic approach* consists of a ligand exchange with monomers or initiator for living polymerization able to be subsequently polymerized over the NP surface. In both cases, they have to be functionalized with the correct functional group able to displace the original ligand because its higher affinity with the inorganic surface and, at the same time, should be provided with the adequate functional group able to polymerize or initiate the polymerization process. Keep in mind that the complete removal of the former ligand by replacement with the monomer it is highly desirable in order to avoid competences between them and to allow the correct polymerization. This approach is also known as "*grafting from*" approach because the polymer grows from the NP surface itself. When an initiator is used during the ligand exchange, the new NP becomes a *macroinitiator* and the polymer chain grows in a *brush-like* performance. However, when a monomer is used in the ligand exchange, the hybrid NP becomes a *multicrosslinker* and the polymerization could produce a dense polymer capping maintaining its "nano" nature (i.e., nanogels) or being part as a bulk polymer which integrates more inorganic NPs attached to the whole polymeric matrix.

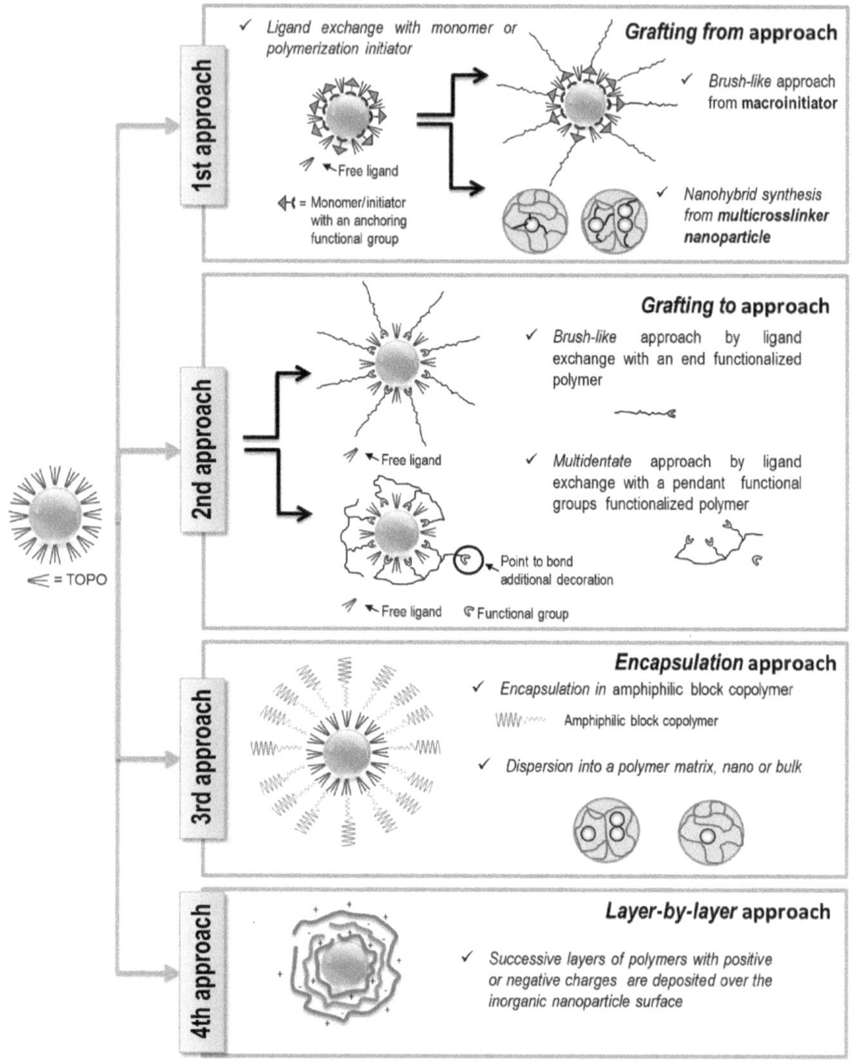

FIGURE 7.2 Different synthetic approaches for capping inorganic nanoparticles. Although the figure illustrates the case of a QD originally capped with trioctylphosphate oxide ligand, these approaches can be translated to any inorganic colloidal nanoparticle and any original ligand.

- The *second approach* is known as "*grafting to*" approach because it is based on ligand exchange using directly a functionalized polymer. Also, in this case the polymeric functionalization has to obey to ligand engineering to guarantee a correct coating. So, an adequate choice of the pendant functional groups is mandatory. With this approach, we can find two scenarios: (1) the first one, when the active functional group is located at the end of the main polymer chain leading a brushed polymer disposition; (2) the second one, when the polymer has hanging functional groups able to perform ligand exchange, thus allowing the multifunctional attach to the inorganic surface. As the ligand is attached/detached in equilibrium with the surface and free nonattached ligands, the multifunctional polymer covers the surface more efficiently than the monofunctional polymers. Therefore the NP stabilization (protection *versus* quencher and so on) is higher using this polymer capping design. Moreover, in the multifunctional polymer ligand exchange process, some of the nonattached functional groups could be used to bind other molecules or agents of interest or could be responsible of new nanohybrid properties as, for instance, pH responsiveness.

- The *third approach* is based on the ligand attraction by Van der Waals forces between the original ligand and the polymer. In general, the original ligand is a hydrophobic molecule with more or less long hydrocarbon chain and this attraction obeys to the interpenetration of the hydrocarbon lateral chain present in the monomer/polymer of choice. The use of amphiphilic block copolymers that allows the functionalization of the inorganic NP by affinity between hydrophobic polymeric blocks and the ligands can also be considered in this approach. In contrast, the incorporation of inorganic NPs into bulk polymer could be also considered. To avoid precipitation or aggregation into the bulk polymer, a good chemical affinity between both the polymer matrix and original ligands must be guaranteed. This approach can be considered as an "*encapsulation process*" because the original ligand remains and the interaction between the polymer capping is with the ligand and not with the inorganic NP surface. In this case, there are no covalent bonds between the NP and the polymer and sometimes the NP can leach from the polymer (e.g., during swelling–deswelling process of hydrogels).
- The *fourth approach* is based on the electrostatic attraction between the polymer and the inorganic surface. The assembly of several polymeric layers over the inorganic surface is possible by layer-by-layer deposition technique. The successive layers covering the NP surface are led by electrostatic interactions with the polymeric cation or anion sites.

On other hand, when the final nanohybrid is aimed to biological applications, the polymer capping nature should meet several requirements: (1) amphiphilicity or hydrophilicity to achieve a nano-display dispersible in physiological media; (2) biocompatibility to avoid cytotoxicity; (3) to have an anchoring functional group to the NP surface; (4) possibility of prefunctionalization or postfunctionalization aiming to bind other interesting functions such as fluorophores, target agents, and drugs; and (5) if possible, it should be able to respond to stimuli, such as temperature and pH.

The control over the polymeric molecular weight is an important issue to have into account in the design of polymeric capping, and even more if we are thinking of ligand exchange with polymer (second approach described above). In order to maintain the colloidal NPs original polydispersity, the polymer must show a low polydispersity index and so, the use of living polymerization techniques is mandatory [29].

In this chapter, we are going to focus our interest on these luminescent inorganic NPs able to be used in the biomedical field as monitoring agents, diagnosis or therapeutic carriers. For instance, Fig. 7.3 depicts the general strategies employed in the use of nanohybrids for cancer diagnosis, imaging, and therapy [30]. QDs are the quintessential highly luminescent NPs and, in fact, they have been broadly studied, mainly as bio-image carriers and for

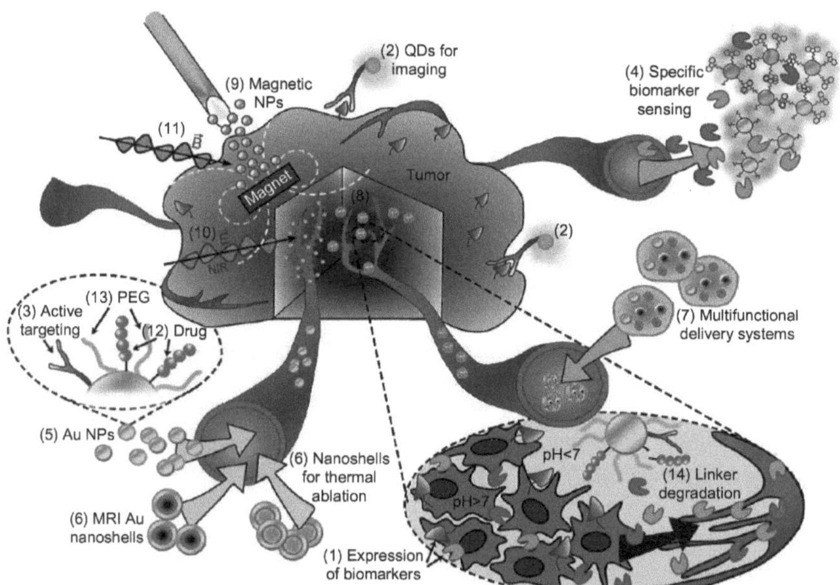

FIGURE 7.3 Nanomaterial-based strategies for tumor diagnosis, imaging, and therapy. Tumor cells may overexpress biomarkers (1) that can be used as anchoring points to direct the binding of NPs used for labeling (2), imaging, or drug delivery (3). Some biomarkers may be released into the bloodstream, from where they can be detected using NP-based systems (4). Gold NPs (5), composite NPs (6), and multifunctional delivery systems (7) circulating in the bloodstream accumulate in tumors (8). Alternatively, magnetic NPs can be directed into the tumor by means of magnetic fields (9). Engineered NPs can act as heat sources when irradiated, for example, with NIR light (10) and magnetic fields (11), causing tumor cell death. NPs can also be used to deliver drugs and image contrast agents into the tumor (12). NP surface PEGylation is commonly used to improve NP circulation time in blood (13). Drug release can be triggered by environmental factors, such as the acidic pH found in tumor cells (14). *Reprinted with permission from C. Minelli, S.B. Lowe, M.M. Stevens, Engineering nanocomposite materials for cancer therapy, Small. 6 (2010) 2336–2357 [30]. Copyright (2010) by Wiley and Son.*

monitoring as sensors; equally important, upconversion nanoparticles (UCNPs) are relevant not only for their luminescence but also because they are active upon near-infrared (NIR) light excitation, which offers a high tissue penetration and therefore they have potential in photodynamic therapy.

7.2 Coating quantum dots with polymers

Without any doubt, QDs have been the fluorescent NPs with more applications not only in the biomedical field [31] but also in optoelectronic and photovoltaics [32]. They are semiconductor colloidal NPs with size-dependent physical properties originating from quantum confinements, show a broad excitation spectra coupled with a narrow size-tunable luminescence emission, high brightness and high photo-stability, as compared with conventional organic dyes (Fig. 7.4) [33].

The inorganic core is composed of a blend of elements drawn from groups III and V, or II and VI of the periodic table elements such as cadmium, lead, zinc, arsenic, indium, tellurium, selenium, sulfur, and phosphorus, which are used in pairs or as alloys. Derived from their synthetic method, the original organic ligands are in general olefins ended with amino groups (i.e., dodecylamine), carboxylates (i.e., oleic acid), or phosphates (i.e., TOPO or TOP) that allow the use of high temperature ($>300°C$) needed during the nanocrystallization process [34]. To increase stability or to

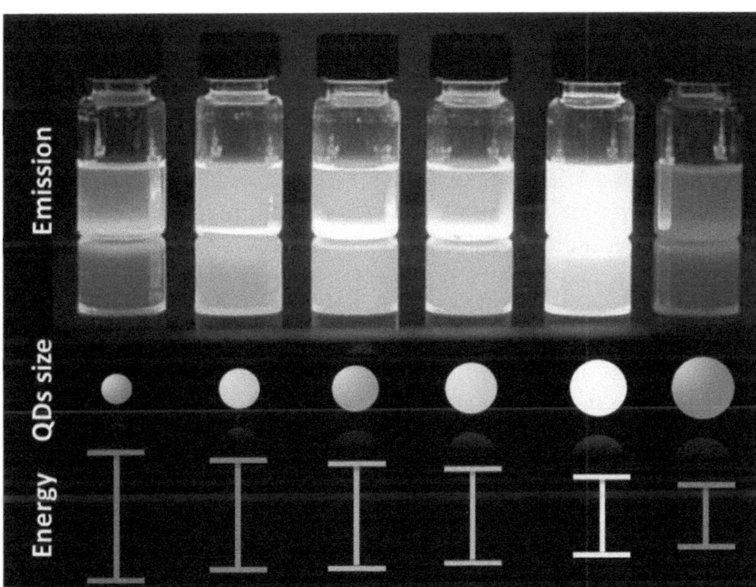

FIGURE 7.4 Photograph showing different QDs colloidal nanoparticles and illustration of the relative particle sizes and the corresponding band-gap energy.

modify the properties of the QDs, TOPO is replaced with another ligand of choice by the ligand exchange methodology [35].

In the literature, the examples of polymer capping of QDs surface are really extensive (for a valuable revision of the topic, see reference [36]). In this chapter, we are going to focus on both the importance of thiol pendant polymers and the multidentate approach versus the brush-like manner in a *grafting to* approach (see Fig. 7.2).

The best organic functionality able to displace the original ligands is the thiol moiety because possess better affinity for the QDs surface than the original ligands used during their synthesis. In fact, the use of aliphatic thiols functionalized with carboxylic acid in a ligand exchange process is the habitual protocol to achieve water soluble QDs [37]. For this reason, the design of polymers with thiol functionalities as end anchoring group or pendant group was crucial to be able to functionalize QDs with a polymeric capping. However, the synthesis of polymers with thiol groups is not easy due to the tendency of free thiols to form disulphide bridges among other reasons. For example, Aldana et al. described photooxidation of the thiol catalyzed for the NP itself when the ligand is mercaptopropionic acid and the nanocrystal are dispersible in water [38] as well as thiol/thiolate pH-dependent equilibrium when mercaptopropanol (MPOH) was used as ligand [39]. In fact, it was not until 2011 when Liras and coworkers published the first strategy to functionalize polymers synthesized from atom transfer radical polymerization (ATRP), a living polymerization technique, by transforming the bromine end group into thiol [40]. To do so, the bromine end group was substituted by a thioacetate group, as a protective group, which was easily hydrolyzed into thiol. The same research group checked the ability of both new acetyl protected thiol and free thiol poly(methyl methacrylate) (PMMA) polymers to undergo ligand exchange in CdSe@TOPO to synthesize a new kind of hybrid QD@polymers in a *brush-like* manner [41]. Remarkably, in this contribution, the strategy allows the release in situ of the thiol group from the acetyl protected thiol under mild conditions. Thus, the QDs@acetyl−PMMA show long stabilities for long periods of time in a low-emissive *standby* mode until the enhanced fluorescence is recovered due to thiol release from the NP surface of QDs@thiol−PMMA.

Surface ligands are in equilibrium with free ligands in solution, although ligands with more than one anchoring group can bind more effectively to the nanocrystal surface. The translation of this concept to a polymeric design was first used by Bawendi group to design of multidentate phosphine oxide oligomers [42], and afterwards with a multidentate phosphine oxide gel using bis(dichlorophosphino)ethane and poly(ethylene glycol) [43] that could be used to transfer various NPs from organic solvents to water. Other multifunctional polymers based on amine functionalities, such as poly(dimethylaminoethyl methacrylate) [44], poly(acrylic acid) (PAA) modified with pendant amine groups [45], poly(imidazole) [46], hyperbranched polyethyleneimides

(PEI) [47], polyethylene glycol (PEG) grafted polyethyleneimine (PEI-g-PEG) [48] or poly(amidoamine) dendrimers [49], have been used successfully to achieve ligand exchange and exchange of the original TOPO ligand.

Meanwhile, the use of multidentate polymer based on thiol has been elusive for two reasons. The first one, due to problems described for thiolated small molecules used as ligands as commented before. The second one, due to the absence of monomers based on thiol groups able to polymerize without side reactions. In spite of this, Weiss et al. described the use of polypeptides with domains based on cysteine (amino acid based on thiol) able to exchange TOPO from the surface [50–53]. With this strategy the polymer thickness was really low (*ca.*1 nm) due to the dense coating for the elevated number of interactions with the surface that also enhance the brightness of the hybrid material. On the other hand, postfunctionalization of denaturalized proteins (albumin) by reduction of the disulphide bridges and introduction of additional PEG chains, amino functionality, and thioctic acid groups (bidentate thiol) by reaction with the free thiol or amine groups conduces to polymer–peptide hybrids able to anchor to the QD surface [54]. In this case the thioctic group acts as the anchoring group while the PEG lateral chain allows for the nanohybrid dispersion in water.

The first methacrylate monomer based on thiol groups described in the literature, named SEMA, comes from its analog with the thiol group protected with an acetyl group AcSEMA [41,55]. AcSEMA is a versatile tool to synthesize metacrylic polymers, even when using living polymerization techniques, because the acetyl groups can be removed in situ using really mild condition without interference with the living polymerization conditions. A series of multidentate-protected thiol methacrylic polymers were synthesized using AcSEMA and, subsequently, used as effective ligands to keep QDs in luminescent *standby* mode (Fig. 7.5, upper side) [41]. This chapter demonstrates the effectiveness of multidentate capping *versus* the brush-like functionalization in terms of protection against external quenchers. As a first approximation, copolymers with MMA were synthesized, due to their transparency, stability and inert chemical properties, but the methodology opened the door to the synthesis of tailor-made thiol-multidentate polymers of different natures to functionalize QDs in an easy "*grafting to*" approach. Thus, a series of thermo-responsive polymers based on AcSEMA and short oligo(ethylene glycol) methyl ether methacrylates (MEO_nMA, $n = 2$, 3, or 8) were synthesized by ATRP and then, used to functionalized QDs (Fig. 7.5, lower side) [56]. The length of the ethylene glycol side chain of the co-monomer MEO_nMA was critical for a successful functionalization; it was established that the shortest MEO_2MA-based copolymers result in a compact coating and higher quantum yield (up to a factor of 6) than that of CdSe@TOPO in dichloromethane. This was probably due to the length of MEO_2MA, which was the closest one to the AcSEMA/SEMA. In addition, the amphiphilic character of the copolymer allows the CdSe@P(MEO_nMA-*co*-SEMA)

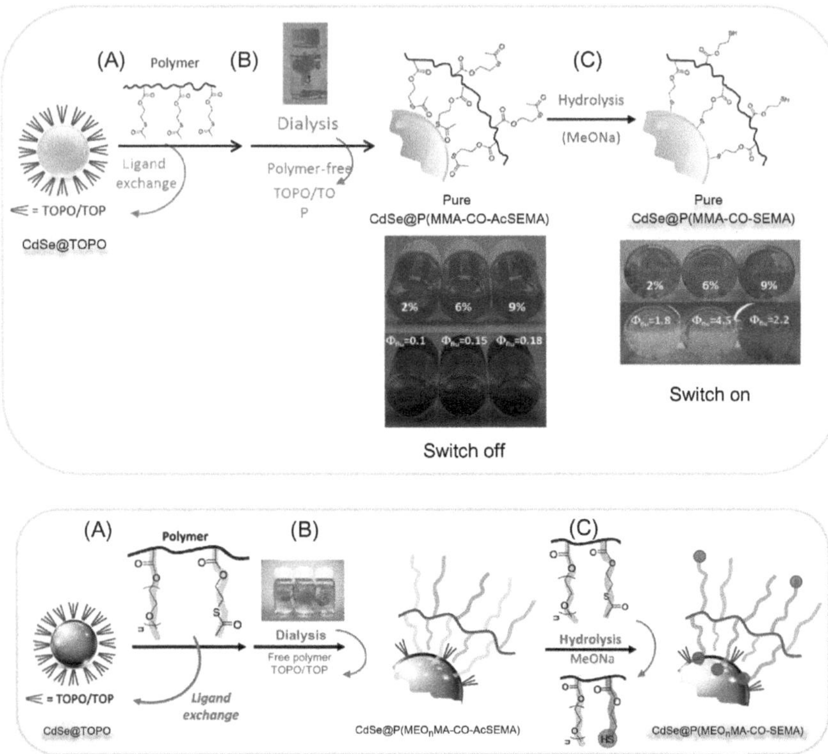

FIGURE 7.5 Upper side: A summary of the procedure for multidentate QD functionalization. (A) ligand exchange with thioacetate multidentate polymer, formation of CdSe@P(MMA-*co*-AcSEMA), (B) purification of CdSe@P(MMA-*co*-AcSEMA) by dialysis of (A); (C) posthydrolysis of CdSe@P(MMA-*co*-AcSEMA) and formation of CdSe@P(MMA-*co*-SEMA). Photographs show purified polymer-coated QDs in dichloromethane solution under visible light (upper) and under UV light (lower). They show switch off and switch on luminesces state. *Lower side*: Schematic representation of the procedure to multidentate QD functionalization. (A) Ligand exchange reaction with thioacetate multidentate polymer, formation of CdSe@P(MMA-*co*-AcSEMA), (B) purification of CdSe@P(MMA-*co*-AcSEMA) by dialysis, and (C) posthydrolysis of CdSe@P(MMA-*co*-AcSEMA) and formation of CdSe@P(MMA-*co*-SEMA). Upper side*: Adapted from M. Liras, I. Quijada-Garrido, M. Palacios-Cuesta, S. Muñoz-Durieux, O. García, Acetyl protected thiol methacrylic polymers as effective ligands to keep quantum dots in luminescent standby mode, Poly. Chem. 5 (2014) 433–442, with permission from The Royal Society of Chemistry (2013) [41].* Lower side*: Reproduced from M. Liras, E. Peinado, P. Cañamero, I. Quijada-Garrido, O. García, Smart photoluminescent nanohybrids based on CdSe quantum dots capped with multidentate thiolated pH-responsive and thermoresponsive polymers for nanosensing, J. Polym. Sci. A. 82 (2014) 3087–3095, [56], with permission from Wiley and Son (2014).*

nanohybrids to disperse in water. It is important to note that all the inherent properties of the polymers, including their thermo-sensitiveness, were translated to the hybrid showing differences in photoluminescence as temperature changes. Besides, the residual ionizable thiol groups do not get attached to

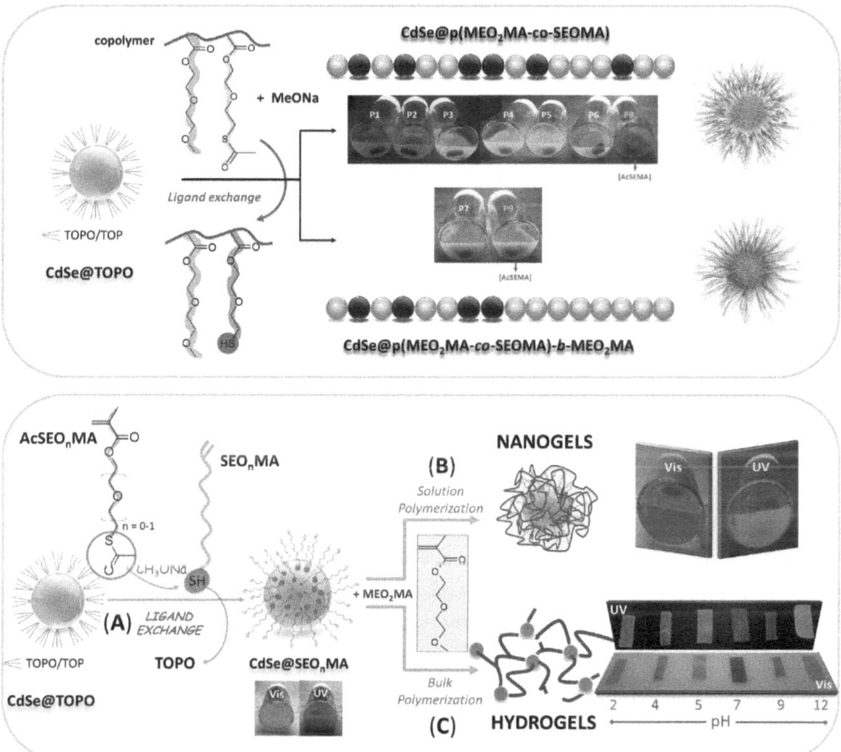

FIGURE 7.6 Upper side: Synthesis of CdSe@p(MEO_2MA-*co*-SEOMA) and CdSe@p (MEO_2MA-*co*-SEOMA)-*b*-MEO_2MA hybrids. *Lower side*: Schematic of (A) QDs (CdSe@TOPO) functionalized with thiolated monomers ($AcSEO_nMA$) to form multicrosslinked QDs (CdSe@SEO_nMA). (B) Synthesis of QD@SEO_nMA@$pMEO_2MA$ nanogels and UV/Vis photographs of the nanogels in an organic medium. (C) Synthesis of hybrid hydrogels of $pMEO_2MA$ with nanocrosslinked QD@SEO_nMA, and images of the hybrid nanocomposite sheets swollen in aqueous solutions at different pH values, illuminated under both visible and UV light. Upper side: *Reprinted with permission from A. Ferrández-Montero, I. Quijada-Garrido, M. Liras, O. García, Thermoresponsive random and block copolymers based on diethylene glycol methacrylate and a novel thiolated methacrylic monomer for the coating of semiconductor nanoparticles, Eur. Polym. J. 84 (2016) 565–576 [57]. Copyright (2016) by Elsevier. Lower side: Adapted and reprinted from M. Liras, I. Quijada-Garrido, O. García, New multicrosslinkers for the synthesis of smart luminescent nanogels and hydrogels, Polym. Chem. 8, (2017), 5317–5326 [58]. Copyright (2017) by Royal Society of Chemistry.*

the QD surface as indicated the dependence of the lower critical solubility temperature (LCST) on pH. Thus, at acidic pH, an abrupt increase in the luminescence emission accompanied the polymer collapse.

A second generation of an acrylic monomer with pendant acetyl-protected thiol but with longer ethylenglycol side chain, AcSEOMA, has been recently described [57]. The new monomer is a better analog of the well-known thermo-responsive monomer, MEO_2MA, than the former monomer AcSEMA, because has an additional ether chain achieving almost identical length. Two different families of polymers based on AcSEOMA and MEO_2MA were synthesized by ATRP: (1) well-controlled random copolymers p(MEO_2MA-*co*-AcSEOMA); and (2) well-controlled diblock copolymers p[(MEO_2MA-*co*-AcSEOMA)-*b*-MEO_2MA] (Fig. 7.6, upper side). The monomeric feed ratio determines the cloud temperature (T_{cp}) values by the modulation of the hydrophilic/hydrophobic balance. After deprotection, the free thiolated random and block polymers also show pH responsiveness. The ability of this kind of polymers to functionalize NPs was probed by the in situ ligand exchange of CdSe QDs. These new multithiolated copolymers easily replaced the hydrophobic ligand at the QDs surface leading to stable amphiphilic photoluminescent NPs. Moreover, QD@polymer photoluminescence was closely related to the thiol content and its distribution along the polymer chains. In addition, functionalization with block copolymers led to a noticeable increase of hybrid water dispersibility (*ca.* 15 mg/mL). The polymer coating triggered a remarkable increase in the nanohybrid fluorescence at acidic pH and high temperature (above the LCST).

Moreover, both monomers AcSEMA and AcSEOMA can be used to exchange the original TOPO ligand of pristine hydrophobic QDs bearing a protected thiol group capable of connecting to the surface of the NPs [58]. Therefore the organic–inorganic hybrid QDs decorated with methacrylic groups are able to connect the inorganic core with growing $pMEO_2MA$ chains during the radical polymerization reaction, acting as multicrosslinker of the NPs. Depending on whether the bottom-up polymerization approach is conducted in solution or in bulk, nanometric core–shell QD@$pMEO_2MA$ nanoparticles or hybrid $pMEO_2MA$ hydrogels crosslinked by QDs are achieved, respectively (Fig. 7.6, lower side).

In the synthesis of QD@nanogels, a thin layer of $pMEO_2MA$, a thermo-responsive polymer, grows over the multifunctional NPs because of their reactive methacrylic groups, under suitable radical solution polymerization conditions. The new hybrid QD@nanogels maintained nanometric size (6–8 nm) and fluorescence emission properties, both in organic solvents and in aqueous media. Increase of temperature eventually causes change of the QD@nanogels size due to the collapse of the polymeric shell, forming larger aggregates, and at the same a remarkable enhancement of the fluorescence emission. This reversible aggregation behavior with increasing temperature is potentially useful for local imaging and drug release.

On the other hand, the QD decorated with thiolmethacrylic-monomer acts as a multicrosslinking agent in the synthesis of new hybrid hydrogels based on $pMEO_2MA$. These hydrogels based on QDs show suitable mechanical properties, maintaining the photoluminescence of the starting NPs, and also response to external stimuli such as the pH and temperature of the medium. As it was observed for QD@nanogels, an important increase in fluorescence was detected with the collapse of the gels at all pH values studied. Also, at very basic pH (pH 12) the hydrogels undergo a color change from red to fully reproducible blue, which makes these gels promising pH sensors. These new multifunctional hybrid materials present very promising expectations for applicability as biosensors in the biomedical field, derived from the change in properties under certain environmental conditions (light, pH, and temperature).

7.3 Upconversion nanoparticles capped with polymers

Lanthanide pairs, which can upconvert low-energy photons into higher-energy photons, are promising for efficient up conversion emission [59]. Lanthanide UCNPs are unique as they can be excited with low-power continuous wave NIR sources and emit at shorter wavelengths due to their intra-configurational $4f^n$ electron transitions [60–62]. This fact is really important from a point of view of their potential use as therapeutic agent in those photodynamic therapies where deeper absorption of NIR light by the tissues is needed. Undoubtedly, hexagonal lanthanide trivalent-doped sodium yttrium fluoride NPs ($NaYF_4:Ln^{3+}$) are the most widely used due to their superior upconversion efficiency. Their emission colors from visible to NIR by single wavelength excitation can be tailored with the choice of lanthanides used as dopants (e.g., Yb^{3+}, Tm^{3+}, and Er^{3+}) in the $NaYF_4$ lattice NPs host; in addition, the emission intensity can be balanced and precisely controlled using different host-activator system and dopant concentrations (Fig. 7.7) [63].

UCNPs are being used in a great variety of applications such as light-emitting diodes, solar cells [59], and bioimaging agents [64], among others. These UCNPs are usually synthesized from the corresponding rare-earth

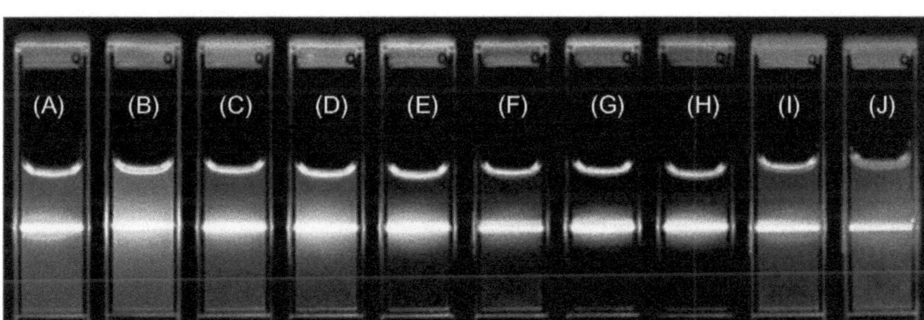

FIGURE 7.7 Compiled luminescent photos showing corresponding colloids of (A) $NaYF_4$: Yb/Tm (20/0.2 mol.%), (B–F) $NaYF_4$: Yb/Tm/Er (20/0.2/ 0.2–1.5 mol.%), and (G–J) $NaYF_4$: Yb/Er (18–60/2 mol.%). The samples were excited at 980 nm with a 600-mW diode laser. The photographs were taken with exposure times of 3.2 s for A–H and 10 s for I and J. *Adapted and reprinted from F. Wang, X. Liu, Upconversion multicolor fine-tuning: visible to near-infrared emission from lanthanide-doped NaYF4 nanoparticles, J. Am. Chem. Soc. 130 (2008) 5642–5643 [63], with permission from American Chemical Society (2008).*

chlorides, using an organic ligand and a non-coordinating solvent, such as oleic acid and 1-octadecene, respectively, and performing the reaction at high temperature [65–67]. However, the NPs must be water-dispersible for their bio-application and several methodologies have been addressed to address their water dispersibility, some of them related to the use of polymers following the same approaches described above (Fig. 7.2). In fact, an exhaustive revision about their use in biological applications and so, about their polymer capping, was published in 2015 by Sedlmeier and Gorris [68].

The most popular polymer used to obtain a good biocompatibility and water dispersibility is PEG. One approach is the use of end functionalized PEGs in a grafting to approach by ligand exchange reaction. This methodology was first used by Veggel's group using PEG–phosphate ligands [69] and afterwards, by the Perez-Prieto's group using amine and carboxylic end functionalized PEGs [70]. In the last case, they obtained a series of noncytotoxic water-dispersible UCNCs, stable in the physiological environment which, at the same time, decorated with amine and carboxylic groups on their periphery, thus allowing rapid, selective, and modular covalent conjugation with other molecules such as dyes [70]. On the other hand, this approach allows for a second ligand exchange with a small molecule well-functionalized without complete removal of the first polymer capping. For instance, this was used with the aim of designing water soluble UCNPs post-functionalized with a diiodo BOPIPY dye (IBDP), a very good singlet oxygen generator, hanging a carboxylic group [71]. The new hybrid, UCNP–IBDP@PEG, showed activity in the killing of cancerous cells by photodynamic therapy inducing cell death under NIR excitation (Fig. 7.8). The in vitro assays demonstrated that the UCNP–IBDP@PEG nanohybrid was taken up by the SH-SY5Y human neuroblastoma-derived cells and showed low cytotoxicity. In fact, *ca.* 50% cancer cell death was observed only upon NIR irradiation (45 minutes, 239 mW). To understand the resonant energy transfer processes between the acceptor IBDP dye and the emissive NP donor ($NaYF_4$: Yb^{3+}, Er^{3+}) steady state and time-resolved emission studies were performed and show that the Yb^{3+} metastable state, formed after absorption of a NIR photon, can decay via two competitive energy transfer processes: sensitization of IBDP after absorption of a second NIR photon and population of Er^{3+}-excited states [72].

Of course, as described for other luminescent NPs, a better coverage of the surface is obtained when a multidentate polymer is used in the ligand exchange process. In fact, and due to the affinity of the carboxylate group for the UCNPs surface, commercial available poly(acrylic acid), PAA (molecular weight, $M_w = 1800$) was one of the polymers first used to carry out the ligand exchange reaction in a grafting to approach [73–75]. Thus, each multidentate PAA can bind tightly to the UCNPs surface through multiple $-COO^-$ groups, while an abundance of uncoordinated carboxyl groups on the polymer chain extend into the solution, making the UCNPs water

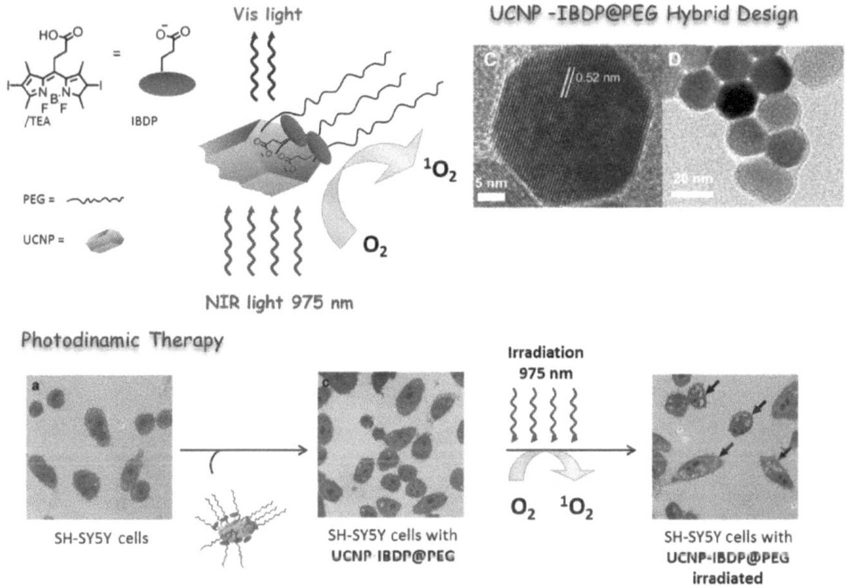

FIGURE 7.8 Oversimplified image illustrating how UCNP@PEG (hexagonal nanoprism with polymeric shell) transfers their energy (wavy green arrows) to IBDP (gray color ovals: see structure on the left) coordinated to the UCNP surface after being excited with a 975nm CW diode. Triplet–triplet energy transfer from IBDP to oxygen leads to singlet oxygen generation. High-resolution transmission electron microscopy (HRTEM) images of a hexagonal nanoprism of UCNP@oleate (left) and transmission electron microscopy (TEM) image of UCNP@PEG with a shell thickness of 2.5 nm. Schematic illustration about photodynamic therapy over SH-SY5Y cells by UCNP–IBD@PEG is shown. *Adapted from M. Bejar, M. Liras, L. Francés-Soriano, V. Voliani, V. Herranz-Perez, M. Duran-Moreno, et al., NIR excitation of upconversion nanohybrids containing a surface grafted Bodipy induces oxygen-mediated cancer cell death, J. Mater. Chem. B. 2 (2014) 4554–4563 [71], with permission from Royal Society of Chemistry (2014).*

dispersible. With this methodology, a complete exchange of the original oleate was achieved with the subsequent dispersibility in water but caused a relative loss of the emissive properties. The presence of free carboxyl groups on the surface also facilitates the conjugation of the UCNPs with various biomolecules, which paves the way for further bio-applications. With the aim of improving the water dispersibility and taking advantage of this good results, a block polymer based on PAA and PEG (PAA-*co*-PEG) was employed [76]. In this case the PAA block acts as binder with the UCNP surface meanwhile the PEG moiety is responsible for the water dispersibility.

A different design of UCNP nanohybrids with multidentate polymer based on acid groups was described by Pérez-Prieto's group last year [77]. They prepared dispersible UNCPs (β-$NaYF_4$: Yb^{3+}, Er^{3+}, UCNP) coated with a thin shell of a biocompatible copolymer comprising 2-hydroxyethylmethacrylate and 2-acrylamido-2-methyl-1-propanesulphonsulphonic acid, which they will term COP, by multidentate grafting. This capping shows remarkable resistance to strong acidic conditions (as low as pH 2) and also enables the UCNP@COP nanohybrid to bind electrostatically to cationic molecules, which can be progressively released when the pH decreases. This additional functionality of the smart UCNP@COP nanosystem was proved by its association to a well-known photosensitizer (namely, methylene blue, MB). The green-to-red emission ratio of the UC@COP@MB nanohybrid exhibited excellent linear dependence in the pH 7−2 range as a consequence of the release of the dye as the pH decreases.

Other multidentate and commercial available polymer used in ligand exchange reaction is the polyvinylpyrrolidone (PVP), a polymer with a high biocompatibility and low toxicity. The polymer binds to the UCNP surface across its carbonyl groups yielding hybrid materials dispersible in water as well as in several organic solvents. Besides, and thanks to the affinity of PVP to silica a thin and uniform shell of silica around the NPs was grown by Stöber-like synthesis. However, the tendency of PVP to crosslink produces UCNPs aggregation [78].

An alternative polymer able to undergo the ligand exchange of the oleate-capped UCNPs is the copolymer based on AcSEMA and MEO_2MA, P (MEO_2MA-*co*-ACSEMA), previously used in the capping of QDs by Liras et al. (see Fig. 7.6, lower side). The strategy used in this case was an in situ hydrolysis of the acetyl-protected group to thiol under mild condition which allows the effective ligand exchange (Fig. 7.9) and provides a thin amphiphilic-capped nanohybrids, UCNP@P(MEO_2MA-*co*-SEMA) [79]. They exhibited an enhanced emission by up to a factor of 10, as compared to that of their hydrophobic precursors in dichloromethane and even in water

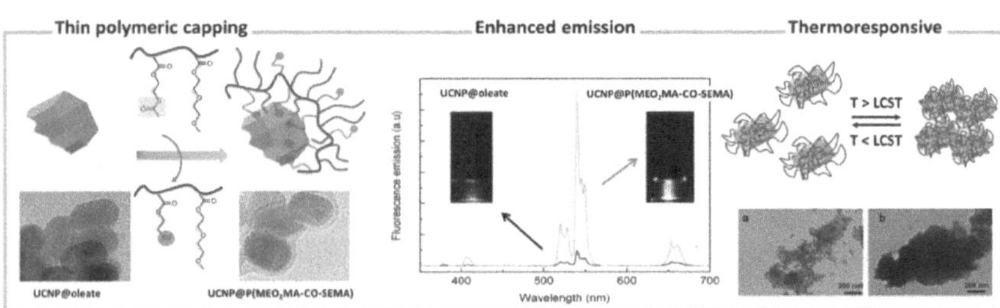

FIGURE 7.9 Oversimplification illustrating the synthesis of UCNP@P(MEO_2MA-*co*-SEMA) (hexagonal nanoprism as UCNP) together with their respective HRTEM images. Note that in the UCNP@P(MEO_2MA-*co*-SEMA) TEM image the thin polymeric capping can be appreciated. Emission spectrum comparison of both UCNP@oleate and UCNP@P(MEO_2MA-*co*-SEMA) as well as photograph of both water solution samples taken in the same irradiation conditions. The reversibility self-assembly of the nanohybrid controlled by the lower critical solution temperature is illustrated at right. *Reprinted from M. Liras, M. González-Béjar, E. Peinado, L. Francés-Soriano, J. Pérez-Prieto, I. Quijada-Garrido, et al., Thin amphiphilic polymer-capped upconversion nanoparticles: enhanced emission and thermoresponsive properties, Chem. Mater. 26 (2014) 4014–4022 [79], with permission of American Chemical Society (2014).*

(a factor of 2). Moreover, the presence of MEO_2MA as part of the copolymer endows thermo-responsiveness to the nanohybrid, which was modulated by the pH due to the presence of some thiol groups at the nanohybrid periphery. Temperature changes provoke reversible self-assembly, while retaining their upconversion emission properties. Remarkably, the nanohybrid stability was almost independent of the aggregation state (in the basic−acid and temperature range studied). The formation of stable water-dispersible UCNPs with enhanced emission, together with their amphiphilic and temperature-responsive polymer coating, is promising for building multifunctional nanostructures for intracellular imaging, therapy, and drug delivery (Fig. 7.9).

7.4 Conclusions

In summary, in this chapter we have highlighted the importance of the design and synthesis of polymers to provide functional nanohybrids based on inorganic NPs. The polymer capping can be directly responsible for the final properties of the nanohybrid, such as water dispersibility, biocompatibility, low cytotoxicity, possibility of postfunctionalization with biomarkers and others interesting molecules and so on. In addition, synergistic effect between both inorganic core and the polymer capping can also play a key role in their performance. When the relevance of these NPs is associated to the emission of light (e.g., QDs) or its transfer (e.g., UCNPs), the polymers used as coverage acquire greater importance. A good polymeric capping choice can be responsible of the future behavior of the nanohybrid system. Thus the emissive properties can be enhanced, the accessibility of quenchers of luminescence can be avoided, and new smart properties such as thermo and pH responsiveness can emerge.

Acknowledgments

This work received financial support from Ministerio de Economía y Competitividad (MINECO) from Spain through the Project MAT2014−57429-R. M. Liras also thanks MINECO and European Social Fund for a Ramón y Cajal contract (RyC-2015−18677).

References

[1] a) T.L. Doane, C. Burda, The unique role of nanoparticles in nanomedicine: imaging, drug delivery and therapy, Chem. Soc. Rev. 41 (2012) 2885−2911. Available from: https://doi.org/10.1039/C2CS15260F. b) R. Jin, C. Zeng, M. Zhou, Y. Chen, Atomically precise colloidal metal nanoclusters and nanoparticles: fundamentals and opportunities, Chem. Rev. 116 (2016) 10346−10413. Available from: https://doi.org/10.1021/acs.chemrev.5b00703.

[2] M. Nguyen, N. Felidj, C. Mangeney, Looking for synergies in molecular plasmonics through hybrid thermoresponsive nanostructures, Chem. Mater. 28 (2016) 3564−3577. Available from: https://doi.org/10.1021/acs.chemmater.6b00230.

[3] K. Baranes, M. Shevach, O. Shefi, T. Dvir, Gold nanoparticle-decorated scaffolds promote neuronal differentiation and maturation, Nano. Lett. 16 (2016) 2916–2920. Available from: https://doi.org/10.1021/acs.nanolett.5b04033.

[4] R. Mout, D.F. Moyano, S. Rana, V.M. Rotello, Surface functionalization of nanoparticles for nanomedicine, Chem. Soc. Rev. 41 (2012) 2539–2544. Available from: https://doi.org/10.1039/C2CS15294K.

[5] J. Nam, N. Won, J. Bang, H. Jin, J. Park, S. Jung, et al., Surface engineering of inorganic nanoparticles for imaging and therapy, Adv. Drug Deliv. Rev. 65 (2013) 622–648. Available from: https://doi.org/10.1016/j.addr.2012.08.015.

[6] J. Kao, K. Thorkelsson, P. Bai, B.J. Rancatore, T. Xu, Toward functional nanocomposites: taking the best of nanoparticles, polymers, and small molecules, Chem. Soc. Rev. 42 (2013) 2654–2678. Available from: https://doi.org/10.1039/C2CS35375J.

[7] N. Hildebrandt, C.M. Spillmann, W. Russ Algar, T. Pons, M.H. Stewart, E. Oh, et al., Energy transfer with semiconductor quantum dot bioconjugates: a versatile platform for biosensing, energy harvesting, and other developing applications, Chem. Rev. 117 (2017) 536–711. Available from: https://doi.org/10.1021/acs.chemrev.6b00030.

[8] S. Her, D.A. Jaffray, C. Allen, Gold nanoparticles for applications in cancer radiotherapy: mechanisms and recent advancements, Adv. Drug Deliv. Rev. 109 (2017) 84–101. Available from: https://doi.org/10.1016/j.addr.2015.12.012.

[9] Y. Cui, B. Li, H. He, W. Zhou, B. Chen, G. Qian, Metal–organic frameworks as platforms for functional materials, Acc. Chem. Res. 49 (2016) 483–493. Available from: https://doi.org/10.1021/acs.accounts.5b00530.

[10] C. He, D. Liu, W. Lin, Nanomedicine applications of hybrid nanomaterials built from metal–ligand coordination bonds: nanoscale metal–organic frameworks and nanoscale coordination polymers, Chem. Rev. 115 (2015) 11079–11108. Available from: https://doi.org/10.1021/acs.chemrev.5b00125.

[11] a) Y. Tao, M. Li, J. Ren, X. Qu, Metal nanoclusters : novel probes for diagnostic and therapeutic applications, Chem. Soc. Rev (2015). Available from: https://doi.org/10.1039/c5cs00607d. b) I. Chakraborty, T. Pradeep, Atomically precise clusters of noble metals: emerging link between atoms and nanoparticles, Chem. Rev., 117 (2017) 8208–8271. Available from: https://doi.org/10.1021/acs.chemrev.6b00769.

[12] V. Stavila, A.A. Talin, M.D. Allendorf, MOF-based electronic and opto-electronic devices, Chem. Soc. Rev. 43 (2014) 5994–6010. Available from: https://doi.org/10.1039/c4cs00096j.

[13] Y. Wang, D. Zhao, Beyond equilibrium: metal-organic frameworks for molecular sieving and kinetic gas separation, Cryst. Growth Des. 17 (2017) 2291–2308. Available from: https://doi.org/10.1021/acs.cgd.7b00287.

[14] S. Razzaque, S. Hussain, I. Hussain, B. Tan, Design and utility of metal/metal oxide nanoparticles mediated by thioether end-functionalized polymeric ligands, Polymers. 8 (2016) 156. Available from: https://doi.org/10.3390/polym8040156.

[15] G. De Crozals, R. Bonnet, C. Farre, C. Chaix, Nano today nanoparticles with multiple properties for biomedical applications : a strategic guide, Nano Today. 11 (2016) 435–463. Available from: https://doi.org/10.1016/j.nantod.2016.07.002.

[16] X. Xue, F. Wang, X. Liu, Emerging functional nanomaterials for therapeutics, J. Mater. Chem. 21 (2011) 13107. Available from: https://doi.org/10.1039/c1jm11401h.

[17] R. Ladj, A. Bitar, M.M. Eissa, H. Fessi, Y. Mugnier, R. Le Dantec, et al., Polymer encapsulation of inorganic nanoparticles for biomedical applications, Int. J. Pharm. 458 (2013) 230–241. Available from: https://doi.org/10.1016/j.ijpharm.2013.09.001.

[18] R. Francis, N. Joy, E.P. Aparna, R. Vijayan, Polymer grafted inorganic nanoparticles, preparation, properties, and applications: a review, Polym. Rev. 54 (2014) 268–347. Available from: https://doi.org/10.1080/15583724.2013.870573.

[19] R.B. Grubbs, Roles of polymer ligands in nanoparticle stabilization, Polym. Rev. 47 (2007) 197–215. Available from: https://doi.org/10.1080/15583720701271245.

[20] A. Quarta, A. Curcio, H. Kakwere, T. Pellegrino, Polymer coated inorganic nanoparticles: tailoring the nanocrystal surface for designing nanoprobes with biological implications, Nanoscale. 4 (2012) 3319–3334. Available from: https://doi.org/10.1039/C2NR30271C.

[21] F. Zhang, E. Lees, F. Amin, P. Rivera-Gil, F. Yang, P. Mulvaney, et al., Polymer-coated nanoparticles: a universal tool for biolabelling experiments, Small. 7 (2011) 3113–3127. Available from: https://doi.org/10.1002/smll.201100608.

[22] M.V. Zyuzin, T. Honold, S. Carregal-Romero, K. Kantner, M. Karg, W.J. Parak, Influence of temperature on the colloidal stability of polymer-coated gold nanoparticles in cell culture media, Small. 12 (2016) 1723–1731. Available from: https://doi.org/10.1002/smll.201503232.

[23] S. Ehlert, S.M. Taheri, D. Pirner, M. Drechsler, H.-W. Schmidt, S. Förster, Polymer ligand exchange to control stabilization and compatibilization of nanocrystals, ACS Nano. 8 (2014) 6114–6122. Available from: https://doi.org/10.1021/nn5014512.

[24] L.J. Hill, N. Pinna, K. Char, J. Pyun, Colloidal polymers from inorganic nanoparticle monomers, Progress Polym. Sci. 40 (2015) 85–120. Available from: https://doi.org/10.1016/j.progpolymsci.2014.08.003.

[25] B. Sarkar, P. Alexandridis, Block copolymer–nanoparticle composites: structure, functional properties, and processing, Progress Polym. Sci. 40 (2015) 33–62. Available from: https://doi.org/10.1016/j.progpolymsci.2014.10.009.

[26] B. Sierra-Martin, A. Fernandez-Barbero, Multifunctional hybrid nanogels for theranostic applications, Soft Matter. 11 (2015) 8205–8216. Available from: https://doi.org/10.1039/C5SM01789K.

[27] S. Ehlert, C. Stegelmeier, D. Pirner, S. Förster, A general route to optically transparent highly filled polymer nanocomposites, Macromolecules. 48 (2015) 5323–5327. Available from: https://doi.org/10.1021/acs.macromol.5b00565.

[28] M. Karg, Functional materials design through hydrogel encapsulation of inorganic nanoparticles: recent developments and challenges, Macromol. Chem. Phys. 217 (2016) 242–255. Available from: https://doi.org/10.1002/macp.201500334.

[29] O.W. Webster, Living polymerization methods, Science 251 (1991) 887–893. Available from: https://doi.org/10.1126/science.251.4996.887.

[30] C. Minelli, S.B. Lowe, M.M. Stevens, Engineering nanocomposite materials for cancer therapy, Small. 6 (2010) 2336–2357. Available from: https://doi.org/10.1002/smll.201000523.

[31] G. Lv, W. Guo, W. Zhang, T. Zhang, S. Li, S. Chen, et al., Near-infrared emission cuins/zns quantum dots: all-in-one theranostic nanomedicines with intrinsic fluorescence/photoacoustic imaging for tumor phototherapy, ACS Nano. 10 (2016) 9637–9645. Available from: https://doi.org/10.1021/acsnano.6b05419.

[32] O.E. Semonin, J.M. Luther, M.C. Beard, Quantum dots for next- generation photovoltaics, Mater. Today. 15 (2012) 508–515. Available from: https://doi.org/10.1016/S1369-7021(12)70220-1.

[33] J.A. Smyder, T.D. Krauss, Coming attractions for semiconductor quantum dots, Mater. Today. 14 (2011) 382–387. Available from: https://doi.org/10.1016/S1369-7021(11)70182-1.

[34] S.B. Brichkin, V.F. Razumov, Colloidal quantum dots : synthesis, properties and applications, Russ, Chem. Rev. 85 (2016) 1297–1312. Available from: https://doi.org/10.1070/RCR4656.

[35] J. Zhou, Y. Liu, J. Tang, W. Tang, Surface ligands engineering of semiconductor quantum dots for chemosensory and biological applications, Mater. Today. 20 (2017) 360–376. Available from: https://doi.org/10.1016/j.mattod.2017.02.006.

[36] N. Tomczak, R. Liu, J.G. Vancso, Polymer-coated quantum dots, Nanoscale. 5 (2013) 12018–12032. Available from: https://doi.org/10.1039/c3nr03949h.

[37] W.W. Yu, E. Chang, R. Drezek, V.L. Colvin, Water-soluble quantum dots for biomedical applications, Biochem. Biophys. Res. Commun. 348 (2006) 781–786. Available from: https://doi.org/10.1016/j.bbrc.2006.07.160.

[38] J. Aldana, Y.A. Wang, X. Peng, Photochemical instability of cdse nanocrystals coated by hydrophilic thiols, J. Am. Chem. Soc. 123 (2001) 8844–8850. Available from: https://doi.org/10.1021/ja016424q.

[39] J. Aldana, N. Lavelle, Y. Wang, X. Peng, Size-dependent dissociation pH of thiolate ligands from cadmium chalcogenide nanocrystals, J. Am. Chem. Soc. 127 (2005) 2496–2504. Available from: https://doi.org/10.1021/ja047000.

[40] M. Liras, O. Garcia, I. Quijada-Garrido, R. París, Transformation of the bromine end group into thiol in (meth)acrylic polymers synthesized by atom transfer radical polymerization, Macromolecules. 44 (2011) 1335–1339. Available from: https://doi.org/10.1021/ma102829s.

[41] M. Liras, I. Quijada-Garrido, M. Palacios-Cuesta, S. Muñoz-Durieux, O. García, Acetyl protected thiol methacrylic polymers as effective ligands to keep quantum dots in luminescent standby mode, Polym. Chem. 5 (2014) 433–442. Available from: https://doi.org/10.1039/C3PY00987D.

[42] S. Kim, M.G. Bawendi, Oligomeric ligands for luminescent and stable nanocrystal quantum dots, J. Am. Chem. Soc. 125 (2003) 14652–14653. Available from: https://doi.org/10.1021/ja0368094.

[43] S. Kim, S. Kim, J.B. Tracy, A. Jasanoff, M.G. Bawendi, Phosphine oxide polymer for water-soluble nanoparticles, J. Am. Chem. Soc. 127 (2005) 4556–4557.

[44] X. Wang, T.E. Dykstra, M.R. Salvador, I. Manners, G.D. Scholes, M.A. Winnik, Surface passivation of luminescent colloidal quantum dots with poly (dimethylaminoethyl methacrylate) through a ligand exchange process, J. Am. Chem. Soc. 126 (2004) 7784–7785. Available from: https://doi.org/10.1021/ja0489339.

[45] I. Potapova, R. Mruk, C. Hübner, R. Zentel, T. Basché, A. Mews, CdSe/ZnS nanocrystals with dye-functionalized polymer ligands containing many anchor groups, Angew. Chem. Int. Edit. 44 (2005) 2437–2440. Available from: https://doi.org/10.1002/anie.200462236.

[46] W. Liu, A.B. Greytak, J. Lee, C.R. Wong, J. Park, L.F. Marshall, et al., Compact biocompatible quantum dots via raft-mediated synthesis of imidazole-based random copolymer ligand, J. Am. Chem. Soc. 132 (2010) 472–483. Available from: https://doi.org/10.1021/ja908137d.

[47] T. Nann, Phase-transfer of CdSe@ZnS quantum dots using amphiphilic hyperbranched polyethylenimine, Chem. Commun. (2005) 1735–1736. Available from: https://doi.org/10.1039/B414807J.

[48] H. Duan, S. Nie, Cell-penetrating quantum dots based on multivalent and endosome-disrupting surface coatings, J. Am. Chem. Soc. 129 (2007) 3333–3338. Available from: https://doi.org/10.1021/ja068158s.

[49] C. Zhang, S.O. Brien, L. Balogh, Comparison and stability of cdse nanocrystals covered with amphiphilic poly(amidoamine) dendrimers, J. Phys. Chem. B 106 (2002) 10316–10321. Available from: https://doi.org/10.1021/jp014241k.

[50] J. Xu, P. Ruchala, Y. Ebenstain, J.J. Li, S. Weiss, Stable, compact, bright biofunctional quantum dots with improved peptide coating, J. Phys. Chem. B 116 (2012) 11370–11378. Available from: https://doi.org/10.1021/jp306453y.

[51] G. Iyer, X. Michalet, Y.-P. Chang, F.F. Pinaud, S.E. Matyas, G. Payne, et al., High affinity scFv – Hapten Pair as a tool for quantum dot labeling and tracking of single proteins in live cells, Nano. Lett. 8 (2008) 4618–4623. Available from: https://doi.org/10.1021/nl8032284.

[52] G. Iyer, F. Pinaud, J. Tsay, S. Weiss, Solubilization of quantum dots with a recombinant peptide *Escherichia coli*, Small. 3 (2007) 793–798. Available from: https://doi.org/10.1002/smll.200600516.

[53] F. Pinaud, D. King, H.-P. Moore, S. Weiss, Bioactivation and cell targeting of semiconductor cdse/zns nanocrystals with phytochelatin-related peptides, J. Am. Chem. Soc. 126 (2004) 6115–6123. Available from: https://doi.org/10.1021/ja031691c.

[54] Y. Wu, S. Chakrabortty, R.A. Gropeanu, J. Wilhelmi, Y. Xu, K.S. Er, et al., pH-responsive quantum dots via an albumin polymer surface coating, J. Am. Chem. Soc. 132 (2010) 5012–5014. Available from: https://doi.org/10.1021/ja909570v.

[55] M. Liras, O. García, N. Guarrotxena, M. Palacios-Cuesta, I. Quijada-Garrido, Versatile thiolated thermosensitive polymers synthesized by ATRP of MEO_2MA and AcSEMA, a new methacrylic monomer with a protected thiol group, Polymer Chem. 4 (2013) 5751. Available from: https://doi.org/10.1039/c3py00773a.

[56] M. Liras, E. Peinado, P. Cañamero, I. Quijada-Garrido, O. García, Smart photoluminescent nanohybrids based on CdSe quantum dots capped with multidentate thiolated pH-responsive and thermoresponsive polymers for nanosensing, J. Polym. Sci. A. 82 (2014) 3087–3095. Available from: https://doi.org/10.1002/pola.27358.

[57] A. Ferrández-Montero, I. Quijada-Garrido, M. Liras, O. García, Thermoresponsive random and block copolymers based on diethylene glycol methacrylate and a novel thiolated methacrylic monomer for the coating of semiconductor nanoparticles, Eur. Polym J. 84 (2016) 565–576. Available from: https://doi.org/10.1016/j.eurpolymj.2016.09.054.

[58] M. Liras, I. Quijada-Garrido, O. García, Polymer chemistry new multicrosslinkers for the synthesis of smart luminescent nanogels and hydrogels, Polym. Chem. 8 (2017) 5317–5326. Available from: https://doi.org/10.1039/c7py00954b.

[59] R. Naccache, F. Vetrone, J.A. Capobianco, Lanthanide-doped upconverting nanoparticles: harvesting light for solar cells, ChemSusChem. 6 (2013) 1308–1311. Available from: https://doi.org/10.1002/cssc.201300362.

[60] A. Gnach, A. Bednarkiewicz, Lanthanide-doped up-converting nanoparticles : merits and challenges, Nano Today. 7 (2012) 532–563. Available from: https://doi.org/10.1016/j.nantod.2012.10.006.

[61] A. Bagheri, H. Arandiyan, C. Boyer, M. Lim, Lanthanide-doped upconversion nanoparticles : emerging intelligent light-activated drug delivery systems, Adv. Sci. 3 (2016) 1500437. Available from: https://doi.org/10.1002/advs.201500437.

[62] C. Chen, C. Li, Z. Shi, Current advances in lanthanide-doped upconversion nanostructures for detection and bioapplication, Adv. Sci. 3 (2016) 1600029. Available from: https://doi.org/10.1002/advs.201600029.

[63] F. Wang, X. Liu, Upconversion multicolor fine-tuning : visible to near-infrared emission from lanthanide-doped $nayf_4$ nanoparticles, J. Am. Chem. Soc. 130 (2008) 5642–5643. Available from: https://doi.org/10.1021/ja800868a.

[64] Z. Li, Y. Zhang, G. Han, Lanthanide-doped upconversion nanoparticles for imaging-guided drug delivery and therapy BT – advances in nanotheranostics I. Chapter 4 in: Z. Dai (Ed.), Design and Fabrication of Theranosic Nanoparticles, Springer Berlin Heidelberg, Berlin, Heidelberg, 2016, pp. 139–164. Available from: http://dx.doi.org/10.1007/978-3-662-48544-6_4.

[65] J. Boyer, F. Vetrone, L.A. Cuccia, J.A. Capobianco, Q.C. Hb, I. National, et al., Synthesis of colloidal upconverting nayf$_4$ nanocrystals doped with Er^{3+}, Yb^{3+}, and Tm^{3+}, Yb^{3+} via thermal decomposition of lanthanide trifluoroacetate precursors, J. Am. Chem. Soc. 128 (2006) 7444–7445. Available from: https://doi.org/10.1021/ja061848b.

[66] H.S. Qian, H.C. Guo, P.C. Ho, R. Mahendran, Mesoporous-silica-coated up-conversion fluorescent nanoparticles for photodynamic therapy, Small. 5 (2009) 2285–2290. Available from: https://doi.org/10.1002/smll.200900692.

[67] H. Mai, Y. Zhang, R. Si, Z. Yan, L. Sun, L. You, et al., High-quality sodium rare-earth fluoride nanocrystals: controlled synthesis and optical properties, J. Am. Chem. Soc. 128 (2006) 6426–6436. Available from: https://doi.org/10.1021/ja060212h.

[68] A. Sedlmeier, H.H. Gorris, Photon-upconverting nanoparticles for bioanalytical applications, Chem. Soc. Rev. 44 (2015) 1526–1560. Available from: https://doi.org/10.1039/C4CS00186A.

[69] J. Boyer, M. Manseau, J.I. Murray, F.C.J.M. Van Veggel, P.O. Box, B. Columbia, Surface modification of upconverting $NaYF_4$ nanoparticles with PEG - phosphate ligands for NIR (800 nm) biolabeling within the biological window, Langmuir. 26 (2010) 1157–1164. Available from: https://doi.org/10.1021/la902260j.

[70] V. Voliani, M. González-Béjar, V. Herránz-Pérez, M. Duran-Moreno, G. Signone, J.M. García-verdugo, et al., Orthogonal functionalisation of upconverting $NaYF_4$ nanocrystals, Chem. Eur. J. 19 (2013) 13538–13546. Available from: https://doi.org/10.1002/chem.201301353.

[71] M. Bejar, M. Liras, L. Francés-Soriano, V. Voliani, V. Herranz-Perez, M. Duran-Moreno, et al., NIR excitation of upconversion nanohybrids containing a surface grafted bodipy induces oxygen-mediated cancer cell death, J. Mater. Chem. B. 2 (2014) 4554–4563. Available from: https://doi.org/10.1039/C4TB00340C.

[72] L. Francés-Soriano, M. Liras, A. Kowalczyk, A. Bednarkiewicz, M. González-Béjar, J. Pérez-Prieto, Energy transfer in diiodobodipy-grafted upconversion nanohybrids, Nanoscale. 8 (2016) 204–208. Available from: https://doi.org/10.1039/C5NR07229H.

[73] R. Naccache, F. Vetrone, V. Mahalingam, L.A. Cuccia, J.A. Capobianco, Controlled synthesis and water dispersibility of hexagonal phase $NaGdF_4$: Ho^{3+}/Yb^{3+} nanoparticles, Chem. Mater. 21 (2009) 717–723. Available from: https://doi.org/10.1021/cm803151y.

[74] S.J. Budijono, J. Shan, N. Yao, Y. Miura, T. Hoye, R.H. Austin, et al., Synthesis of stable block-copolymer-protected $NaYF_4$: Yb^{3+}, Er^{3+} up-converting phosphor nanoparticles, Chem. Mater. 22 (2010) 311–318. Available from: https://doi.org/10.1021/cm902478a.

[75] Z. Wang, C. Liu, L. Chang, Z. Li, One-pot synthesis of water-soluble and carboxyl-functionalized β-$NaYF_4$: Yb, Er(Tm) upconversion nanocrystals and their application for bioimaging, J. Mater. Chem. 22 (2012) 12186–12192. Available from: https://doi.org/10.1039/c2jm31572f.

[76] M. Kamimura, D. Miyamoto, Y. Saito, K. Soga, Y. Nagasaki, Design of poly(ethylene glycol)/streptavidin coimmobilized upconversion nanophosphors and their application to fluorescence biolabeling, Langmuir 24 (2008) 8864–8870. Available from: https://doi.org/10.1021/la801056c.

[77] I. Recalde, N. Estebanez, L. Francés-Soriano, M. Liras, M. González-Béjar, J. Pérez-Prieto, Upconversion nanoparticles with a strong acid-resistant capping, Nanoscale. 8 (2016) 7588–7594. Available from: https://doi.org/10.1039/C5NR06653K.

[78] N.J.J. Johnson, N.M. Sangeetha, J. Boyer, F.C.J.M. Van Veggel, Facile ligand-exchange with polyvinylpyrrolidone and subsequent silica coating of hydrophobic upconverting β-$NaYF_4$: Yb^{3+}/ Er^{3+} nanoparticles, Nanoscale. (2010) 771–777. Available from: https://doi.org/10.1039/b9nr00379g.

[79] M. Liras, M. González-Béjar, E. Peinado, L. Francés-Soriano, J. Pérez-Prieto, I. Quijada-Garrido, et al., Thin amphiphilic polymer-capped upconversion nanoparticles: Enhanced emission and thermoresponsive properties, Chem. Mater. 26 (2014) 4014–4022. Available from: https://doi.org/10.1021/cm501663n.

Chapter 8

Photosensitive materials for constructing on-demanded drug-release systems

Feng Zhang[a] and Hélder A. Santos[a,b,*]

*[a]Drug Research Program, Division of Pharmaceutical Chemistry and Technology, Faculty of Pharmacy, University of Helsinki, Helsinki, Finland, [b]Helsinki Institute of Life Science (HiLIFE), University of Helsinki, Helsinki, Finland, *Corresponding author. email address: helder.santos@helsinki.fi*

Chapter Outline

8.1 Introduction

Controlled drug-delivery technology has emerged as a frontier science to overcome the limitations associated with conventional drug formulations, especially related to the issues of pharmacokinetics and biodistribution of free drugs [1,2]. As a multi-disciplinar technology, the drug-delivery system (DDS) can be based on nanotechnology, integrating the developments of chemistry, physics, material science, and medicine to treat the diseases, for example, cancer. The discovery and development of the enhanced permeability and retention effect in various pathologies set the stage of nanomedicine [3–6], which means that the nanoparticle-based DDSs can accumulate at the site of activity within the body by passive targeting. Thus, the behavior of the encapsulated theranostics, such as chemotherapeutics and bio-imaging

Photoactive Inorganic Nanoparticles. DOI: https://doi.org/10.1016/B978-0-12-814531-9.00008-7
© 2019 Elsevier Inc. All rights reserved.

tags can be guided, manifesting itself as the improved therapeutic efficacy and reduced systemic toxicity [1,2,7,8]. Various nanostructures can be used as the cargo carriers, which include liposomes, micelles, carbon tubes, inorganic nanoparticles made of gold, iron, lanthanides, or quantum dots, and biomolecules (*e.g.*, DNA nanostructures). The therapeutic efficiency of DDS through systemic administration has been investigated in numerous studies [9,10].

Although nanomedicines have many benefits, such as prolonged half-life in systemic circulation and reduced side effects compared with the conventional drugs, they modestly improved the overall survival of patients, presenting a long distance to clinical transition of nanomedicines [11]. More importantly, the need of stimuli–responsive surface is not only to improve the release efficiency in the active site, but also a requirement to overcome the inherent bio-barriers, such as the clearance by mononuclear phagocyte system cells, and the dense interstitial matrix that can prevent the particle penetration throughout the entire tumor [12,13]. To solve these problems, the design of stimuli–responsive systems that recognize a special signal in the microenvironment or exogenous stimulus and response, are paramount.

To achieve the on-demand delivery, various stimuli–responsive materials, such as pH [14], enzymes [15], temperature [16], ultrasound [17], magnetic field [18], and light [19], have been extensively employed to achieve the purpose by changing their structural composition and conformation. As an extracorporeal physical stimulus, light has gained especial attention, in view of its clinical relevance, non-invasiveness, and precise spatiotemporal controllability [20]. In addition, light wavelength, light intensity, and duration of exposure to modulate, brings great potential to manipulate the drug carriers' behaviors and drug-release profiles. In recent years, a large number of light-responsive materials have been developed for constructing smart DDS, which can be classified into two regions: ultraviolet (UV)-sensitive materials and near-infrared (NIR)-responsive materials (Fig. 8.1) [20–22].

In this chapter, we discuss the mainstream mechanisms of the stimulus, the well-established strategies and most significant progresses made in the past ten years in the field of photoresponsive materials for drug-delivery applications.

8.2 Ultraviolet light-sensitive material-based drug-release systems

In view of the sufficient energy of single UV-photon photochemical reactions, such as photoisomerization and photodegradation, it can be achieved on the materials cross-linked with metastable photochromic groups [21]. Generally, the photoisomerization molecules, azobenzene (Azo), for example, whose process of the hydrophobic–hydrophilic transition is accompanied

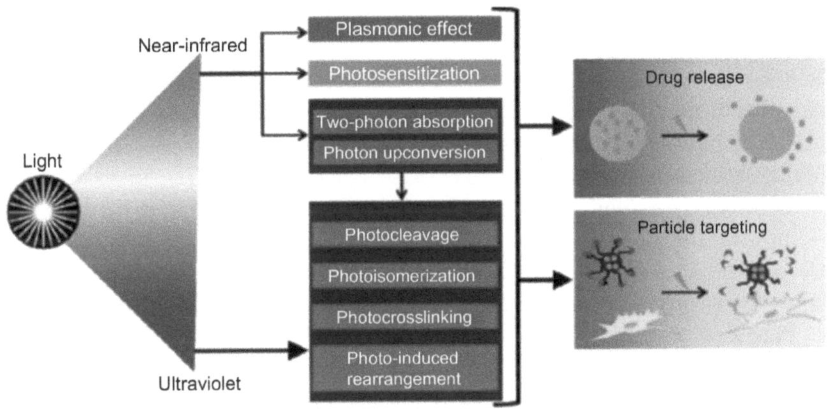

FIGURE 8.1 Mechanisms of UV-induced or NIR-induced photoresponsiveness for targeting delivery and drug release. *Reprinted with permission from A.Y. Rwei, W. Wang, D.S. Kohane, Photoresponsive nanoparticles for drug delivery, Nano Today 10 (4) (2015) 451–467. Copyright © 2015 Elsevier Ltd.*

with the *trans–cis* photoisomerization, are used to construct photoswichable dynamic systems that can be reversibly "turned on" and "off" under irradiation at different wavelengths [23]. In contrast, the photodegradations are usually achieved by photocleavable groups, as represented by *o*-nitrobenzyl (ONB) alcohol derivatives [24]. Once these species are fractured from the polymer scaffold, the chain structure is unable to be recovered, leading to the changes in electrical property and hydrophobicity of the scaffold. Thus, photocleavage can be used to reconstruct the particle surface, affect the hydrophilic–hydrophobic balance, by removing the function moieties, and adjust the particle interactions ("caging" groups).

Although most of the photochemical reactions rely on UV-triggering and the reactions are highly efficient, the UV resource suffers from many drawbacks. The major disadvantage of UV light is the low penetration depth [25]. UV light cannot penetrate into tissue deeper than 1 cm, as the existing hemoglobin, oxy-hemoglobin, and water can cause strong scattering and absorption. In addition, UV light is harmful to the tissues and cells, which restricts the clinical applications as stimulus for drug release [26]. However, the development of the two-photon technology and upconversion technology based on lanthanide-doped nanoparticles are promising to break the limitations of clinical application of UV-sensitive materials.

8.2.1 Photoisomerization of the azobenzene derivatives

Azo and its derivatives are classical photoisomerization molecules that have been used widely to construct photoswitches with high quantum yields. The *trans* conformation of Azo is more stable than the *cis* isomer in the dark

(Fig. 8.2A). The transition of *trans*–*cis* conformation can be achieved by UV irradiation in the range of 300–400 nm, whereas the *cis*-Azo can revert back to the *trans*-form under the light with a wavelength above 400 nm. The transition from *trans*-form to *cis*-form will increase the dipole moment, leading to the hydrophobic–hydrophilic transition (Fig. 8.2B) [8,27]. In addition, as the self-assembly block together with cyclodextrin (CD), the *trans*-Azo can form complex with CD by the host–guest effect, which can be disrupted by the transition to the hydrophilic *cis*-form [28]. In view of this reversible interaction, Azo–CD complex can act as a kind of photocleavable group to adjust the surface function or the conformation of nanostructures.

Based on the photolabile hydrophobic–hydrophilic transition, numerous Azo-based photosensitive DDSs have been developed [29]. For example, Zhao et al. conjugated the *trans*-Azo on the poly(ethyleneoxide)-*block*-poly (*N*-isopropylacrylamide) (PEO-*b*-PNIPAM) to form an amphiphilic copolymer [21]. The *trans*–*cis* isomerization can shift the lower critical solution temperature (LCST) of the polymer, thus adjusting the assembly or disassembly of copolymer to reorganize the formation of micelles and release of encapsulated cargos. Azo and the derivatives can also act both as the gatekeepers and impellers of mesoporous silica particles to regulate cargo release. The driven force is the continuous photoisomerization of the Azo acting as an impeller and expeller of cargos out of the pore [30]. The coassembly of Azo–CD system is another widely explored strategy for drug delivery. Kros et al. used the dextran that separately modified with *trans*-Azo and *beta*-CD to form a hydrogel system for protein-controlled release. After UV irradiation, the isomerized *cis* configurations resulted in the

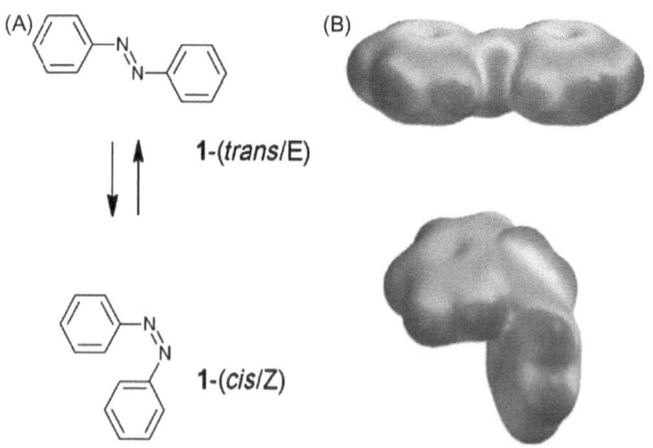

FIGURE 8.2 (A) Configuration transition of *trans* and *cis* isomers of azobenzene. (B) Space filling models are represented by electrostatic potential. *Reprinted with permission from A.A. Beharry, G.A. Woolley, Azobenzene photoswitches for biomolecules, Chem. Soc. Rev. 40(8) (2011) 4422–4437. Copyright © 2011, Royal Society of Chemistry.*

disruption of the Azo–CD assembly and dissociation of the gel network. As a consequence, the hydrogel converted into a sol and then released the proteins [31]. Zhao et al. introduced a piston-like gatekeeper based on the [2] rotaxane–Azo–*alpha*-CD system [32]. The *alpha*-CD was hanged on the surface of mesoporous silica particles dragged by a linear [2]rotaxane axle with a preattached Azo stopper at the end. The *trans–cis* isomerization of the Azo moiety caused the changing of the distance between CD and particle surface, leading to the photoswitchable gate blocking and unblocking.

Huang et al. first used the pillar[7]-arene (WP7) as novel macrocyclic host of Azo to form a supraamphiphiles that had dual thermo- and photoresponsiveness (Fig. 8.3) [33]. In this system, the Azo moieties showed a photoswitchable interaction with WP7, which is similar as the pattern of Azo–CD system. However, the complexation of *trans*-Azo-WP7 can be reversed when the temperature is above the LCST of WP7. This LCST behavior of WP7 is achieved by the heating/cooling controlled dethreading/rethreading process. After forming vesicles in water using Azo-WP7-based supra-amphiphilic polypseudorotaxane, both the UV-induced *trans*-to-*cis*-transition of Azo and heating caused due to

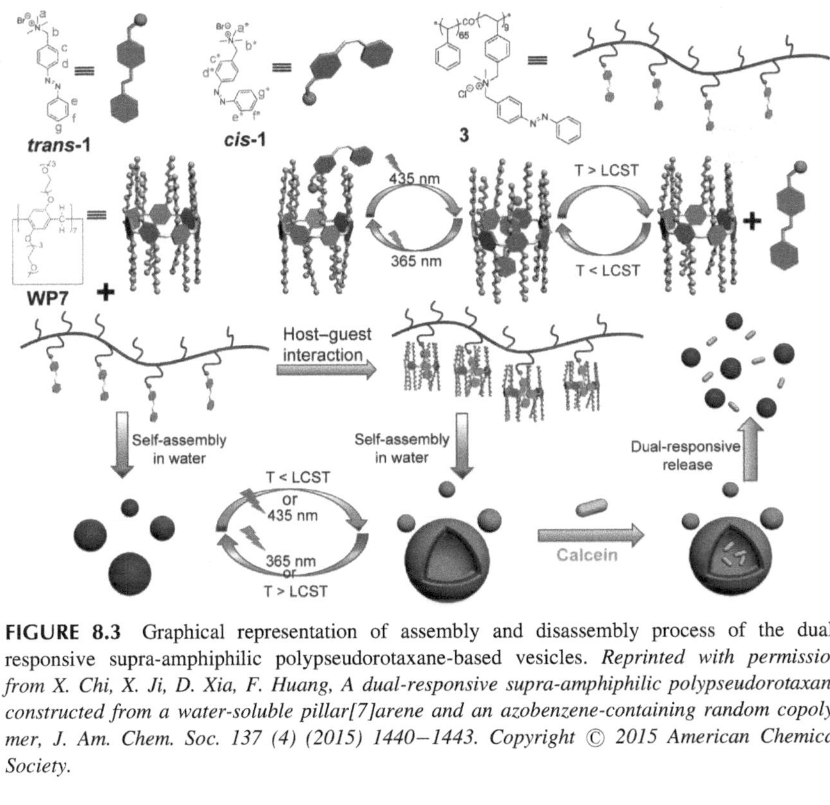

FIGURE 8.3 Graphical representation of assembly and disassembly process of the dual-responsive supra-amphiphilic polypseudorotaxane-based vesicles. *Reprinted with permission from X. Chi, X. Ji, D. Xia, F. Huang, A dual-responsive supra-amphiphilic polypseudorotaxane constructed from a water-soluble pillar[7]arene and an azobenzene-containing random copolymer, J. Am. Chem. Soc. 137 (4) (2015) 1440–1443. Copyright © 2015 American Chemical Society.*

dethreading of WP7 can lead to the vesicles disassembly and trigger the drug-release process.

Recently, many photomanipulated systems based on Azo biomolecules have been developed, offering new platforms to construct more powerful photosensitive materials. For example, Tan et al. grafted the Azo into the DNA sequence to achieve a photoswitchable dehybridization–hybridization of a hairpin structure [34]. The driven force was from the stereo-isomerization of Azo moiety. Then they immobilized a double-stranded Azo-tethered DNA sequence onto the mesoporous silica particles as gatekeeper (Fig. 8.4) [35]. The complementary and dissociation of the double-stranded structure was reversibly regulated by *cis–trans* isomerization of the Azo upon UV–vis light irradiation, causing uncapping/capping of pore gates of mesoporous silica.

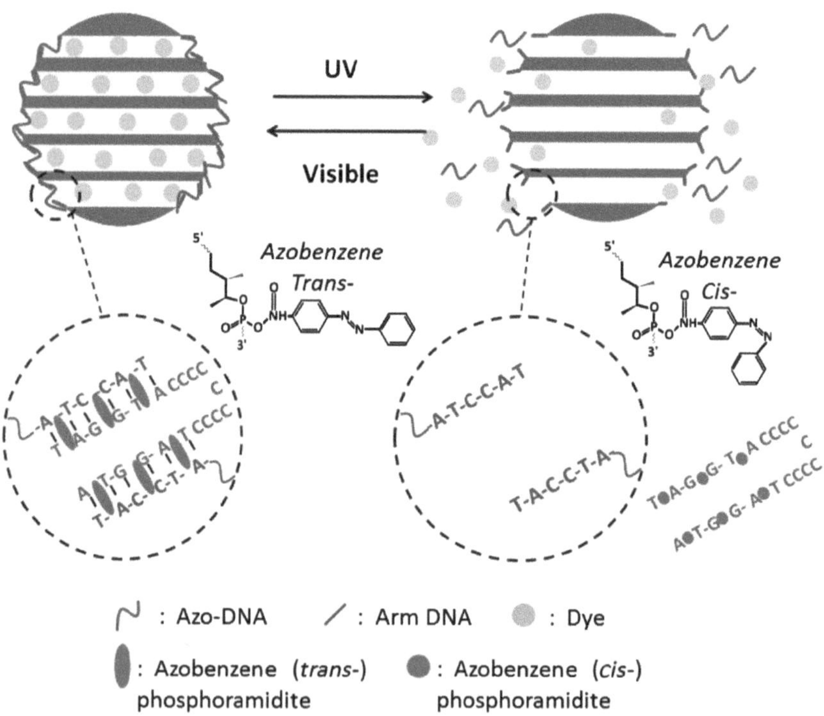

FIGURE 8.4 Schematic of azobezene-modified DNA-controlled reversible release system. Visible irradiation at 450 nm (azobenzene *trans*) leads to hybridization of the linker and the complementary DNA arm. Irradiation with UV (365 nm) converts azobenzene to the *cis* form, leading to dehybridization and pore opening. *Reprinted with permission from Q. Yuan, Y. Zhang, T. Chen, D. Lu, Z. Zhao, X. Zhang, et al., Photon-manipulated drug release from a mesoporous nanocontainer controlled by azobenzene-modified nucleic acid, ACS Nano 6 (7) (2012) 6337–6344. Copyright © 2012, American Chemical Society.*

8.2.2 Photocleavable systems based on *o*-nitrobenzyl derivatives

As one of the most representative photocleavable groups, ONB alcohol derivatives were first described by Schofield et al. [36]. The ONB-based linkers can usually be cleaved within several minutes when exposed to 300–365 nm light [37]. This photocleavage undergoes a Norrish II-type intramolecular rearrangement and generates the nitrobenzaldehyde species irreversibly. The cleavage of the ONB moieties from the polymer backbone lead to the hydrophobic–hydrophilic transition, as consequence of the leaving of the hydrophobic benzyl moieties and generating hydrophilic terminus (carboxyl group or amino group). After photolysis, the shift in hydrophilicity can increase the LCST in this system [38]. This hydrophobic core-forming function of OBN moieties is usually adopted to modify the side-chain units of the polymer to further increase the LCST after photolysis. The other option to construct the OBN-based photocleavable skeleton is by inserting into the main chain of the polymer, resulting in the main-chain degradation. Through this strategy, fast photo-induced change of surface properties or degradation of micelles can be achieved to facilitate the cargo-release process [39].

In addition to the application in photo-induced reconstructing the polymers for drug release, ONB derivatives can act as a kind of efficient photosensitive caging group to achieve inactive-to-active targeting conversion [40]. Kohane et al. used 4,5-dimethoxy-2-nitrobenzyl as caging group to mask the tyrosine unit of the targeting peptide YIGSR, which can nonspecifically target the integrin beta 1. The caged YIGSR lost the targeting capacity, resulting in the attenuated off-target effect. The targeting capacity was recovered after UV irradiation (Fig. 8.5). This light-induced uncaging strategy can minimize the off-target effect of the delivery system.

8.2.3 Spiropyran–merocyanine–photoisomerization

Compared with the Azo derivatives and other kinds of photoswitchable molecules, such as stilbenes and fulgides, spiropyran owns two isomers with vastly different properties [41]. The spiropyrans are a type of photochromic compounds with a hydrophobic and ring-closed structure. Under UV irradiation, this non-polar structure with spiro-form (SP) transformed into a zwitterionic merocyanine (MC)-form with extended π-electron delocalization. The change during this hydrophilic–hydrophobic balance is larger than that with Azos (Fig. 8.6). This property endows spiropyran-based materials with valuable building block features for reversible light-induced transformation. Matyjaszewski et al. took advantage of this capacity and designed a poly(ethylene oxide)-block-poly(methacrylate) whose methacrylate block is bearing SP side chains (PEO-*b*-PMSP) [42]. Micelle disruption was observed upon irradiation of the PEO-*b*-PMSP micelles in water due to the photo-induced

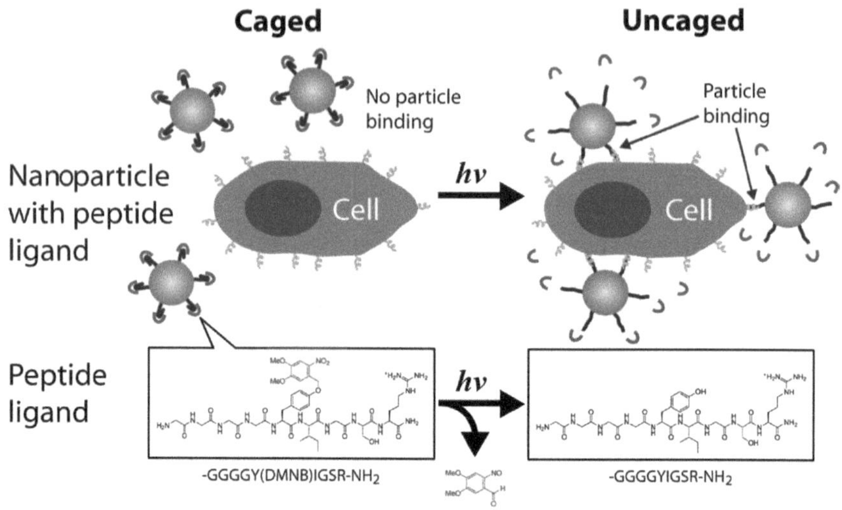

FIGURE 8.5 Schematic of the light-triggered uncaging reaction on the targeting peptide and the phototriggered activation targeting delivery. The peptide is caged on tyrosine with a 4,5-dimethoxy-2-nitrobenzyl group (DMNB). After irradiation, the caging group is released and the targeter becomes active. *Reprinted with permission from T. Dvir, M.R. Banghart, B.P. Timko, R. Langer, D.S. Kohane, Photo-targeted nanoparticles, Nano Lett. 10 (1) (2010) 250–254. Copyright © 2009 American Chemical Society.*

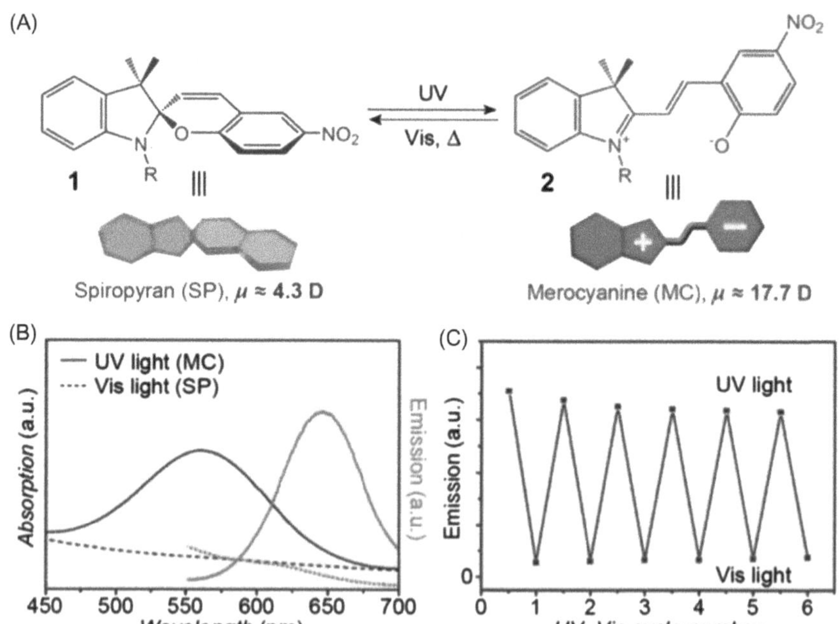

FIGURE 8.6 (A) Reversible transformations between the SP state 1 to MC state 2. (B) Typical absorption (solid line, left) and emission (solid line, left) profiles of SP (dashed lines) and MC (solid lines). (C) Reversible fluorescence ($\lambda_{em} = 645$ nm) switching in a spiropyran-decorated polymer upon exposure to UV and V is light. *Reprinted with permission from R. Klajn, Spiropyran-based dynamic materials, Chem. Soc. Rev. 43 (1) (2014) 148–184. Copyright © 2013, Royal Society of Chemistry.*

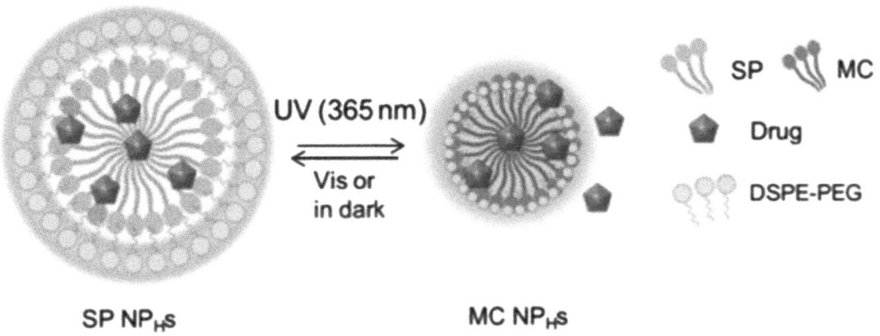

FIGURE 8.7 Scheme of the photo-induced particle shrink and drug release. *Reprinted with permission from R. Tong, H.H. Chiang, D.S. Kohane, Photoswitchable nanoparticles for in vivo cancer chemotherapy, Proc. Natl. Acad. Sci. 110 (47) (2013) 19048–19053. Copyright © 2013, National Academy of Sciences.*

conversion of neutral SP to charged MC. Moreover, this system was successfully used for the encapsulation and release of molecules.

More recently, spiropyran photoisomerization was used as the basis of a photoshrinkable nanoparticle (Fig. 8.7) [43–45]. Nanoparticles were formed by self-assembly of alkyl derivatives of the chromophore spiropyran, hydrophobic drugs, and lipid polyethylene glycol chains. Upon UV light irradiation, hydrophobic spiropyran underwent isomerization to zwitterionic MC, which induced nanoparticle shrinkage from 150 to 40 nm and triggered drug release. Irradiated particles were able to penetrate deeper into the cornea ex vivo, and into mouse tumors in vivo, enhancing the chemotherapeutic efficacy.

8.3 Near-infrared light-trigged drug-release systems

Compared with UV light, NIR light has many advantages for biomedical applications. Due to the minimal attenuation by soft tissues, blood, and water, NIR light allows deeper tissue penetration than UV light (up to a couple of centimeters) [25,46,47]. In addition, NIR light is less harmful to cells and tissues than UV light [25]. There are many successful clinical applications that use NIR systems, such as blood oxygenation levels monitoring [48], deep-tissue fluorescent imaging [49], and cancer therapy by hyperthermia [50].

NIR light cannot directly achieve most of the photo-initiated reactions in view of the limited photon energy. Although the emerged simultaneous two-photon technology offers great opportunity to replace UV/vis light by NIR light to achieve photocleavage, the limitations due to the time-consuming spot-by-spot process and the required high-intensity restricts their clinical transition [51,52]. Recently, most of the preclinical NIR-triggered DDSs are based on photothermal nanostructures, such as gold nanoparticles, that can convert the absorbed NIR light into heat by coherent oscillations of metal conduction band electrons [53]. By grating thermosensitive materials onto the photothermal nanoparticles, the converted thermal signal can be used to trigger drug release [22,54]. For example, by incorporating gold nanoparticles into the poly(nisopropyl methacrylamide) (pNIPAm) hydrogels, the generated photothermal effect can heat the polymer beyond a critical temperature and collapse the gel structure [55]. The other fast-emerging strategy is upconverting the NIR light into UV light by using lanthanide-doped upconverting nanoparticles (UCNPs) [56]. Typically, UCNPs codoped with Tm^{3+} and Yb^{3+} can efficiently convert NIR light into high-energy UV emission [57]. Thus, the limitations of UV light can be avoided and the UV-induced photoreactions can be achieved by NIR triggering [58,59].

8.3.1 Thermal-sensitive materials for near-infrared light-triggered drug release

The thermosensitive materials are the responsive segments in the photothermal release system. During NIR irradiation, this kind of materials can respond to change in temperatures, resulting in the change of the physical features, such as solubility, conformation, and especially hydrophilic–hydrophobic balance, which can be quantitatively characterized by the LCST [60]. Most of the thermal-sensitive materials are amphiphilic copolymers. The LCST is majorly influenced by the ratio of hydrophilic and hydrophobic moieties on the backbone or side chains [61,62]. Generally, these polymers undergo a volume phase transition from miscible in aqueous solutions to insoluble when the temperature is above LCST, resulting in the collapse and aggregation of polymer chains.

There are comprehensive reviews about the synthesis and development of the thermal-sensitive materials [54,63]. When grafting the photothermal nanocarriers with these polymers, the NIR-irradiation-induced local heating can change or disrupt the morphology of the polymer surface, leading to the subsequent release of the encapsulated drugs [64,65]. For example, poly(*N*-isopropyl acrylamide) (PIPAAm) is widely used to construct thermal-sensitive drug carriers, as the LCST of PIPAAm is near the physiological temperature and can be adjusted by introducing other hydrophilic unite, such as dimethylacrylamide or acrylic acid [66]. The other representative example is the poly(ethylene oxide)-poly(propylene oxide)-poly(ethylene oxide) triblock copolymer featured with the properties of thermal-responsive micellization and gelation [67]. However, the low biodegradability is a major drawback that restricts the relevant biomedical applications. To solve this problem, the biodegradable segments, such as poly(3-carprolactone) and poly (L-lactic acid), have been used to accelerate the degradation process in vivo [68,69].

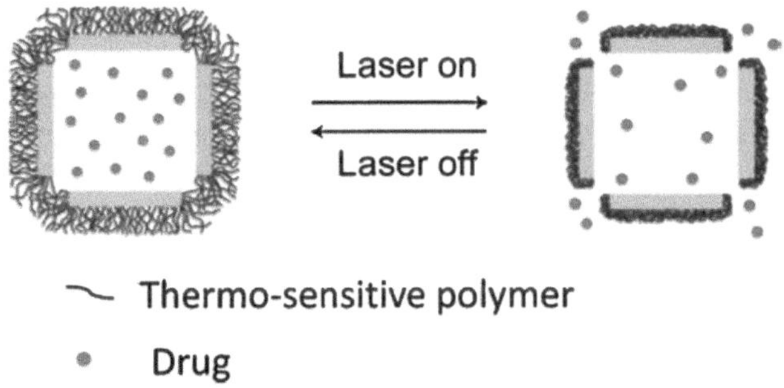

FIGURE 8.8 Gold nanocage-controlled release system. The nanocage can efficiently adsorb the NIR irradiation and covert into heat, triggering the thermos-sensitive polymer to collapse and release the drug payload. After turning off the laser, the polymer chains revert to the extended conformation and suspend drug release. *Adapted with permission from M.S. Yavuz, Y. Cheng, J. Chen, C.M. Cobley, Q. Zhang, M. Rycenga, et al., Gold nanocages covered by smart polymers for controlled release with near-infrared light, Nat. Mater. 8 (2009) 935. Copyright © 2009, Nature Publishing Group.*

Recently, various plasmonic nanoparticle-based NIR-triggered photothermal DDSs have been developed. An example is the gold nanocages coated with a monolayer of poly(*N*-isopropylacrylamide-*co*-acrylamide) polymer as the thermosensitive gatekeeper, with a LCST of 39°C (Fig. 8.8) [70]. The polymer is extended under body temperature, so that can block the pores of the nanocage and prevent the drug release. The NIR irradiation can fast heat the structure temperature beyond the LCST, resulting in the polymer collapse and drug release. This transforming process is reversible by manipulating the laser power, enabling the species control of the drug-release profile.

8.3.2 Lanthanide-doped upconversion system

As discussed above, UV resource suffered many difficulties for clinical application, considering the limited penetration depth and toxicity to the tissue. However, the UV light-induced chemical reactions are of great value in clinical medicine and usually irreplaceable by other extracorporeal physical stimulus, such as ultrasound and magnetic field. In this case, increasing efforts have been made to achieve efficacious photo-induced reactions for clinical application. Among various strategies, the utilization of lanthanide-doped UCNPs is considered to be the most promising one. UCNPs can absorb several photons of NIR sequentially and then emit a single photon with higher frequency, such as UV or vis photon. Especially, UCNPs codoped with Yb^{3+} and Tm^{3+} are capable of converting NIR light into high-energy UV emission [57,59,71]. Unlike the two-photon absorption, which relies on high-intensity pulsed lasers ($> 10^6$ W/cm^2), excitation intensity around several to several hundred watt per square centimeter is sufficient to trigger the UCNP-assisted photoreaction [51,52,72–74]. Therefore, UV light-responsive drug carriers can be utilized with UCNPs for phototriggered DDSs under NIR irradiation, avoiding the possibility of tissue damage from UV light. Recently, combined technology of light-sensitive drug carriers and UCNPs has been intensively developed to achieve low-energy NIR-triggered drug delivery, which is suitable for biomedical applications [75].

To achieve the phototriggered targeting, Yeh et al. conjugated the OBN-caged folic acid onto the UCNPs [76]. After NIR irradiation, the targeting capacity of folic acid was recovered by cleavage of OBN moieties. Zhao et al. encapsulated the $NaYF_4$:TmYb UCNPs inside micelles and proved that the 980 nm excitation can efficiently cleave the ONB-based crosslinker and trigger drug release [77]. Zhang et al. also demonstrated that the ONB-caged plasmid DNA and siRNA can also recover the function after upconversion of NIR signal [78]. Beside the ONB-based photocleavage reaction, the Azo- and SP-based photoisomerization can also be achieved by the process of NIR–UV upconversion [79,80].

Li et al. first reported the application of UCNPs-based DDS in animal tumor tissues [81]. The heterogeneous core–shell nanocrystals ($NaYF_4$: Tm^{3+}, Yb^{3+} @$NaLuF_4$) were encapsulated with the hollow $mSiO_2$ to form yolk-shell structured UCNPs, termed as YSUCNPs. The amino-coumarin unit was conjugated on the anticancer drug chlorambucil as the photocleavable cage. By modifying the coumarin segment with octyl groups, the formed prodrug was highly hydrophic and showed strong physical adsorption with mesopores cavity, with no premature drug release of drug under physiological conditions. After NIR irradiation at 980 nm, the unconverted UV luminescence caused the uncaging of the amino-coumarin and initiated the drug release in the on–off pattern.

Although the upconversion-based UV-triggered photoreactions require much lower NIR excitation intensity than that of two-photon irradiation, the laser intensity is still a safety problem [82]. Compared with the UV emission (<400 nm), which is easily vanished under low NIR excitation intensities (0.19–0.81 W/cm^2), blue emissions (470 nm) are still emitted. Wu for the first time demonstrated that the ruthenium (Ru) complexes with the metal-to-ligand charge transfer band at 453 nm can achieve the blue emissions-triggered photocleavage under ultralow-intensity NIR excitation (0.35 W/cm^2). In this system, the *beta*-phase $NaYF_4$:TmYb@$NaYF_4$ UCNPs were coated with mesoporous silica with the Ru complexes, acting as gatekeepers. The phototriggered drug release can be achieved by the cleavage of the Ru complexes (Fig. 8.9) [83].

8.3.2.1 Outlook

Photo-based theranostic technologies have increased extensively in the past decade; the multiphoton laser scanning [84], photoacoustic tomography [85], and even the optogenetics [86] reveal tremendous potential in medical

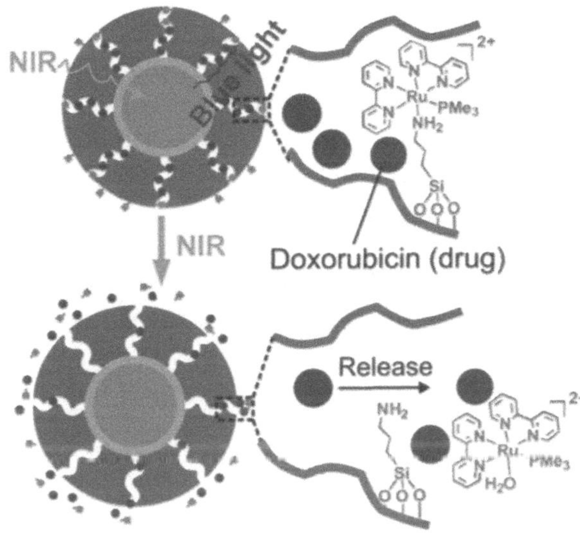

FIGURE 8.9 Schematic illustration of upconverted blue luminescence induces cleavage of Ru complexes and release of doxorubicin from nanoparticles. *Adapted with permission from S. He, K. Krippes, S. Ritz, Z. Chen, A. Best, H.-J. Butt, et al., Ultralow-intensity near-infrared light induces drug delivery by upconverting nanoparticles, Chem. Commun. 51 (2) (2015) 431–434. Copyright © 2014, Royal Society of Chemistry.*

applications. In the development of stimulus-triggered release systems, photoresources also play a crucial role. Especially, the photochemistry-induced reconstruction of system surface and regulation of the particles' behavior owns great potential to overcome the bio-obstacles during drug delivery, achieve the targeting delivery and precise delivery on-demand. The efficiency and capacity of photochemistry for this purpose cannot be matched by other stimulus. In this case, the potentially optimized drug-release kinetics can minimize the in vivo drug toxicity and improve the therapeutic index. In addition, by combining with other stimuli, such as pH, enzyme, or magnetic field, the capacity for species spatial and temporal drug release can be further improved.

The multiphoton technology and UCNPs development can promote the clinical transition of the phototriggered DDS, especially the photochemistry-based delivery system. However, further improvements and better understanding are needed as to the quantum yield in vivo, the post-administration behavior, and the metabolism pathways. Furthermore, the improvement of the biocompatibility and sensitivity of photosensitive polymers is also urgent. Great potential of the photochemistry for reengineering the delivery system on-demand is unexplored, such as the reasonable design of the secondary form after the transition for better signal response, which owns significant means to improve the performance of the DDS and enhance the clinical safety.

8.4 Acknowledgments

Prof. H.A. Santos acknowledges financial support from the Academy of Finland (Decision No. 304844), the University of Helsinki Research Funds, the HiLIFE Research Funds, Sigrid Jusélius Foundation (Decision No. 4704580), and the European Research Council under the European Union's Seventh Framework Programme (FP/2007–2013, Grant No. 310892).

References

[1] V.P. Chauhan, R.K. Jain, Strategies for advancing cancer nanomedicine, Nat. Mater. 12 (2013) 958.

[2] D. Peer, J.M. Karp, S. Hong, O.C. Farokhzad, R. Margalit, R. Langer, Nanocarriers as an emerging platform for cancer therapy, Nat. Nanotechnol. 2 (2007) 751.

[3] J. Fang, H. Nakamura, H. Maeda, The EPR effect: unique features of tumor blood vessels for drug delivery, factors involved, and limitations and augmentation of the effect, Adv. Drug Deliv. Rev. 63 (3) (2011) 136–151.

[4] E.A. Azzopardi, E.L. Ferguson, D.W. Thomas, The enhanced permeability retention effect: a new paradigm for drug targeting in infection, J. Antimicrob. Chemother. 68 (2) (2013) 257–274.

[5] C.N. Marti, M. Gheorghiade, A.P. Kalogeropoulos, V.V. Georgiopoulou, A.A. Quyyumi, J. Butler, Endothelial Dysfunction, Arterial Stiffness, and Heart Failure, J. Am. Coll. Cardiol. 60 (16) (2012) 1455–1469.

[6] S. Goel, D.G. Duda, L. Xu, L.L. Munn, Y. Boucher, D. Fukumura, et al., Normalization of the vasculature for treatment of cancer and other diseases, Physiol. Rev. 91 (3) (2011) 1071–1121.

[7] J. Shi, P.W. Kantoff, R. Wooster, O.C. Farokhzad, Cancer nanomedicine: progress, challenges and opportunities, Nat. Rev. Cancer 17 (2016) 20.

[8] J. Shi, A.R. Votruba, O.C. Farokhzad, R. Langer, Nanotechnology in drug delivery and tissue engineering: from discovery to applications, Nano Lett. 10 (9) (2010) 3223–3230.

[9] K. Riehemann, S.W. Schneider, T.A. Luger, B. Godin, M. Ferrari, H. Fuchs, Nanomedicine—challenge and perspectives, Angew. Chem. Int. Edit. 48 (5) (2009) 872–897.

[10] J. Li, D.J. Mooney, Designing hydrogels for controlled drug delivery, Nat. Rev. Mater. 1 (2016) 16071.

[11] R.K. Jain, T. Stylianopoulos, Delivering nanomedicine to solid tumors, Nat. Rev. Clin. Oncol. 7 (2010) 653.

[12] E. Blanco, H. Shen, M. Ferrari, Principles of nanoparticle design for overcoming biological barriers to drug delivery, Nat. Biotechnol. 33 (2015) 941.

[13] S. Nie, Understanding and overcoming major barriers in cancer nanomedicine, Nanomedicine 5 (4) (2010) 523–528.

[14] Z. Deng, Z. Zhen, X. Hu, S. Wu, Z. Xu, P.K. Chu, Hollow chitosan–silica nanospheres as pH-sensitive targeted delivery carriers in breast cancer therapy, Biomaterials. 32 (21) (2011) 4976–4986.

[15] H. Kim, S. Kim, C. Park, H. Lee, H.J. Park, C. Kim, Glutathione-induced intracellular release of guests from mesoporous silica nanocontainers with cyclodextrin gatekeepers, Adv. Mater 22 (38) (2010) 4280–4283.

[16] K.-J. Chen, H.-F. Liang, H.-L. Chen, Y. Wang, P.-Y. Cheng, H.-L. Liu, et al., Liposomal system for triggering localized extracellular drug delivery, ACS Nano. 7 (1) (2013) 438–446.

[17] C.-H. Wang, S.-T. Kang, Y.-H. Lee, Y.-L. Luo, Y.-F. Huang, C.-K. Yeh, Aptamer-conjugated and drug-loaded acoustic droplets for ultrasound theranosis, Biomaterials. 33 (6) (2012) 1939–1947.

[18] L. Zhang, T. Wang, L. Yang, C. Liu, C. Wang, H. Liu, et al., General route to multifunctional uniform yolk/mesoporous silica shell nanocapsules: a platform for simultaneous cancer-targeted imaging and magnetically guided drug delivery, Chem-Eur. J. 18 (39) (2012) 12512–12521.

[19] J.L. Vivero-Escoto, I.I. Slowing, C.-W. Wu, V.S.Y. Lin, Photoinduced intracellular controlled release drug delivery in human cells by gold-capped mesoporous silica nanosphere, J. Am. Chem. Soc. 131 (10) (2009) 3462–3463.

[20] A.Y. Rwei, W. Wang, D.S. Kohane, Photoresponsive nanoparticles for drug delivery, Nano Today 10 (4) (2015) 451–467.

[21] J.-F. Gohy, Y. Zhao, Photo-responsive block copolymer micelles: design and behavior, Chem. Soc. Rev. 42 (17) (2013) 7117–7129.

[22] H.J. Cho, M. Chung, M.S. Shim, Engineered photo-responsive materials for near-infrared-triggered drug delivery, J. Ind. Eng. Chem. 31 (Suppl. C) (2015) 15–25.

[23] A.A. Beharry, G.A. Woolley, Azobenzene photoswitches for biomolecules, Chem. Soc. Rev. 40 (8) (2011) 4422–4437.

[24] J. Jiang, X. Tong, D. Morris, Y. Zhao, Toward photocontrolled release using light-dissociable block copolymer micelles, Macromolecules 39 (13) (2006) 4633–4640.

[25] R. Weissleder, A clearer vision for in vivo imaging, Nat. Biotechnol. 19 (2001) 316.

[26] G.J. Clydesdale, G.W. Dandie, H.K. Muller, Ultraviolet light induced injury: immunological and inflammatory effects, Immunol. Cell Biol. 79 (2001) 547.

[27] Y.-C. Liu, A.-L.M. Le Ny, J. Schmidt, Y. Talmon, B.F. Chmelka, C.T. Lee, Photo-assisted gene delivery using light-responsive catanionic vesicles, Langmuir 25 (10) (2009) 5713–5724.

[28] A. Harada, Cyclodextrin-based molecular machines, Acc. Chem. Res. 34 (6) (2001) 456–464.

[29] F.D. Jochum, P. Theato, Thermo- and light responsive micellation of azobenzene containing block copolymers, Chem. Commun. 46 (36) (2010) 6717–6719.

[30] J. Lu, E. Choi, F. Tamanoi, J.I. Zink, Light-activated nanoimpeller-controlled drug release in cancer cells, Small. 4 (4) (2008) 421–426.

[31] K. Peng, I. Tomatsu, A. Kros, Light controlled protein release from a supramolecular hydrogel, Chem. Commun. 46 (23) (2010) 4094–4096.

[32] H. Yan, C. Teh, S. Sreejith, L. Zhu, A. Kwok, W. Fang, et al., Functional mesoporous silica nanoparticles for photothermal-controlled drug delivery in vivo, Angew. Chem. Int. Ed. 51 (33) (2012) 8373–8377.

[33] X. Chi, X. Ji, D. Xia, F. Huang, A dual-responsive supra-amphiphilic polypseudorotaxane constructed from a water-soluble pillar[7]arene and an azobenzene-containing random copolymer, J. Am. Chem. Soc. 137 (4) (2015) 1440–1443.

[34] H. Kang, H. Liu, J.A. Phillips, Z. Cao, Y. Kim, Y. Chen, et al., Single-DNA molecule nanomotor regulated by photons, Nano Lett. 9 (7) (2009) 2690–2696.

[35] Q. Yuan, Y. Zhang, T. Chen, D. Lu, Z. Zhao, X. Zhang, et al., Photon-manipulated drug release from a mesoporous nanocontainer controlled by azobenzene-modified nucleic acid, ACS Nano. 6 (7) (2012) 6337–6344.

[36] J.A. Barltrop, P.J. Plant, P. Schofield, Photosensitive protective groups, Chem. Commun. (London) 22 (1966) 822–823.

[37] C.P. Holmes, D.G. Jones, Reagents for combinatorial organic synthesis: development of a new o-nitrobenzyl photolabile linker for solid phase synthesis, J. Org. Chem. 60 (8) (1995) 2318–2319.

[38] X. Jiang, C.A. Lavender, J.W. Woodcock, B. Zhao, Multiple micellization and dissociation transitions of thermo- and light-sensitive poly(ethylene oxide)-b-poly(ethoxytri(ethylene glycol) acrylate-co-o-nitrobenzyl acrylate) in water, Macromolecules 41 (7) (2008) 2632–2643.

[39] D. Han, X. Tong, Y. Zhao, Fast photodegradable block copolymer micelles for burst release, Macromolecules 44 (3) (2011) 437–439.

[40] T. Dvir, M.R. Banghart, B.P. Timko, R. Langer, D.S. Kohane, Photo-targeted nanoparticles, Nano Lett. 10 (1) (2010) 250–254.

[41] R. Klajn, Spiropyran-based dynamic materials, Chem. Soc. Rev. 43 (1) (2014) 148–184.

[42] H.-i Lee, W. Wu, J.K. Oh, L. Mueller, G. Sherwood, L. Peteanu, et al., Light-induced reversible formation of polymeric micelles, Angew. Chem. Int. Edit. 46 (14) (2007) 2453–2457.

[43] R. Tong, H.D. Hemmati, R. Langer, D.S. Kohane, Photoswitchable nanoparticles for triggered tissue penetration and drug delivery, J. Am. Chem. Soc. 134 (21) (2012) 8848–8855.

[44] H.-J. Li, J.-Z. Du, X.-J. Du, C.-F. Xu, C.-Y. Sun, H.-X. Wang, et al., Stimuli-responsive clustered nanoparticles for improved tumor penetration and therapeutic efficacy, Proc. Natl Acad. Sci. 113 (15) (2016) 4164–4169.

[45] R. Tong, H.H. Chiang, D.S. Kohane, Photoswitchable nanoparticles for in vivo cancer chemotherapy, Proc. Natl. Acad. Sci. 110 (47) (2013) 19048–19053.

[46] D.E. Hudson, D.O. Hudson, J.M. Wininger, B.D. Richardson, Penetration of laser light at 808 and 980 nm in bovine tissue samples, Photomed. Laser. Surg. 31 (4) (2013) 163–168.

[47] P. Juzenas, A. Juzeniene, O. Kaalhus, V. Iani, J. Moan, Noninvasive fluorescence excitation spectroscopy during application of 5-aminolevulinic acid in vivo, Photochem. Photobiol. Sci. 1 (10) (2002) 745–748.

[48] R. Boushel, H. Langberg, J. Olesen, J. Gonzales-Alonzo, J. Bülow, M. Kjær, Monitoring tissue oxygen availability with near infrared spectroscopy (NIRS) in health and disease, Scand. J. Med. Sci. Sport. 11 (4) (2001) 213–222.

[49] S. Gioux, H.S. Choi, J.V. Frangioni, Image-guided surgery using invisible near-infrared light: fundamentals of clinical translation, Mol. Imaging 9 (5) (2010) 7290.2010.

[50] L.C. Kennedy, L.R. Bickford, N.A. Lewinski, A.J. Coughlin, Y. Hu, E.S. Day, et al., A new era for cancer for cancer treatment: gold-nanoparticle-mediated thermal therapies, Small. 7 (2) (2011) 169–183.

[51] T. Furuta, S.S.-H. Wang, J.L. Dantzker, T.M. Dore, W.J. Bybee, E.M. Callaway, et al., Brominated 7-hydroxycoumarin-4-ylmethyls: Photolabile protecting groups with biologically useful cross-sections for two photon photolysis, Proc. Natl. Acad. Sci. 96 (4) (1999) 1193–1200.

[52] M. Álvarez, A. Best, S. Pradhan-Kadam, K. Koynov, U. Jonas, M. Kreiter, Single-photon and two-photon induced photocleavage for monolayers of an alkyltriethoxysilane with a photoprotected carboxylic ester, Adv. Mater. 20 (23) (2008) 4563–4567.

[53] S. Link, M.A. El-Sayed, Shape and size dependence of radiative, non-radiative and photothermal properties of gold nanocrystals, Int. Rev. Phys. Chem. 19 (3) (2000) 409–453.

[54] P. Shao, B. Wang, Y. Wang, J. Li, Y. Zhang, The application of thermosensitive nanocarriers in controlled drug delivery, J. Nanomater. 2011 (2011) 1–12.

[55] S.R. Sershen, S.L. Westcott, N.J. Halas, J.L. West, Temperature-sensitive polymer–nanoshell composites for photothermally modulated drug delivery, J. Biomed. Mater. Res. 51 (3) (2000) 293–298.

[56] F. Wang, X. Liu, Recent advances in the chemistry of lanthanide-doped upconversion nanocrystals, Chem. Soc. Rev. 38 (4) (2009) 976–989.

[57] M.L. Viger, M. Grossman, N. Fomina, A. Almutairi, Low power upconverted near-IR light for efficient polymeric nanoparticle degradation and cargo release, Adv. Mater. 25 (27) (2013) 3733–3738.

[58] B. Yan, J.-C. Boyer, D. Habault, N.R. Branda, Y. Zhao, Near infrared light triggered release of biomacromolecules from hydrogels loaded with upconversion nanoparticles, J. Am. Chem. Soc. 134 (40) (2012) 16558–16561.

[59] S. Wu, H.-J. Butt, Near-infrared-sensitive materials based on upconverting nanoparticles, Adv. Mater. 28 (6) (2016) 1208–1226.

[60] J. Kost, R. Langer, Responsive polymeric delivery systems, Adv. Drug Deliv. Rev. 46 (1) (2001) 125–148.

[61] A.P. Goodwin, J.L. Mynar, Y. Ma, G.R. Fleming, J.M.J. Fréchet, Synthetic micelle sensitive to IR light via a two-photon process, J. Am. Chem. Soc. 127 (28) (2005) 9952–9953.

[62] Y.H. Bae, T. Okano, R. Hsu, S.W. Kim, Thermo-sensitive polymers as on-off switches for drug release, Die Makromolekulare Chemie, Rapid Commun. 8 (10) (1987) 481–485.

[63] F.D. Jochum, P. Theato, Temperature- and light-responsive smart polymer materials, Chem. Soc. Rev. 42 (17) (2013) 7468–7483.

[64] W.-L. Chiang, T.-T. Lin, R. Sureshbabu, W.-T. Chia, H.-C. Hsiao, H.-Y. Liu, et al., A rapid drug release system with a NIR light-activated molecular switch for dual-modality photothermal/antibiotic treatments of subcutaneous abscesses, J. Control. Release 199 (Suppl. C) (2015) 53–62.

[65] M. Liras, M. González-Béjar, E. Peinado, L. Francés-Soriano, J. Pérez-Prieto, I. Quijada-Garrido, et al., Thin amphiphilic polymer-capped upconversion nanoparticles: enhanced emission and thermoresponsive properties, Chem. Mater. 26 (13) (2014) 4014–4022.

[66] M. Nakayama, T. Okano, T. Miyazaki, F. Kohori, K. Sakai, M. Yokoyama, Molecular design of biodegradable polymeric micelles for temperature-responsive drug release, J. Control. Release 115 (1) (2006) 46–56.

[67] C. He, S.W. Kim, D.S. Lee, In situ gelling stimuli-sensitive block copolymer hydrogels for drug delivery, J. Control. Release 127 (3) (2008) 189–207.

[68] X.J. Loh, Y.X. Tan, Z. Li, L.S. Teo, S.H. Goh, J. Li, Biodegradable thermogelling poly (ester urethane)s consisting of poly(lactic acid) – thermodynamics of micellization and hydrolytic degradation, Biomaterials. 29 (14) (2008) 2164–2172.

[69] X.J. Loh, K.B. Colin Sng, J. Li, Synthesis and water-swelling of thermo-responsive poly (ester urethane)s containing poly(ε-caprolactone), poly(ethylene glycol) and poly(propylene glycol), Biomaterials. 29 (22) (2008) 3185–3194.

[70] M.S. Yavuz, Y. Cheng, J. Chen, C.M. Cobley, Q. Zhang, M. Rycenga, et al., Gold nanocages covered by smart polymers for controlled release with near-infrared light, Nat. Mater. 8 (2009) 935.

[71] S. Gai, C. Li, P. Yang, J. Lin, Recent progress in rare earth micro/nanocrystals: soft chemical synthesis, luminescent properties, and biomedical applications, Chem. Rev. 114 (4) (2014) 2343–2389.

[72] J. Liu, W. Bu, L. Pan, J. Shi, NIR-triggered anticancer drug delivery by upconverting nanoparticles with integrated azobenzene-modified mesoporous silica, Angew. Chem. Int. Ed. 52 (16) (2013) 4375–4379.

[73] J.-C. Boyer, C.-J. Carling, B.D. Gates, N.R. Branda, Two-way photoswitching using one type of near-infrared light, upconverting nanoparticles, and changing only the light intensity, J. Am. Chem. Soc. 132 (44) (2010) 15766–15772.

[74] D. Yang, Pa Ma, Z. Hou, Z. Cheng, C. Li, J. Lin, Current advances in lanthanide ion (Ln3 +)-based upconversion nanomaterials for drug delivery, Chem. Soc. Rev. 44 (6) (2015) 1416–1448.

[75] A. Bagheri, H. Arandiyan, C. Boyer, M. Lim, Lanthanide-doped upconversion nanoparticles: emerging intelligent light-activated drug delivery systems, Adv. Sci. 3 (7) (2016). 1500437-n/a.

[76] Y.-H. Chien, Y.-L. Chou, S.-W. Wang, S.-T. Hung, M.-C. Liau, Y.-J. Chao, et al., Near-infrared light photocontrolled targeting, bioimaging, and chemotherapy with caged upconversion nanoparticles in vitro and in vivo, ACS Nano. 7 (10) (2013) 8516–8528.

[77] B. Yan, J.-C. Boyer, N.R. Branda, Y. Zhao, Near-infrared light-triggered dissociation of block copolymer micelles using upconverting nanoparticles, J. Am. Chem. Soc. 133 (49) (2011) 19714–19717.

[78] M.K.G. Jayakumar, N.M. Idris, Y. Zhang, Remote activation of biomolecules in deep tissues using near-infrared-to-UV upconversion nanotransducers, Proc. Natl. Acad. Sci. 109 (22) (2012) 8483–8488.

[79] Q. Xing, N. Li, Y. Jiao, D. Chen, J. Xu, Q. Xu, et al., Near-infrared light-controlled drug release and cancer therapy with polymer-caged upconversion nanoparticles, RSC Adv. 5 (7) (2015) 5269–5276.

[80] C. Yao, P. Wang, X. Li, X. Hu, J. Hou, L. Wang, et al., Near-infrared-triggered azobenzene-liposome/upconversion nanoparticle hybrid vesicles for remotely controlled drug delivery to overcome cancer multidrug resistance, Adv. Mater. 28 (42) (2016) 9341–9348.

[81] L. Zhao, J. Peng, Q. Huang, C. Li, M. Chen, Y. Sun, et al., Near-infrared photoregulated drug release in living tumor tissue via yolk-shell upconversion nanocages, Adv. Funct. Mater. 24 (3) (2014) 363–371.

[82] L.I.o.A. American National Standard for Safe Use of Lasers, Orlando, FL, 2000.

[83] S. He, K. Krippes, S. Ritz, Z. Chen, A. Best, H.-J. Butt, et al., Ultralow-intensity near-infrared light induces drug delivery by upconverting nanoparticles, Chem. Commun. 51 (2) (2015) 431–434.

[84] J. Paoli, M. Smedh, A.-M. Wennberg, M.B. Ericson, Multiphoton laser scanning microscopy on non-melanoma skin cancer: morphologic features for future non-invasive diagnostics, J. Invest. Dermatol. 128 (5) (2008) 1248–1255.

[85] O. Liba, A. de la Zerda, Photoacoustic tomography: breathtaking whole-body imaging, Nat. Biomed. Eng. 1 (2017) 0075.

[86] K. Deisseroth, Optogenetics, Nat. Methods 8 (2010) 26.

Chapter 9

State-of-the-art strategies for the biofunctionalization of photoactive inorganic nanoparticles for nanomedicine

María Antonietta Parracino[1,*], Beatriz Martín[2] and Valeria Grazú[2,*]

*[1]Italian Institute of Technology, Genova, Italy, [2]Institute of Materials Science of Aragon, University of Zaragoza, CSIC and CIBER-BBN, Zaragoza, Spain, *Corresponding authors. email address: antonietta.parracino@iit.it; vgrazu@unizar.es*

Chapter Outline

Photoactive Inorganic Nanoparticles. DOI: https://doi.org/10.1016/B978-0-12-814531-9.00009-9
© 2019 Elsevier Inc. All rights reserved.

9.1 An overview of the photoactive nanoparticles most used in nanomedicine

Photoactive nanomaterials are one of the most important classes of nanomaterials. They are currently being applied in several fields but especially in the medical field due to their unique properties and the wide range of methodologies that could be used for their biofunctionalization. They can achieve unconventional photophysical phenomena as they provide significant strong and photostable optical signals. For example, they can absorb UV–visible light or near-infrared (NIR) irradiation and transform light energy into thermal energy with a very high efficiency, almost of the 100% [1], or they can act as UV or visible light sources to activate photoactive moieties in the immediate (nanoscale) vicinity of the nanoparticles (NPs) that cannot be directly excited by NIR light [2]. These intrinsic properties have triggered their use in a myriad of applications in biomedicine ranging from obtaining more sensitive biosensors to the development of more efficient therapies, therapeutics, and theragnostic agents.

Among photoactive NPs, gold NPs (AuNPs) are one of the most used in nanomedicine for their intrinsic optical and chemical properties. In this sense, they enable strong absorption light in the visible and NIR range, with high absorption coefficients (10^7-10^9/M/cm), which are much larger than those of most small dye molecules. This advantageous optical property is derived from the localized surface plasmon resonance of AuNPs, which differs from the light-absorbing mechanism of organic dyes [3]. As their optical properties change dramatically depending on their size, shape, and geometry [4,5], various asymmetric gold nanostructures with absorption peaks in the NIR range have been developed for photothermal therapy such as nanoshells [6], nanorods [7,8], and nanocages [9–12]. Their high efficiency in transforming light into heat has been employed in the development of thermal-responsive drug-delivery systems. As it has been extensively reviewed, in those cases thermoresponsive polymers linked to specific drugs are attached onto gold nanostructure surfaces [13], and the drug release is triggered in response to light-induced changes in the structure of the polymer/gold assemblies. In particular, the photothermal properties of functionalized AuNPs, used as nanocarriers, induce the raising of temperature in their immediate environment with the concomitant formation of micro-bubbles, leading to a localized drug release. Besides the use of light for photothermal release of drugs, AuNPs are considered as a good candidate to substitute commonly used contrast agents for biomedical diagnosis in advanced light-triggered technologies such aspolarized resonance scattering [14], optical coherence tomography [15], two-photon luminescence [16], surface-enhanced Raman scattering imaging [17], and photoacoustic techniques [18,19]. In contrast, gold nanopopcorn, nanorods, nanoshells and nanoprisms are also employed in sensing applications [20]. Their high absorption coefficient permits their use, when linked to other active biomolecules, as excellent sensors for applications including genomics [21], immunoassays [22,23], detection of microorganisms [24], clinical chemistry [25], etc. More in detail, AuNPs of different size and shape properly functionalized with biomolecules such as proteins, DNA/RNA, antibodies, and dyes molecules are applied, for example, to report the presence of trace amounts of analytes in the diagnostic field.

Some analytes with very high medical interest, that is, glucose in diabetes and/or cancer biomarkers, drove the developments of ultrasensitive biosensors that employ plasmonic NPs as target to monitor the presence and concentration of such analytes. For example, enzyme-responsive Ag/Au core−shell NPs, constitute an innovative sensing platform for monitoring the presence of glucose in blood samples. This sensing platform allows to improve the detection limits of glucose in 1−2 orders of magnitude [26]. In addition, photoactive asymmetric NPs are widely used to sense the presence of cancer biomarkers [133]. For example, Polo et al, reported the use of gold nanoprisms to develop an innovative thermal-sensing platform to detect the biomarker carcinoembryonic antigen. The sensor, developed by Polo et al., is based on the use of antibodies labeled with asymmetric gold nanomaterial able to release thermal energy upon illumination with NIR light beam. The use of asymmetric metal-nanomaterial, leads to an improvement of sensitivity up to the femto-molar concentration without the need of additional complex and time-consuming amplification steps [133].

Among inorganic photoactive nanostructures are AgNPs and TiO_2NPs. AgNPs also play an important role, especially as photoactive labels in a variety of biosensing systems and for their antibacterial properties. As AuNPs, AgNPs also have strong size-dependent optical properties and UV−visible extinction bands that depending on the size and shape give unique optoelectrical properties [2]. In the case of TiO_2NPs, they have been studied and widely used in pharmaceutical fields for their antimicrobial assets. The antimicrobial properties of TiO_2 are attributed to the high redox potential of reactive oxygen species generated by the photo-excitation. Several authors reported the photocatalytic disinfection properties of TiO_2 that are used to kill viruses including poliovirus-1 [27], hepatitis B virus [28], herpes simplex virus, and MS2 bacteriophage 2005 [29]. Moreover TiO_2NPs decorated with other functionalized NPs provide an increase in the antimicrobial properties, as it is reported by Mura et al. [30].

Other new class of photoactive metallic NPs used for photothermal therapy consist of aqueous clusters of copper sulfide (CuSNPs). These NPs have a surface plasmon absorption band in the NIR region (900−1100 nm) similar to that of gold asymmetric nanostructures, but they are much smaller ($<$15 nm). This makes them more likely to reach their targets and more readily cleared by the renal system. Cu-labeled CuSNPs can be used for photothermal therapy and also permit positron emission tomographic imaging and radiotherapy [31].

Together with metallic NPs, mesoporous silica nanoparticles (MSNPs) are also listed as photoactive nanomaterials when loaded with photoactive molecules [32]. They have a honeycomb-like porous structure with hexagonal channels (diameters vary ~2–30 nm), which enables physical adsorption or encapsulation of chromogenic, contrast agents, and therapeutic agents. In this sense they could also be used to entrap photoactive NPs, as it is the case of CuSNPs. In particular, these hybrid nanostructures have been used for imaging of small tumors that is especially crucial for cancer diagnosis. MSNPs are also optically transparent, which is advantageous for photic control and spectroscopic monitoring of encapsulated chromogenic species. It has also been reported that light is used to modulate drug release from MSNPs by opening and closing their pores. This has been achieved by grafting the pores with molecules that can be photoactivated such as photoswitching azobenzene groups [33], and coumarin derivatives that could photodimerize reversibly. Photocaging *o*-nitrobenzyl derivatives have also been used as photo-removable "caps" to regulate pore opening [34].

Other NPs with an important role as photoactive NPs are known as "upconverting NPs." These are usually $NaYF_4$ NPs doped with trivalent rare-earth ions (Yb^{3+}, Tm^{3+}, Er^{3+}, Ho^{3+}, etc.). These NPs are able to be excited with continuous-wave (power is constant over time, in contrast to pulsed lasers) NIR light (900–1000 nm) to emit at shorter wavelengths such as visible and UV light [35,36]. Such upconversion NPs properly modified with biomolecules have attracted considerable attention in bioimaging applications due to high resistance to photobleaching, stable emission, and the ability to be detected deep within tissue (using NIR light) [37]. While these NPs have great potential in medical application, the luminescence imaging systems for its visualization are still laboratory-based techniques [38–40], none commercially available yet. Preliminary studies using $NaYF_4$ upconversion NPs coated with polyethylene glycol (PEG) or polyacrylic acid reported no apparent toxicity or adverse effects in vitro and in vivo in mice [41–44]. There are other studies that show the use of upconverting NPs coated with other inorganic nanocomposites (e.g., mesoporous silica) [45] or polymeric materials (e.g., PEG or polylactide-PEG) [46] containing photosensitizers for photodynamic therapy. Those NPs could be irradiated with NIR and then made to emit light at a wavelength equal to that of the excitation band of the photosensitizers incorporated in the polymer or silica nanocomposites. In some reports, upconverting NPs have been coated with Ag nanocomposites so that NIR "upconverted" to shorter wavelengths. Thus they are able to trigger heating from the silver shell, which has surface plasmonic resonance absorption in the visible spectrum region (500–600 nm) and can also be employed in phototherapy [47]. It is also worth mentioning that gold-coated magnetic NPs (MNPs) could also be classified as upconverting NPs. These NPs also are utilized in medicine with an appropriate functionalization. For example, Fe_2O_3/AuNPs, properly functionalized, are ideal probes for magnetomotive photoacustic imaging, which is a powerful noninvasive diagnostic tool. Gold-coated MNPs, give a remarkable contrast enhancement, and improve the specificity, compared with conventional NPs used as contrast agents in this imaging technique [48].

Other important classes of photoactive NPs are quantum dots (QDs), containing heavy metals such as cadmium and selenium. The composition and size of QDs allow them to be endowed with certain physically tailorable chemical properties, such as a tunable fluorescence emission wavelength and a strong resistance against photobleaching. Therefore they are considered to be one of the best substitutes for conventional organic fluorescent materials to be used in vivo and in vitro studies. Several studies describe the QD functionalization with polymer brushes in a controlled manner, as it is the case of QD tethered with liposomes containing drugs for the generation of one-pot diagnostic and therapeutic agent [49]. Moreover, the functionalization of QD with antibodies has been widely applied for specific targeting of antigens.

More recently in development timeline of photoactive NPs, it has also been reported that the combination of multiple discrete photoactive components, such as QDs, MNPs, and plasmonic Au/AgNPs, into a single and com pact NP has been proved useful in a variety of applications, as molecular diagnosis, imaging, and therapeutics. Such kind of multimodality-imaging nanoprobes have an advantage over single-modality ones as the complementary ability of different imaging modalities could be harnessed to great effect by using them in a row.

9.2 Influence of the functionalization strategy on the final bio-application of the nanoparticle

For the use in medicine of all the abovementioned photoactive NPs, it is not enough to have these novel physicochemical properties. Their adequate mono and/or multibiofunctionalization with different biomolecules (drugs, peptides, nucleic acids, antibodies, enzymes, etc.) is also a critical aspect. This allows adding to the NPs biological properties of interest such as specificity, capacity of overcoming biological barriers, therapeutic action, etc. But biofunctionalization of nanomaterials, and especially their multifunctionalization, is not an easy task. Proof of this is the low number of articles that could be found indexed by ISIWeb of Knowledge when searching for the topic "functionalization of NPs" versus "synthesis of NPs" (Fig. 9.1).

There are several factors that contribute to their functionalization being not straightforward: (1) big differences among NPs in terms of size, surface area, colloidal stability, density, and type of reactive groups; (2) presence of surfactants used to transfer NPs that are synthesized in organic solvents to water; and (3) big differences among the biomolecules to be joined to NPs in terms of size, chemical composition, 3D complexity, location of its biological active site, and so forth. Thus no universal methodologies exist to

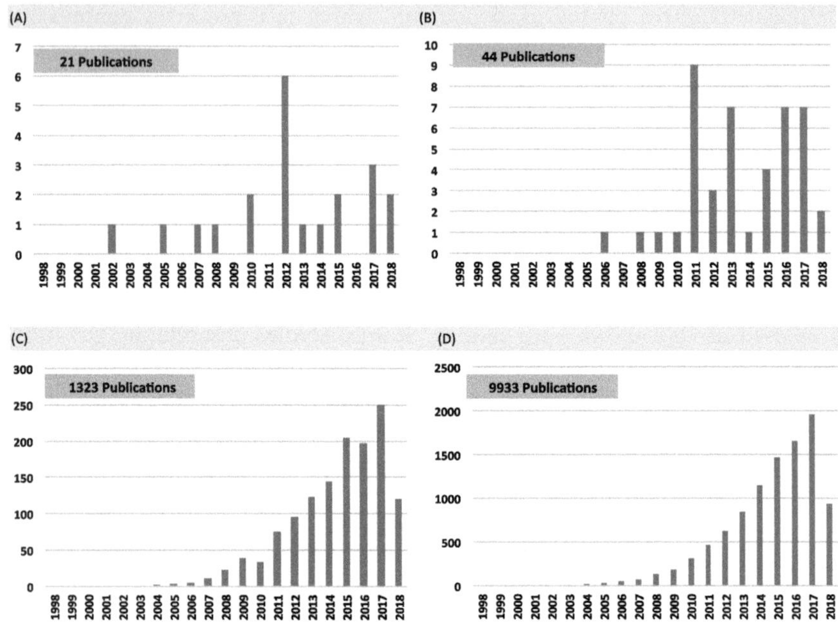

FIGURE 9.1 Articles indexed by ISIWeb of Knowledge when searching for topics: (A) NPs + biofunctionalization, (B) NPs + biomolecules + functionalization, (C) NPs + functionalization, and (D) NPs + synthesis.

cover the wide variety of NPs and biomolecules available for this purpose. Even if we reduce the population of NPs to only inorganic photoactive ones, each particular case (NP + biomolecule) requires optimization in order to find the most adequate functionalization protocol (Scheme 9.1).

It is already well known, from the experience achieved from the functionalization of microstructured materials, that the methodology selected for the binding of the biomolecules could affect their biological function. In particular, in the case of large biomolecules such as proteins (e.g., antibodies, enzymes), their orientation on the NP's surface could hamper or improve their biological activity. For example, in the case of enzymes it is well known that certain immobilization methodologies that enable the rigidification of the enzyme structure could greatly improve the stability of monomeric or multimeric enzymes. Stabilization factors as high as 60,000 times, defined as the ratio between half-lives of the immobilized versus the soluble enzyme have been reported. Moreover, enantioselectivity of different enzymes (e.g., lipases) may also be dramatically improved (from $E = 1$ to >100) by performing different immobilization protocols with the same particle as support [50].

However, not only the orientation of the bounded biomolecule is critical for the biotechnological or biomedical NPs' outcome, but also its density

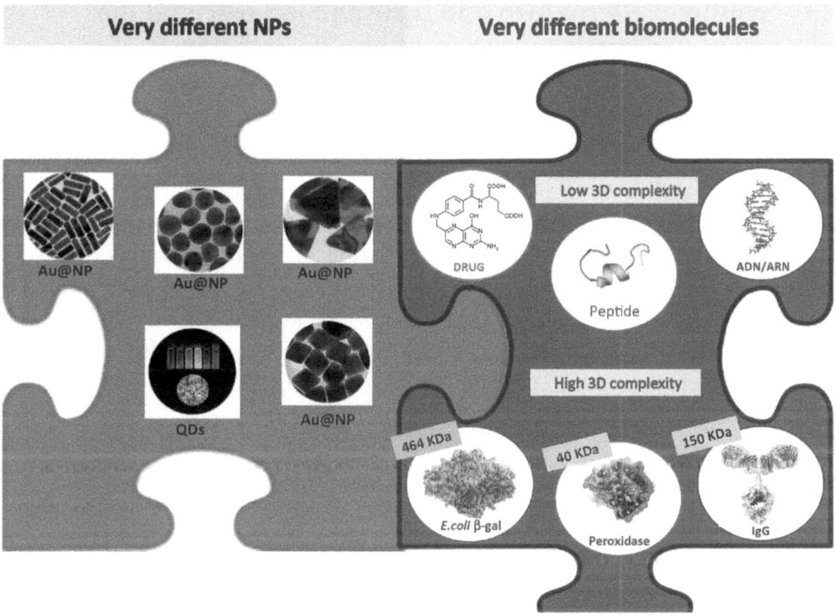

SCHEME 9.1 Nonuniversality of NPs' biofunctionalization protocols due to the big differences among NPs to be functionalized and biomolecules to be linked.

and even its spatial localization in the case of hybrid nanostructures. In this sense, a well-known system used as nanocarrier for drug-delivery is the tethering of gold nanoshells to liposomes via a lipid–PEG–thiol linker. This particular spatial localization of the AuNPs within the nanohybrid showed an enhancement in the drug release in response to a photic stimulus compared to that obtained for AuNPs free outside the liposomes or coencapsulated with the drug inside them [51].

Although all these factors are usually widely taken into account when functionalizing microstructured materials, more fundamental and detailed control of the NP–biomolecule interface is required for a sustained and general success of biofunctionalized NPs in the biomedical field. At the nanoscale, the interaction between biomolecules and surfaces has been shown to induce size-dependent effects on protein conformation upon binding that do not occur with microstructured materials [52]. Thus this chapter will focus on giving some tips to select/find the best functionalization methodology of photoactive inorganic NPs depending on the biomolecule(s) to be attached, and the intrinsic physicochemical properties of the NPs to be functionalized. Besides, a section explaining different techniques for the characterization of the functionalization process and/or the functionalized NPs is also included. In this sense, we want to highlight the importance of performing a good characterization of the NP's functionalization in order to understand or subsequently improve its biomedical application.

9.3 Synthetic alternatives and their and influence on the subsequent biofunctionalization of photoactive nanoparticles

Different synthetic protocols are used to create core–shell photoactive NPs depending on their composition and expected final application. However, whatever the chosen protocol, it will have its influence on the subsequent biofunctionalization of the obtained NPs. In particular, the ligands that are bound to the NP surface (shell) will dictate not only the type and density of chemical functional groups for the conjugation of biomolecules, but also the stability of the NPs. In this sense, they stabilize NPs suspensions by (1) blocking physical and chemical access to the NP surface, (2) favoring dispersion of the NPs in the solvent, (3) inhibiting particle–particle interactions by steric and/or electrostatic repulsion, and (4) slowing down the loss of surface atoms. Due to the wide variety of NPs, there is no single ligand structure that stabilizes any NP in every application. However, the most successful ones have in common that strong bind to the NP's surface, self-interact usually via solvophobic segments, and have charged and/or polymer segments that provide electrostatic and steric stabilization. A good colloidal stability is a critical aspect in the selection of the biofunctionalization methodology to be used, as the NPs must be compatible with the reaction conditions. Besides, it could also affect the final biomolecule@NP dispersibility and long-term stability [53].

In the case of AuNPs, variants of the classical Turkevich/Frens reaction based on the reduction of tetra-choloauric acid ($HAuCl_4$) by sodium citrate allow the synthesis of quasi-spherical NPs ranging from 5 to 300 nm with low coefficient of variations (<10%) [54–57]. However, this AuNPs, which are stabilized with citric acid, can go through irreversible aggregation during its biofunctionalization. That is why the organic shell of citric acid is usually replaced by place exchange reaction using different types of ligands. Mostly thiolated ligands are used because of the strong interaction between the Au surface and thiol functional groups. Besides, place exchange reaction could be done including or not a surfactant (Tween 20) or even using thioctic acid as an intermediate through a two-step functionalization [58,59].

Although the water-based synthesis of core–shell AuNPs makes easier their subsequent biofunctionalization for their use in nanomedicine, the biphasic synthetic strategy reported by Brust–Shiffrin allows obtaining monolayer-protected clusters from 1.5 to 5 nm with low dispersity. This methodology uses tetraoctylammonium bromide (TOAB) as phase-transfer reagent, and sodium borohydride ($NaBH_4$) as the reducing agent. These AuNPs possess higher stability than others due to their alkanethiol-protecting organic layer, making possible its drying and redispersion without any aggregation. This is due to a synergic effect of the strong thiol–gold interactions and van der Waals attractions between the neighboring ligands [60,61]. However, place exchange reactions with one or several thiolated ligands could not ensure the complete removal of TOAB. This could raise problems during their biofunctionalization (e.g., short- or long-time denaturation of the biomolecule once it is attached), or even induction of cytotoxicity [62].

In the case of AuNPs of less than 10 nm, other strategies to obtain them are based on the use of thiol-containing polymers as stabilizing ligands. In this sense, Hussain and coworkers have developed a simple one-step protocol for the preparation of highly monodisperse AuNPs in aqueous solution with diameters of less than 5 nm, using thioether or thiol-functionalized polymer ligands, such as alkyl thioether end-functionalized poly (methacrylic acid). In this approach, the particle size and size distribution is controlled by subtle variation of the polymer structure and concentration [63]. By systematically varying the polymer-to-gold ratio, the size of the NPs can be finely tuned and a transition from nonfluorescent to fluorescent NPs is observed for core diameters between 1.7 and 1.1 nm [63]. The beneficial effect of using this type of polymers over monothiol ligands is that they are bound to the inorganic core through several stable bonds. This multivalence of the ligand renders AuNPs more stable than AuNPs protected with monothiol ligands to harsh experimental conditions encountered in biotechnological/biological protocols, such as high ionic strength, high temperatures, presence of competing thiols or oxidizing agents, and solvents where the ligand is extremely well solvated [64,65]. This higher colloidal stability allows both the use of a wider range of methodologies for their biofunctionalization and their compatibility for their use in biotechnological and biomedical applications.

Other physical techniques, such as vacuum deposition, electrical circuitry dispersion, and laser ablation method, are also used to produce monodisperse AuNPs. Employing physical techniques allows the easy control of the shape; nevertheless, those methods are based on the use of expensive equipment, and/or complex preparation processes that have several potential drawbacks such as being expensive and time-consuming [66–69].

For the synthesis of asymmetric AuNPs it is worth mentioning that minimization of the surface energy concurs with the decrease of the surface area, and this is why the formation of spherical objects is favored. In order to control the formation of nonspherical AuNPs, several methodologies have been reported. They are usually based on the use of templates to give shape to the NPs, the use of adequate ligands to kinetically control the growth of certain facets of the seeds, and even the assembly of performed spherical NPs [70]. Among these methodologies the use of mesophase structures self-assembled from surfactants is very popular to promote the formation of AuNPs of various shapes. However, it has been reported that it is not so easy and reproducible to achieve a complete removal of the surfactants used during the synthesis. As it was mentioned previously, this could raise problems for NP's biofunctionalization and even cytotoxicity issues. This has led to a significant increase of methodologies that avoid the use of surfactants for the synthesis of anisotropic AuNPs. It is the case of the method described by

Pelaz et al. [1] for the production of gold nanoprisms that avoids the use of the most widely used surfactant cetyltrimethylammonium bromide (CTAB). In this synthesis, potassium iodide and thiosulphite are used to reduce the $HAuCl_4$. The created asymmetric NPs are then stabilized directly with bifunctoinal PEG polymers having a thiol on one end and a functional group (amine, carboxyl, etc.) at the other end. As most of the reported synthetic methods have been developed recently, there is a lack of information about important factors such as reproducibility, purity of the product, potential for scaling-up, cost-effectiveness, etc. Besides, most of them lack the ability to afford particles of varied shapes.

As it was mentioned in the previous section, superparamagnetic Fe_3O_4/Au core–shell NPs represent another type of next-generation NPs with particular promise as theragnostic agents for clinical applications. They combine the advantages of both superparamagnetic and plasmonic NPs. The Au coating of the magnetic core not only stabilizes more efficiently the NP in corrosive biological conditions, but also introduces NIR response and the well-developed functionalization chemistry of AuNPs. Direct coating of MNPs with Au, however, is a difficult task due to the dissimilar nature of the two surfaces. Usually, direct deposition of Au on MNP surface by iterative hydro-xylamine seeding [71] or by reducing $HAuCl_4$ in a chloroform solution of oleylamine [72] results in either poor shell morphology or failure in producing small-sized NPs with NIR response. A common way to make core–shell NPs is still reliant upon the modified seed-mediated growth approach, resulting in large NPs of 100–200 nm in diameter with uneven surfaces [73,74]. Recently, it was reported a new strategy to obtain this MNP–gold core–shell NPs with NIR response but smaller size (31–33 nm). The key aspect consisted in the creation of a 3-nm gap between the core and the shell. This way the resultant NP not only was superparamagnetic with both NIR and magnetic resonance imaging responses but also enabled a new modality of imaging (magnetomotive photoacoustic, mmPA). To achieve this, monodispersed MNPs with oleic acid as surface ligand were first transferred to water using phospholipid-PEG terminated with carboxylic acid (PL–PEG–COOH). The hydrophobic PL segment interdigitates with oleic acids through hydrophobic interactions, whereas the PEG block facing outward renders MNPs water-soluble and negatively charged due to the terminal carboxylic groups. To create anchor points for gold shell growth, a layer of positively charged peptide, poly-L-histidine, was adsorbed onto the outer surface of MNP–PEG through charge–charge interaction at pH 5–6. As histidine-containing peptides have the capacity of metal ion chelation, Au^{3+} ions could be immobilized on the NP surface and thus their further reduction leads to the formation of the Au shell [48].

In the case of other class of photoactive NPs such as AgNPs, they are usually synthetized with "bottom-up" chemical methods including chemical or electrochemical reduction, thermal decomposition, sono-decomposition, laser irradiation, laser ablation, and lithography [75–84]. Among them, chemical reduction is one of the most used. As in the case of AuNPs, it involves two stages (a nucleation and a subsequent growth stage), and employs three main components (metal precursors, reducing agents, and stabilizing/capping agents). Besides, they could be also obtained by "top-down" methods, as it is the case of mechanical grinding of bulk metals with subsequent stabilization using colloidal protecting agents [71,85,86]. Both in the case of AuNPs and AgNPs, the advantage of the chemical synthetic methods are the ease of production, lower cost, and higher yields; however, the use of chemical-reducing agents is not only harmful to living organisms [87] but also could affect the biofunctionalization of the photoactive particles. Therefore AgNPs produced with this method need further steps of purification to excise out these chemicals. To overcome the shortcomings of chemical methods, biological methods have emerged as viable options. Recently, biologically mediated synthesis of NPs has been shown to be simple, cost effective, and environmentally friendly approach. Thus, much attention has been given to the development of high-yield production strategies of AgNPs of defined size using various biological systems including bacteria, fungi, plant extracts, and small biomolecules like vitamins and amino acids as an alternative method to chemical methods. This alternative has been explored not only for AgNPs, but also for the synthesis of several other NPs, such as gold and graphene [87–94]. The major advantage of biological methods is the availability of amino acids, proteins, or secondary metabolites present in the synthesis process, which allows to eliminate the extra step required for the prevention of particle aggregation in chemical synthetic processes. However, the synthesized nanomaterials are usually obtained as aggregates and not as monodisperse and uniform NPs [95]. If you want to go deeper into the subject, we suggest a recent article of Nadeem et al. that explained a detailed account of synthetic methods, properties, and bio-application of AgNPs [96].

QDs are also widely used in nanomedicine. Although traditional QDs have toxicity issues due to the presence of heavy metals, the recent development of heavy metal-free/cadmium-free biocompatible QDs has renewed the expectation on their potential as nanotheragnostics platforms for simultaneous sensing, imaging, and therapy. Besides, it has been also shown that in the case of QD containing heavy metal, functionalizing the QD surface with biocompatible molecules could ensure minimal toxicity [97].

Similarly to other photoactive NPs, they are synthetized also utilizing chemical and physical methods. Physical methods usually involve nucleation and growth of particles in the vapor phase or in solution. However, the most feasible way for physical preparation of semiconductor QDs is molecular beam epitaxy. This approach involves deposition of atomic or molecular beams produced in some sources on crystalline substrates, with atomically smooth surface, in ultrahigh vacuum conditions. However, the method is not reliable for high semiconductor QD production. It requires sophisticated equipment and high-purity starting materials.

Regarding the chemical methods, there are several described. For example, in the case of CdSe/ZnS QDs, mercaptopropionic acid is widely used for displacing the original trioctylphosphine oxide layer. This approach takes advantage of the fact that mercapto groups can bind easily to zinc atoms. Moreover, it introduces carboxyl groups that could be directly used for their functionalization with aminated biomolecules, or further modified in order to provide reactivity with other functional groups. This is the case for 7-octenyl dimethyl-chlorosilane, which introduces octenyl double bonds able to react with alkylidene groups. Not only mercaptopropionic acid but also small cysteine-rich proteins have been used for their functionalization as they can bind up to monovalent or up to seven divalent heavy metal ions [98,99]. Besides the use of thiol-containing molecules, other approaches have been reported for introducing and controlling the composition of the ligand shell surrounding the QD's surface. A reported example consists in growing an organic polymer corona onto the QD surface using the bottom-up approach of surface-initiated ring-opening metathesis polymerization with Grubbs catalyst [72].

Other class of photoactive NPs is represented by MSNPs loaded with photosensitizers as described in the previous section. The synthesis of SiNPs is based on the use of the surfactant CTAB together with a silica precursor TEOS and sodium hydroxide, which is used to modify the pH of solution for triggering nucleation. The stabilized sample obtained after synthesis is extracted with an alcoholic solution of NH_4NO_3 and annealed in air at 600°C for several hours in order to remove the surfactant from the pores. After this step, the clean surface of SiNPs can be chemically modified usually by silanization in order to introduce chemical groups that enable binding of biomolecules and photoactive dyes (e.g. porphyrins, phthalocyanines) or NPs [48,100].

Liposome-templated gold nanoshells nanocontainers represent another useful "soft" core material for preparing functional photoactive core–shell NPs for biomedical applications. Currently there are two general formats for preparing cargo-loaded plasmonic vesicles. In one of them, small-sized AuNPs are combined with liposomes (e.g., by adsorbing on, forming large aggregates with, or embedding inside liposomes) loaded with cargo molecules. In another one, the liposome-templated AuNS nanocontainers are assembled from amphiphilic AuNPs with mixed polymer brush coatings. The integration of AuNPs with two types of chemically different polymer grafts, which are analogs to block copolymers as a whole, creates a new type of hybrid building block inheriting the amphiphilicity-driven self-assembly of block copolymers to form vesicular structures, and the plasmonic properties of the AuNPs. A key limitation, however, of such satellite AuNP-vesicle assemblies is that leakage or degradation of cargo molecules is unavoidable.

The sizes of the nanocontainers are easily tuned from 10 to 70 nm by varying the liposome size. This design has several important features that benefit for bioapplications. For example, drug/gene release can be spatially confined to targeted sites and help reduce toxicity to other cells or organs; and the release profile can be precisely controlled for optimal dosage using laser illumination at various powers [101]. The drawback of the system at the present stage lies in its scale up due to purification issues of the liposomal system, although it might be addressed by combined use of, for example, a microfluidic method [102].

9.4 Bioconjugation of photoactive nanoparticles: how to select the most adequate methodology?

Selecting the most adequate biofunctionalization strategy is no mean feat, as no universal methodologies exist to cover the wide variety of NPs and biomolecules available for this purpose. In this sense, a functionalization protocol that works well for a particular NP may not work for another as they could be very different in terms of size, surface area, colloidal stability, density, and type of reactive groups, etc. Furthermore, the biomolecules to be joined to the surface of the material could vary significantly in terms of size, chemical composition, 3D complexity, location of its biological active site, and so forth. Consequently, there is not a universal methodology for the immobilization of any biomolecule. But despite this chaotic scenario, there are some useful tips that could help to select the most adequate functionalization protocol of the nanomaterial if we (1) focus on the biomolecule to be linked and (2) take advantage of the existing experience previously acquired when optimizing the biofunctionalization of microstructured materials.

Clearly, the simplest immobilization method for linking biomolecules and nanomaterials is nonspecific adsorption, which is mainly based on physical adsorption or ionic binding [103,104]. When favoring physical adsorption, the biomolecules are attached to the material through hydrogen bonding, van der Waals forces, or hydrophobic interactions. However, when favoring ionic binding, the biomolecules are bound through ionic linkages. It is true that hydrophobic adsorption is energetically more favored than ionic binding because water molecules are released from the surface and from the biomolecule, which leads to a large entropy gain. The facility and versatility of these adsorption processes is based on the fact that it is not necessary to chemically modify the biomolecule to be attached or to activate the functional groups of the NP. In the case of antibodies, it has been recently showed that favoring their ionic adsorption could ensure even an oriented binding to the NP surface [105]. However, as both are noncovalent immobilization processes, the interactions between the biomolecules and the nanomaterial are relatively weak. In fact, they could be reversed by changing the conditions that influence their strength (e.g., pH, ionic strength, temperature, or polarity of the solvent) or even other molecules in biological samples can replace them [53]. In this sense, it is important to highlight that it is possible to tune the strength of protein–NP ionic adsorption interaction by grafting multicharged polymers on the NP surface [106]. It is well known that a multicharged flat surface only permits an interaction with a protein surface area not larger than 15%–20%. However, grafting multicharged polymers on surfaces allows the interaction with a larger percentage of the protein surface. This is a consequence of the promotion of a very intense 3D multipoint ionic adsorption as the surface is not planar and the protein may penetrate within the grafted polymer matrix. This strategy not only allows to increase the binding strength but also allows the adsorption of proteins that could not be adsorbed on conventional supports as they have many charged groups dispersed around their surface instead of having a very rich charged area [107]. This is why coating NPs with multinegative or positive polymers is so popular [108–110]. Other weakness of adsorption strategies is the need of high concentration of the selected biomolecule for the preparation of the biomolecule–NP conjugate. This could be an important disadvantage when the biomolecule to be linked is expensive (e.g., antibodies) [111]. Besides, in the case of hydrophobic adsorption, extensive denaturation usually occurs when the biomolecule to be linked is a protein. This is because hydrophobic interactions force a thermodynamic change in the protein native 3D- structure to allow the hydrophobic portion of the protein to interact with the substrate surface. As a consequence, there is a time dependent and irreversible loss of the biological activity of the attached biomolecule. Therefore although these methods are very simple and economically attractive, they should be selected whenever biomolecule leakage from the nanomaterial must be avoided. As an example, Conde et al. have shown how small interference RNA (siRNA) bounded to the AuNPs ionically has less efficiency on eliciting *c*-myc downregulation in vivo than AuNPs to which a thiolated-siRNA was directly bound to their surfaces (35% vs 65% of downregulation) [112].

Noncovalent, affinity-based biofunctionalization strategies allow the establishment of a rapid and extremely stable bond with the NP, as reported for the well-known avidin–biotin systems [53]. Though the established bond is not covalent, its dissociation constant is of the order of 10^{15} M and thus it is extremely stable, resisting harsh chemical conditions and even high temperatures. This strategy has been also widely employed by researchers to develop NPs for application in medicine. Naturally, the use of this biomolecule binding strategy is necessary to functionalize the NP with the biotin binding protein followed by additional functionalization of the biomolecule with the binding partner (biotin). Usually, the biotin binding protein (avidin or its derivatives such as streptavidin, neutravidin) is coupled directly to the NP surface by adsorption or by covalent binding via its carboxylic or amine groups. As for the biotinylation of the biomolecules, a large variety of biotinylation reagents and even biomolecules like DNA oligomers, peptides, antibodies, and fluorescent dyes already biotynilated could be used and are commercially available. However, although the (strept) avidin–biotin system is simple to set up and use, it does have certain limitations. Because biotin is a biological molecule, endogenous biotin can cause background and specificity issues when the target tissue or extracts are rich in biotin (e.g., brain, liver, etc.). Besides, when using this strategy for the attachment of antibodies, the tetrameric nature of (strept) avidin becomes a problem if the Ab stoichiometry needs to be controlled. To overcome this problem, it is possible to use recombinant monomeric forms of these proteins, but taking into account that the affinity for biotin is much lower (K_d around 10^{-7} M) [113]. A similar well-known system is the binding of recombinant proteins via a polyhistidine tag, commonly introduced by six or more histidine residues to one of the terminal ends of a protein. Polyhistidine binds strongly and specifically to nitrilotriacetic acid (NTA) via a chelation complex with Ni^{2+} or other bivalent metal ions. Originally, this strategy has focused on its application for purification of recombinant proteins. However, its use has also been extended for biofunctionalization of NPs [114,115], as it is quite simple to implement as His-tagged proteins and peptides are routinely prepared using well-established molecular engineering techniques. Even it has been used for site-specific binding of AuNPs to the active site of enzymes in order to improve the electrochemical communication between gold electrode and the metal active site of a redox metalloenzyme (galacto oxidase) [116]. More recently, it was also shown that the polyhistidine tag can directly bind to CdSe/ZnS QDs that were not functionalized with NTA or carboxylic groups on their surface [117]. This categorized this strategy as a self-assembly approach for certain types of NPs.

Other noncovalent binding strategy for biomolecules containing thiol groups is their direct attachment to the surface of the inorganic particle core usually known as "ligand-like binding." It is usually driven by chemisorption of, for example, thiol groups. It is a relatively straightforward one-step functionalization strategy for biomolecules of low 3D complexity (nucleic acids, peptides, sugars, etc.) as they can be synthesized or are commercially available with a large variety of functional end groups, including thiol groups [118–121]. However, even biomolecules with large 3D complexity such as proteins containing cysteine residues at their surface have been attached with this strategy [122]. Although the thiol–Au interaction is stronger than nonspecific adsorption, its dissociation is possible allowing an exchange with thiol-containing molecules [123]. This has been exploited for the release of biomolecules inside cells due to the reducing intracellular environment [124]. This intracellular release of thiotethered ligands from the surface of AuNPs is mediated by glutathione, an abundant thiol-containing molecule found in the millimolar range in the cytoplasm of healthy cells [125]. This has to be taken into account as it could be exploited for the release of therapeutic agents (drugs, siRNA, etc.) but if this is not the final aim of the functionalized NP, this substitution of the NP's functionality could lead to the loss of the NP biological activity. Note that in order to be able to control the number of the bound biomolecules, it is necessary to dilute the biomolecules to be attached with other thiolated molecules that act as "dilutant" (e.g., thiolated PEG, etc.) [126]. Nevertheless, these "dilutant" molecules could have an effect of steric hindrance on the availability of the biomolecule for its cleavage into the intracellular environment if this was needed [125]. Other aspect to highlight, especially for photoactive NPs, is that temperature increases or laser illumination associated with the use of AuNPs for phototherapy can also disrupt thiol bonds, leading to the loss of NP activity [127].

Thus covalent functionalization of NPs whose surface has already been derivatized, significantly expands the range of their potential uses under physiological conditions, compared with traditional noncovalent or thiol-based approaches. However, covalent coupling methodologies are not as straightforward as physical or ionic adsorption. First, it is necessary to have adequate functional groups at the NP surface that can be used for the conjugation of the biomolecule of choice. These functional groups could be present on the NP surface as a consequence of their synthetic procedure, or be introduced in a second step via ligand exchange reactions with small bifunctional ligands or polymer coatings. Then an additional activation step is needed to generate electrophilic groups on the NP or in the protein, which in the coupling step react with strong nucleophiles in the protein or in the nanomaterial, respectively. In this sense a large number of crosslinking and modification methods are available. The most used ones allow (1) the introduction of carbonyl reactive groups which reacts with hydrazide or alkyoxyamine to form hydrazone or oxime bond; (2) the activation of carboxylate groups in order to react with primary amine groups of the biomolecule to form amide bonds;(3) the activation of thiol groups to generate sulfhydryl reactive groups (maleimide, haloacetyl, pyridyl disulfide) to form thioester, or disulfide bonds; and (4) the introduction of azide groups that via an orthogonal reaction known as click chemistry reacts with phosphine or alkyne to form an amide bond or a triazole ring (Table 9.1) [128].

However, not all chemistries ensure the binding of the biomolecule with a correct orientation and desired surface density. In particular, this has been clearly reported when the biomolecule to be attached is a protein (antibodies, enzymes, etc.) due to their greater structural complexity. All proteins are built from series of up to 20 different amino acids with side chains of different sizes, shapes, charges, and chemical reactivity. The side chains of the following amino acids are the most frequently used for covalent binding to NPs: lysine (ε-amino group), cysteine (thiol group), aspartic and glutamic acid (carboxylic group), and tyrosine (phenolate anion). Hence to be able to control the orientation of the attached biomolecule, it is important (if possible) to know the number and location on the protein's surface of the

TABLE 9.1 Type of covalent conjugation chemistries mostly used for nanoparticles biofunctionalization.

Type of covalent conjugation	Linkage	Stability under physiological conditions
Hydrazide–aldehyde	Hydrazone	Stability under physiological conditions
Amine–carboxyl	Amide bond	Stable
Thiol–maleimide	Thioether bond	Stable
Thiol–Thiol	Disulfide bond	Cleaved under reduction conditions
Click chemistry	Triazole ring	Stable

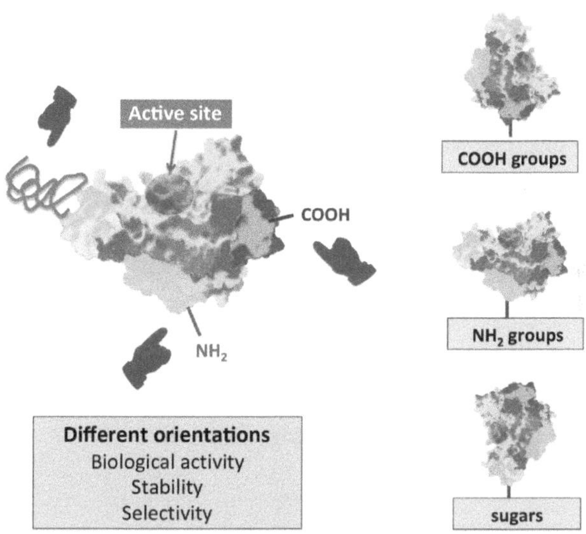

SCHEME 9.2 Representation showing how the attachment through functional groups localized at different regions of the protein surface could lead to different orientations of the attached protein.

abovementioned amino acids as well as the free amino and carboxyl groups of the N- and C-terminal residues. This will ultimately depend on the specific amino acid composition of the protein which determines the sequence location of the individual amino acids in the three-dimensional (3D) structure of the protein. Depending on the functional group of the biomolecule used for the binding, different molecular orientations of the protein would be attained (Scheme 9.2). This has been the focus of a great amount of research in the field of biofunctionalization of microstructured materials for biotechnological and medical applications. Thus it is well known that this not only could affect the biological activity of the protein (antigen binding capacity in the case of antibodies, selectivity in the case of enzymes) but also its stability against different denaturing agents (temperature, pH, organic cosolvents, etc.) [129,130].

In the case of antibodies, their immobilization on nanostructured materials is an essential process for the development of immune-based detection systems and active targeting of nanotherapeutics. The most used class of antibody for NP functionalization is IgG (IgGs, ~150 kDa, 143 × 77 × 40 Å), which consists of two light and two heavy chains, linked by disulfide bonds to form a characteristic Y-shape structure. Two identical antigen binding sites are localized at the end of the arms of the Y. These two antigen binding ends are called the Fab fragments (for antigen binding fragment). The stem of the Y is the Fc fragment (for crystallizable fragment) (Scheme 9.3). It is well known that a bad choice of the immobilization strategy could impair the antigen-recognition capacity of the antibodies attached to the NPs [111,131]. Thus the use of strategies that maintain the antigen binding sites available for antigen capture ensures good selectivity for the target analyte and high sensitivity levels in the case of the use of the antibody functionalized (Ab@NPs) for biosensing [132–134].

Immobilized IgG can adopt four possible molecular orientations: side-on (one Fc and one Fab attached to the surface), end-on (Fc attached to the

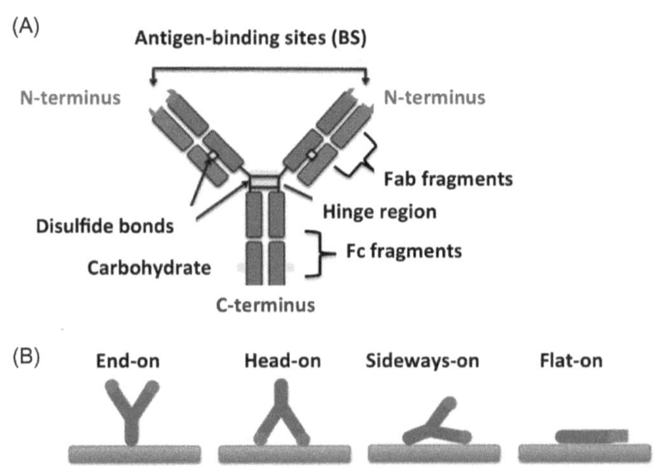

SCHEME 9.3 (A) Schematic cartoon showing the Y-shaped structure of an antibody. The light chains (variable regions) and the heavy chains (constant regions) are colored in light grey and dark grey (violet and red in online version), respectively. The antigen binding sites are drawn in green (online version), while groups that can be used for attachment to NPs are drawn in yellow (online version). (B) Possible orientation of antibodies immobilized on surfaces. *Adapted with permission from R.M. Fratila, S.G. Mitchell, P. Del Pino, V. Grazu, J.M. De La Fuente, Strategies for the biofunctionalization of gold and iron oxide nanoparticles, Langmuir 30 (2014) 15057–15071.*

surface), head-on (both Fabs attached to the surface) or flat-on (all three fragments attached to the surface) (Scheme 9.3). The end-on orientation clearly ensures that the antigen binding sites of the immobilized antibodies will be properly oriented as they will be well exposed to the solution phase. Recently, it has been demonstrated that "flat-on" orientation neither alters the antibody's antigen binding activity [105]. Thus both orientations will exhibit higher analyte binding capacities than randomly oriented antibodies ("head-on" and "side-on" orientations) [135]. In this sense, it has been shown that steric hindrance of antigen-recognition sites when randomly immobilized, could cause even a 1000-fold decrease of the binding affinity of an antibody (in comparison with its soluble counterpart) [136].

For all the abovementioned reasons, it is clear that controlling the orientation is a key aspect for the functionalization of NP with antibodies. Their immobilization via covalent coupling procedures provides advantages over physical adsorption such as higher stability of the bioconjugate, better control of the number and orientation of the attached antibody molecules, and better reproducibility. But above all, thinking on an in vivo use, the clear advantage is that they prevent the competitive displacement of adsorbed antibodies by blood components. The most used covalent binding strategy is its direct coupling through the primary amines of the Ab. Its popularity is based on the fact that this is one of the most abundant functional groups on the Ab surface. Moreover, these functional groups are very reactive with a wide variety of amine-reactive groups on the NP surface without any previous activation (Scheme 9.4). However, this is the less advisable covalent coupling strategy in terms of ensuring a correct orientation of the attached Ab

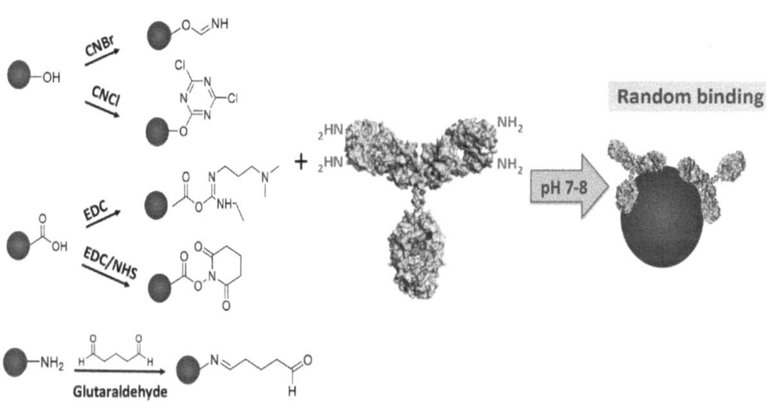

SCHEME 9.4 Different amine-reactive chemistries that could be used for coupling the Ab via its more reactive primary amines to NPs containing either hydroxyl (OH), carboxyl (COOH) or amine (NH_2) functional groups. If the pH of incubation is 7–8, the most reactive primary amines are localized at the Fab region and thus a random binding is promoted. *CNBr*, Cyanogen bromide; *CNCL*, cyanogen chloride; *EDC*, *N*-ethyl-*N*-(3-dimethylaminopropyl)carbodiimide; *NHS*, *N*-hydroxisuccinimide.

molecules. This is because amine-reactive linkers are unstable at alkaline pH values, thus mild pH conditions (pH 7–8) should be used in order to ensure enough stability of these reactive groups so that covalent reaction could take place. At this pH range, the most reactive primary amine present on any protein is the amino-terminal group (pK_a around 7–8) of polypeptide chains [note that the pK_a of the ε-amino of lysine residues (10.5–10.7) is extremely high]. Four polypeptide chains form Abs molecules. Thus they have four amino-terminal moieties and all of them are placed at the antigenbinding sites (Fab regions). Therefore this functionalization chemistry leads to their random immobilization. However, tuning the pH of reaction conditions allow transforming this amine-reactive chemistry from a random functionalization strategy into an oriented one (see Scheme 9.4). In the case of carboxylated NPs, if the pH of incubation with the amino-reactive NPs is lower than the isoelectric point of the antibody, it is possible to exploit the high density of positive charges on the major plane of antibodies. Thus an ionic preadsorption of the antibody takes place, which fixes a flat-on orientation of the antibody before a site-directed covalent reaction occurs with ε-amino of lysine residues now facing the NP surface [105].

Other covalent coupling chemistries that also allow for an oriented binding of antibodies need previous chemical modification of the Ab such as partial reduction of its disulfide bonds or oxidation of its sugar moieties (Scheme 9.5). The disulfides in the hinge region are the most susceptible to

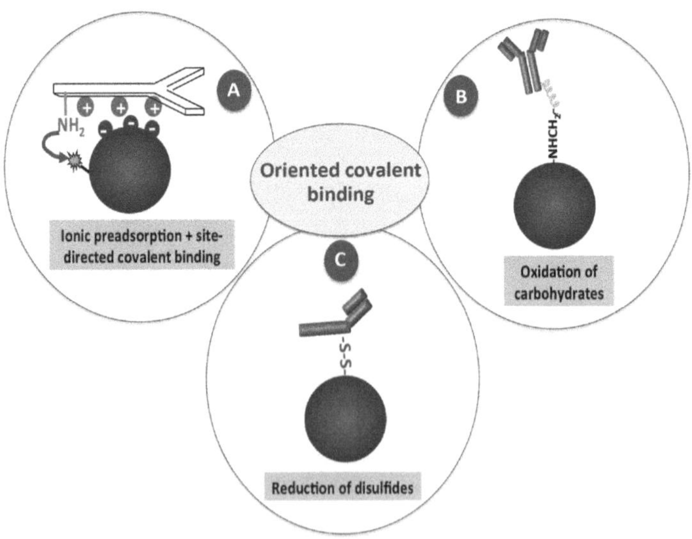

SCHEME 9.5 Different Ab-oriented covalent coupling strategies: (A) a two-step procedure using "amine-reactive" chemistry where first a ionic adsorption of the antibody is promoted, and then site-specifically covalent binding to the amine groups facing the NP surface occurs. (b) Binding through the sugars localized at the Fc region of the antibody; (c) binding through thiols obtained after reduction of disulfide bonds present at the hinge region.

reduction. So, it is possible to selectively cleave only these disulfides, thus splitting the Ab into two monovalent halves without altering their 3D structure and antigen binding efficiency. Thiol reactive groups (maleimide, iodoacetyl, 2-piridyl disulfide, etc.) on the NP surface of the Ab fragments can be selectively immobilized through site-specific localized thiol groups. This strategy has the advantage that it ensures an oriented immobilization ("end-on" orientation), but it has the drawback that requires a previous chemical modification of the Ab. The use of the sugar moieties of the Ab also ensures an oriented Ab immobilization (end-on orientation) taking advantage of the location of these carbohydrate chains at the Fc region far away from the antigen binding sites. However, they have to be first mildly oxidized by the use of sodium periodate to create reactive aldehydes (−CHO) for ensuring covalent coupling. Then, the Ab's aldehyde-activated (oxidized) sugars can be directly coupled to NPs containing primary amines through reductive amination or to NPs that have been activated with hydrazide groups [135].

In the case of enzymes, even although their activity may be quite preserved, it is well known that immobilization by covalent attachment usually leads to structural modifications [137]. This could have a negative effect on the catalytic activity (e.g., by denaturation of the 3D structure or a bad orientation of the enzyme active site) [138,139]. However, it has also been extensively reported, and even an enhancement of the enzyme activity has been observed due to the rigidification of the protein structure by its multipoint covalent attachment [130]. Moreover, it has been clearly established that these observed effects arising from their multipoint covalent attachment depend on the specific involved protein areas. In the case of the usually observed improvements of enzyme stability, even though multipoint covalent attachment produces a global rigidification of the enzyme structure, the stabilization effect is more relevant when the area of the enzyme structure where the inactivation starts is the one directly rigidified. In this sense, site-directed mutagenesis combined with multipoint covalent immobilization showed to be the most-interesting strategy to stabilize an enzyme via site-specific rigidification of the most labile regions of its structure. Following this functionalization strategies, the highest stabilization factors for several enzymes have been achieved (i.e., 1500 times against temperature for penicillin G acylase) [140]. However, to achieve this type of control of the interaction between the enzyme and the support, it is necessary to introduce complex functionalization chemistries involving the introduction of several functionalities on the surface of the material to be functionalized (heterofunctional supports) [141,142]. The use of this functionalization strategy has also showed the importance of a site-directed rigidification to gain control toward the modulation of enzyme selectivity and specificity. This effect has been observed mainly when an enzyme has a flexible active center, like lipases, penicillin G acylase, or multimeric enzymes. Thus this effect could be explained due to a rigidification of certain areas of the enzyme that avoid certain conformational changes which produce severe alterations of the enzyme catalytic properties. This is clearly exemplified with the results obtained by the site-directed immobilization of a thiolated variant of *Geobacillus thermocatenulatus* (BTL2) on tailor-made disulfide-aldehyde agarose microbeads. The enzyme was first site directly attached via disulfide exchange through a cysteine residue genetically introduced in a position ($\theta - 332$) opposite to its lid (active site). Then, an additional rigidification was introduced through the interaction of the reactive amino groups on the protein near to the position of the introduced cysteine with aldehydes on the support. The soluble enzyme was not selective for: (1) the kinetic resolution of rac-2-*O*-butyryl-2-phenylacetic acid to produce (*R*)-mandelic acid, and (2) the asymmetric hydrolysis of phenylglutaric acid dimethyl diester to produce (*S*)-monomethyl-phenylglutaric acid. However, the site-directed rigidification of this region of the lipase structure allowed obtaining a complete selective catalyst ($e > 99\%$) for both enantiomeric biotransformations [143].

In light of the importance of the zone of the enzyme surface that is attached to the support to be biofunctionalized, is this being taken into account for the biofunctionalization of NPs? Immobilization of enzymes onto NPs is a growing field of research [144–150]. When enzymes are immobilized on NPs they often exhibit outstanding stability against high temperature. This is a consequence of a structural confinement of the enzyme due to the size of the support that facilitates multipoint attachment with the NP's surface [151]. Moreover, their immobilization onto NPs also offers improvements on enzyme activity due to the inherent physicochemical properties of NPs: (1) higher enzyme density and higher localized avidity, (2) enhanced mass transport due to the absence of internal diffusion limitations and Brownian motion in solution, and (3) increased surface curvature, which allows for increased center-to-center distances between adjacent immobilized enzymes while limiting unfavorable protein-to-protein interaction [152]. Unfortunately, the previous experience with microstructured materials is not taken into account in most of the reported examples of rationally engineered functionalization protocols. In fact, controlling orientation during NP functionalization is really difficult because not all NPs resist the functionalization conditions needed for the implementation of site-directed immobilization strategies used with microparticles. Thus simple immobilization covalent procedures are usually used where orientation and/or site-specific rigidification of the enzyme molecules is not taken into account. Therefore resulting nanobiocatalysts based on further engineering of the NPs–enzyme interaction will ensure the development of more practical and progressive applications of enzyme technology not only for biotechnological and medical applications (e.g., site-specific prodrug conversion, in vitro and in vivo diagnosis, etc.).

From all the abovementioned information, it is clear that the choice of the conjugation approach is vital regarding successful bioactivity of the biofunctionalized NP. As it has been explained, the orientation of the biomolecule on the NP surface should be a leading factor to select the conjugation methodology. Note that even for the same biomolecule more than one protocol that results in its oriented binding could be available. The right methodology for a NP would depend on the chemical functionalities anchored to the NP surface and its colloidal stability. Once the desired orientation is selected focusing on the type of biomolecule to be attached, it is necessary to ensure the NP stability under the reaction conditions (pH, ionic strength, activation of NP's functional groups with heterofunctional linkers, etc.) needed for a specific functionalization strategy. We could focus on the binding of antibodies through carbodiimide chemistry to show this additional critical aspect. First, it was necessary to promote an ionic adsorption of the antibody for an oriented binding. When using carboxylated NPs, the Ab thus has to have a net positive charge to ensure this ionic attraction. Therefore an incubation buffer with a pH lower than the Ab's isoelectric point (pI), at least 0.5 units below is required. Due to the pK_a of carboxylic moieties ($\sim$4) on the NP surface, these type of NPs lose net negative charge and thus colloidal stability at very acidic pH values. Thus if the Abs pI is alkaline or neutral it will be possible to ensure its ionic adsorption at pH values where the carboxylated NPs are stable. However, if the Ab has a very acidic pI this oriented functionalization strategy could not be used with this type of NPs [105].

In conclusion, outstanding control of the protein orientation at the NP surface can be accomplished by taking into account surface topography and chemical properties of the NPs together with protein structural details.

9.5 Importance of a good characterization

There is no doubt that it is possible to engineer the NP–biomolecule interphase with great precision. Equally important, there is a need of a careful characterization of the biofunctionalized NP to make sure that its biological outcome is the consequence of the selected biofunctionalization strategy. Thus it is mandatory to describe the number of attached biomolecules, their orientation, the effect of the attachment on the biomolecule mobility, long-term effects of the surrounding surface chemistry on the bounded biomolecule, etc. This is the only way to take a step back, if needed, to improve the NPs functionalization, and also to achieve the best control and reproducibility of the bioconjugation process. This is why we considered important to include this section, where different techniques that could be used for following up the biofunctionalization process or the characterization of the biofunctionalized NP are discussed.

9.5.1 Use of microscopies

A wide variety of optical and electronic microscopies are routinely used to characterize the functionalization of NPs focusing, in most cases, on changes in their final size and size dispersion.

9.5.1.1 Electronic microscopies

In particular, transmission electron microscopy (TEM) is usually used for determining the size of NPs. TEM provides direct images and chemical information of nanomaterials at a spatial resolution down to the level of atomic dimensions (<1 nm) [153,154]. The higher resolution is based on beams of accelerated electrons with much shorter wavelengths than visible light photons. The high spatial resolution of TEM enhances the morphological and structural analyses of nanomaterials. A wide variety of analytical techniques can be coupled to TEM for a deeper characterization of the nanomaterials; for example, chemical analyses based on electron energy loss spectroscopy and energy dispersive X-ray spectroscopy can quantitatively investigate the electronic structure and chemical composition of the nanomaterials. However, this microscopy is rather suitable for determining the core size of inorganic NPs but less for organic NPs or the organic layer provided by the biomolecules attached and the protecting shell. This is a consequence of their poor contrast in TEM due to the smaller electron density of organic molecules. A way to overcome this is by the use of staining protocols with heavy elements [155]. Two common widely used contrast agents are uranyl acetate and lead citrate, which enhance the contrast by interaction with lipids and proteins, or proteins and glycogens, respectively. Other important disadvantage of this technique is that the samples need drying for observation under vacuum. This could introduce artifacts that may cause shrinkage of the sample and alter the characteristics of the nanomaterials.

Another electron microscopy widely used is scanning electron microscopy (SEM). Unlike TEM, SEM is a surface-imaging method in which the incident electron beam scans across the sample surface. The interaction with the sample generates signals reflecting the atomic composition and topographic detail of its surface. Similarly to TEM, the size, size distribution, and shape of nanomaterials can be directly acquired from SEM taking into consideration the artifacts that could be introduced due to the process of drying and contrasting samples. In addition, while scanned by an electron beam, many biomolecules that are nonconductive specimens tend to acquire charge, which leads to imaging faults or artifacts. Thus biofunctionalized NPs are usually coated with an ultrathin layer of an electrically conducting material (e.g., gold) during the sample preparation procedure. This could be avoided by using environmental SEM (ESEM). Samples can be imaged in their natural state (without any modification or preparation requirement), as the sample chamber of ESEM is operated in a low-pressure gaseous environment (10–50 Torr) and high humidity. This eliminates the abovementioned charging artifacts and thus the coating of samples with a conductive material is no longer needed [156]. However, this microscopy still presents other disadvantages of electron microscopies, such as its destructive character (prohibiting its analysis by other modalities) and biased statistics of size distribution of heterogeneous samples due to the unavoidable small number of particles observed per analyzed region of the grid containing the sample.

9.5.1.2 Fluorescence microscopy

Among microscopy techniques, also, fluorescence microscopy is used to characterize fluorescent or fluorescent-labeled NPs. Parracino et al. have provided an example of the use of this technique to characterize the functionalization of Fe_3O_4@Au core−shell NPs with fluorescent proteins [157]. They described the immobilization of fluorescent-labeled bovine serum albumin (BSA) onto the NPs, and the characterization of the functionalization also through fluorescence microscopy images. Besides, site-specific fluorescent labeling of certain regions of a protein surface also enables the study of the orientation of the biomolecules onto the NP's surface. In the particular case of BSA, used as a model carrier protein to study the transport of hydrophobic drugs, functionalization protocols that keep exposed its hydrophobic region are preferred. In this sense, Parracino et al. showed an easy way to check if this is ensured by fluorescence microscopy. They have used a fluorescent dye (8-anilinonaphthalene-1-sulfonic acid, ANS) that becomes fluorescent upon binding hydrophobic regions of BSA. Once the functionalization of the NPs was performed, they incubated the NPs-BSA with ANS and monitored the fluorescent of the obtained NPs. If the functionalization strategy did not alter the hydrophobic regions of BSA, ANS has to become fluorescent upon binding to the BSA attached to the NPs.

9.5.1.3 Atomic force microscopy

Other microscopy that is gaining importance for the characterization of biofunctionalized NPs is atomic force microscopy (AFM). It is a very high-resolution type of scanning probe microscopy, consisting of a micromachined cantilever with a sharp tip at one end. Most AFMs use a laser beam deflection system that detects the deflection of the cantilever tip caused by electrostatic and van der Waals repulsion, as well as attraction between atoms at the tip and on the sample's surface. The major difference between AFM and other microscopies (optical and electron microscopy) is that it does not use lenses or beam irradiation. Therefore it does not suffer from a limitation in spatial resolution due to diffraction and aberration that allows generation of images with a vertical resolution of around 0.5 nm (more than 1000 times better than the optical diffraction limit). AFM can also be used for investigating the size, shape, dispersion and aggregation of NPs (as SEM and TEM do) with the advantage that it could be used in aqueous fluids under physiological conditions. This permits its use for the characterization of dynamics between biofunctionalized NPs in biological situations such as their interaction with supported bilayers in real time [153]. However, it is also important to highlight that a major drawback of this technique is that the size of the cantilever tip is generally larger than the dimensions of the analyzed nanomaterial. This leads to unfavorable overestimation of the lateral dimensions of the samples [158]. Besides, the number of particles analyzed by AFM is much smaller than other techniques (e.g., dynamic light scattering, DLS) that could provide better size distribution and polydispersity index.

9.5.2 Dynamic light scattering and fluorescent correlation spectroscopy

DLS is one of the most widely used physicochemical methods to characterize NPs in terms of hydrodynamic size, shape, and aggregation state [159,160]. In this sense, it is able to analyze the size distribution of particles alone and particles modified with polymers, biomolecules and so on, using a monochromatic light source (e.g., a laser) [153,160]. Particles in a colloid dispersion scatter an incident laser and the intensity of the scattered light is detected. The continuously mobile particles within dispersion cause constructive and destructive interferences and hence, the intensity of scattered light fluctuates over time. Thus the principle of DLS is based on monitoring the temporal fluctuation of the elastic scattering intensity of light [161]. The main strengths of DLS include its noninvasive manner, short experiment duration (minutes), and accuracy in determining the hydrodynamic size of monodisperse samples, which are very important when determining the hydrodynamic diameter of functionalized NPs. However, the functions used for correlating fluctuations of intensity over time with NP's size display several disadvantages, such as difficulty in correlating size fractions with a particular composition when aggregates are present. Then it is not possible to distinguish between dust particles and the particles to analyze. Thus an important aspect to be considered when using this technique is that samples to characterize should be pure and not contaminated. Besides, DLS works over a relatively small range of particle sizes (1 nm to 3 μm), although this scale limitation is not really a pitfall for the comparison of naked NPs with the biofunctionalized ones. In this sense, the hydrodynamic diameter of NPs increases during the process of functionalization, due to the presence of other biomolecules at NPs interface that scatter differently the light. For example, Rischitor et al. reported a significant increase in the hydrodynamic diameter of AuNPs when they are simply in contact with cell culture media due to the unspecific adsorption of proteins [162].

Fluorescence correlation spectroscopy (FCS) together with DLS, is an ideal tool for characterizing the size, and, most importantly, the size distribution and polydispersity of a supramolecular nanostructure and for measuring molecular diffusion and under extremely dilute conditions. It is usually used to follow-up QDs or other fluorescent NPs functionalization as they suffer changes in their fluorescence properties upon surface modification. Based on the analysis of the duration of brief bursts of photons from individual diffusing emitters during their brief sojourn in the observation volume of a confocal microscope, FCS can provide quantitative information on NPs functionalized biomolecules, demonstrated the interactions, specifically the increasing NP size due to molecules adsorption, binding affinity, and even the molecules orientation on the NP surfaces. For example, Li Shang et al., have studied the adsorption of human serum albumin onto small carboxylic acid-functionalized CdSe/ZnS QDs with an overall diameter of 16 nm by using FCS, measuring the thickness of the protein corona around QDs, of 3.3 nm, corresponded to a monolayer of proteins adsorbed in a specific orientation [163].

9.5.3 Nanoparticle tracking analysis

Nanoparticle tracking analysis (NTA) is an emerging technique for direct and real-time visualization, sizing, and counting of materials in liquid suspension based on properties of both, light scattering and Brownian motion. It is an innovative system for sizing particles ranging from 10–30 nm up to 1–2 μm in size, with the lower detection limit being dependent on the refractive index of the NP [164,165].

In order to obtain the particle size distribution, this technique combines laser light scattering microscopy with a charge-coupled device camera, which enables the visualization and recording of NPs in solution when a laser beam pass through the sample chamber. Thanks to the NTA software and according to a formula derived from the Stokes–Einstein equation it is possible to identify and track individual NPs moving under Brownian motion and to relate the movement to the particle's size [166].

Several research groups have reported the use of NTA for the characterization of photoactive NPs. For example, Zu et al. described the synthesis of polyethylene-silica particles with a core–shell structure using SEM and NTA determining their spherical morphology and the mean size. Khaydarov et al. used NTA to test the aggregation effect of AgNPs in the development of a novel method of synthesis using cellulose fibers and showing their pronounced antibacterial effect. In the AuNPs' field, Vogel et al. have reported a new way for bulk production of uniform metal NPs in water in which NTA showed that pulsed laser ablation from gold plate in water results in a large amount of NPs [167–169].

Other techniques available for analyzing the particle size distribution, like electron microscopy and AFM, require time-consuming samples preparation and data analysis as compared to DLS and NTA. Moreover, NTA allows for the analysis of polydisperse NPs and the study of the interaction between the particles itself and particles biomolecules, thus complementing the information which DLS provides [148,166].

9.5.4 Zeta potential

For the characterization of NPs, functionalization is also important to analyze the surface charge of the particles. It should be noted that zeta potential (ZP) never measures charge or charge density and rather deals with surface potential. NPs have a surface charge that attracts a thin layer of ions of opposite charge to the NP surface. This double layer of ions travels with the NP as it diffuses throughout the solution. The electric potential at the boundary of the double layer is known as the ZP of the particles. During ZP measurements, an electrical field is applied across the sample, inducing the movement of charged particles. The ratio between the NP velocity and the external applied field, known as electrophoretic mobility, is then measured and converted to a ZP (ζ) value. The usual values typically range from $+100$ to -100 mV. The magnitude of the ZP is predictive of the colloidal stability; typically a value greater than 30 mV indicates a stable NP suspension, whereas lower ZP values indicate lower colloidal stability with high tendency toward aggregation. Besides being an indicator of the colloidal stability of NPs, it could be used to check the functionalization of NPs with biomolecules. The presence of protein or polymers at the NP surface provides changes on their surface charge that could be observed through changes in the ZP before and after their biofunctionalization [156].

9.5.5 X-ray photoelectron spectroscopy and time of flight secondary ion mass spectrometry

To monitor the effectiveness of NPs, functionalization surface analysis spectroscopic methods such as X-ray photoelectron spectroscopy (XPS) and time of flight secondary ion mass spectrometry (ToF-SIMS) are increasingly being used. The combined use of both techniques provides elemental, chemical state, and molecular information from surfaces of NPs. Thus it could be used to analyze the nature of the specific bonds established between the NPs and the biomolecules. For example, Spampinato et al. [170] reported the sequential characterization of the surface chemistry of AuNPs at the different steps needed for their functionalization with 1-ß-D-thio-glucose (TG). Besides, even the interaction of the obtained NPs with a maltose binding protein (MBP) was characterized by its use. ToF-SIMS allowed the identification of some characteristic peaks related to the coordination of thio-glucose with gold substrates. XPS provided quantitative information about the elements content in each functionalization step. In addition, from the XPS high-resolution spectra information about the bonds formed between the different species involved in the functionalization were also inferred. From the ToF-SIMS data it was also possible to determine the reaction of TG with MBP occurred with specific amino acid residues present in the binding pocket of the protein.

9.5.6 UV–visible spectroscopy

The NPs have unique physical properties which depend on the composition, size, and shape. Particularly, metal NPs present an optical property known as surface plasmon resonance (SPR). In this phenomenon, free electrons in the conduction band of the surface of metal NPs are excited in a collective way when a beam of light irradiate the surface of the metal NPs. This generates a delocalized oscillation of the electrons that results in the absorption of electromagnetic radiation by the AuNPs in a certain area of the visible or NIR spectrum giving rise to spectra with resonance bands known as plasmons [171].

The surface functionalization of photoactive NPs results in a modification of the absorption band and its intensity. This is a consequence of a variation in size and shape in comparison with the original nonfunctionalized NPs. Usually this change only implies a displacement of a few nanometers and it may not even be enough to be detected. However, the SPR is also sensitive to the aggregation of NPs in such a way that the SPR of individual particles produce feedback generating a displacement in the absorption spectrum. In that case, the spectrum moves to higher wavelength values or to lower absorbance values due to the loss of AuNPs that are absorbed at this wavelength [172].

Here we report a graph as an example showing NPs aggregation during the functionalization process, demonstrating that the functionalization buffer selected for the functionalization of gold NPs with an adhesion protein (E-cadherin) was not adequate. For these experiments, 40 nm AuNPs provided by Nanoimmunotech SL, were first incubated in a phosphate buffer containing different amounts of calcium chloride. Fig. 9.2 clearly showed the two

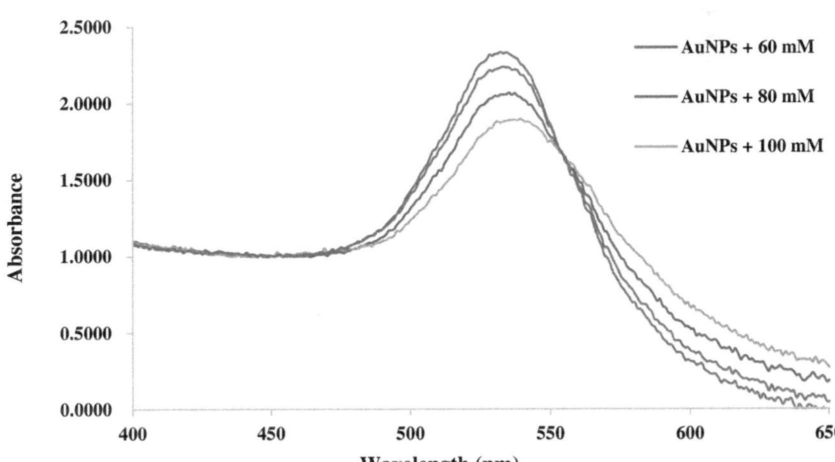

FIGURE 9.2 UV–visible spectra of 40 nm gold nanoparticles incubated with 0, 60, 80, and 100 μg/mL of calcium chloride. The NPs usually have the SPR at 532 nm but this value changes in the presence of different concentrations of calcium chloride; 60 nM (AuNPs + 60 mM), 80 mM (AuNPs + 80 mM), and 100 mM (AuNPs + 100 mM).

effects consequence of NP aggregation mentioned before: (1) the displacement of the absorption spectrum to higher wavelength values due to the formation of aggregates with a bigger size than individual NPs; and (B) a lower absorbance at the original SPR value due to the presence of big aggregates which present a lower absorbance at this wavelength. Thus these UV–visible spectra, showing the particles aggregation at different concentrations of Ca^{+2} ion, clearly indicate the need to change the functionalization conditions by selecting an incubation buffer where AuNPs do not have stability issues.

As UV–visible is a nondestructive technique, it could be used to give information about the colloidal stability of the NPs during the biofunctionalization process. In the case of NPs that do not have an SPR band, DLS or FCS could be used as an alternative to determine their colloidal stability.

9.5.7 Fluorescence spectroscopy

Fluorescence-based spectroscopy as well as imaging techniques has long been used as reliable and quantitative tools for a straightforward bioconjucation characterization of biomolecules and NPs. The fluorescence spectroscopy is used also for the characterization of novel fluorescent nanomaterials including semiconductor QDs, metal nanoclusters, dye-doped silica or polymer NPs, upconverting NPs, and nanodiamonds [173].

Fluorescent nanomaterials frequently exhibit excellent photophysical properties, color tunability as mean of the surface functionalization, and their modification can be monitored by simple steady-state or time-resolved fluorescence spectroscopy. Poderys and coworkers investigated interactions of thioglycolic acid-coated CdTe QDs with proteins by measuring the photoluminescence intensity of QDs upon titration with BSA [174]. A gradual increase in QD photoluminescence (up to 120% of the initial intensity) clearly suggested the adsorption of proteins on the QDs. The enhanced emission efficiency of QDs may result from the recovery of surface defects by the formed protein corona. Recently, it has also been reported the fluorescent characterization of the proteins interactions with ultrasmall gold nanoclusters (AuNCs, diameter 3.2 nm) chemically modified with carboxylic groups. Li Shang et al. demonstrated the substantial increase in the fluorescence intensity of AuNCs upon protein adsorption. Time-resolved fluorescence studies further revealed significantly enhanced long lifetime components of the AuNC luminescence decay curves upon protein association [163].

Parracino et al. also reported the measurement of the fluorescence characterization of native BSA, BSA-AF532, and BSA-ANS coupled to nanoparticles in a RTC 2000 PTI spectrofluorimeter. Intrinsic protein fluorescence spectra were acquired upon 295 nm excitation; meanwhile the extrinsic fluorescence emission spectra of BSA-AF532 were acquired upon 530 nm excitation. ANS fluorescence spectra of BSA-ANS alone and coupled to NPs were acquired upon 350 nm excitation. The emission fluorescence of BSA-AF532 coupled to superparamagnetic gold-coated NPs, together with the fluorescence emission of the ANS, confirm the presence of labeled BSA onto the NPs. This clearly indicates the immobilization of protein onto the particles. The normalized fluorescence emission spectra of free BSA (not immobilized) and BSA onto NPs revealed a 15-nm blue shift of the maximum of intrinsic fluorescence emission after protein immobilization. A blue shift (10 nm) in ANS fluorescence emission is also observed for Fe_3O_4@Au-BSA-ANS versus BSA-ANS. The blue shift is likely to be caused by the lower dielectric constant of the core–shell particles compared to the aqueous environment. Data confirm that ANS as well as the Trp becomes less solvent accessible upon BSA immobilization onto NPs [147]

9.5.8 Electrophoresis

9.5.8.1 Agarose gel electrophoresis

Electrophoresis provides a method for characterization of both bare and bio functionalized NPs as when an electric field is applied, their migration through the agarose gel depends on their shape, size, and charge [175]. The differential migration of NPs biofunctionalized or not is reported in the following example. Link Oriented kit GOLD from Nanoimmunotech was used for the oriented immobilization of a polyhistidine-tagged fragment of E-cadherin (E-EC12). The kit comprised chelate-modified AuNPs of 40 nm that ensure an oriented binding through metal-chelation to histidine-rich tag of the protein. The cadherin fragments have a large size (~25 kDa) and a net negative charge at the pH of the running buffer used for the electrophoresis. The size and the net charge of the functionalized NPs should be different to the nonfunctionalized ones. Thus it is expected to observe a shift in their mobility if cadherin fragments were effectively bounded (Fig. 9.3). The

FIGURE 9.3 1% Agarose gel electrophoresis. 1, Gold nanoparticles without E-cadherin; 2, gold nanoparticles with 45 μg/mL of E-cadherin; 3, gold nanoparticles with 100 μg/mL of E-cadherin.

analysis of the agarose gel shows a different migration of the NPs due to the binding of E-cadherin on their surface. The control (1), AuNPs without E-cadherin, shows the highest electrophoretic mobility while AuNPs functionalized with 45 and 100 μg/mL (2 and 3, respectively) of protein has less mobility toward the anode (positive electrode). As it was expected, the migration of the NPs is altered by the binding of protein on their surface due to the increase in their size. Also, the net superficial charge is modified depending to a greater extent on the isoelectric point of the protein.

This straightforward characterization technique could be used for confirming the effectiveness of the immobilization procedure. Even, for certain biomolecules it could be used to determine the exact number of molecules attached to the NP surface. For example, this technique has been extensively used for the characterization of metal NPs modified by single-stranded DNA short strands with a thiol function. As it can be observed in Fig. 9.4, AuNPs (10 nm) functionalized with different number (one, two, three, etc.) of DNA oligomers (43 pb) have a distinctive shift on their mobility and could be easily discriminated and even separated by gel electrophoresis [176]. This high level of resolution of the technique depends on the size of the final derivative. Although it is possible to play with the percentage of agarose used to control the pore size and thus the mobility of the samples, sometimes the difference in size is not enough as the original NPs or the biomolecule-

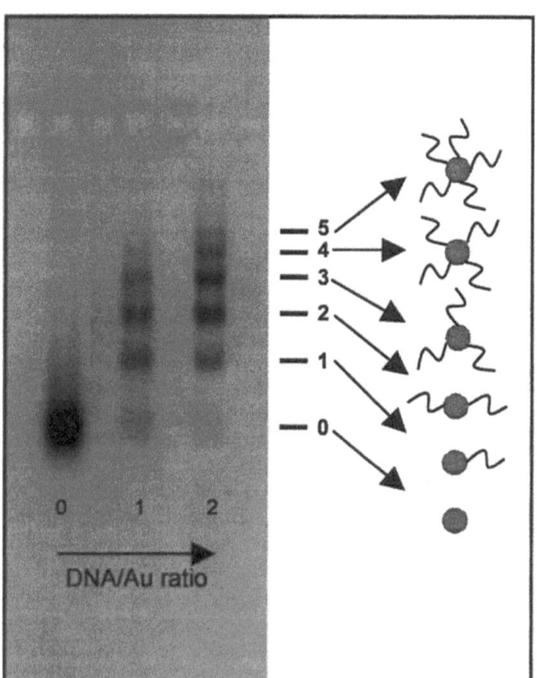

FIGURE 9.4 Electrophoretic mobility of 5 nm Au/100b HS-ssDNA conjugates (3% gel). The first lane (left to the right) corresponds to 5-nm particles (single band). When 1 equiv. of DNA is added to the Au particles (second lane), discrete bands appear (namely, 0, 1, 2, 3, ...). When the DNA amount is doubled (third lane), the intensity of the discrete bands change and additional retarded bands appear (4, 5). Because of the discrete character, each band can be directly assigned to a unique number of DNA strands per particle. *Reproduced with permission from D. Zanchet, C.M. Micheel, W.J. Parak, D. Gerion, A.P. Alivisatos, Electrophoretic isolation of discrete Au nanocrystal/DNA conjugates. Nano. Lett. 1 (2001) 32–35.*

derivative size is too large. As an example, the level of resolution of the number of DNA strands attached to the AuNPs was not possible to be achieved with lager NPs (20 nm) [177].

9.5.8.2 Polyacrylamide gel electrophoresis

Polyacrylamide gel electrophoresis (PAGE) is one of the most useful methods to separate biomolecules, such as proteins and peptides, according to their electrophoretic mobility. Thus when the molecule used to functionalize the AuNPs is a peptide or a protein, this method is an indirect way to analyze the efficacy of the process.

In this type of electrophoresis, similarly to the agarose gels, the electrophoretic mobility is dependent on the size, conformation, and charge of the molecule. The separation of molecules only according to their size can be achieved running them in denaturing conditions. In this case, 2-mercaptoethanol aided by high temperatures removes the disulfide bridges that link polypeptide chains causing the loss of the quaternary and tertiary structure of proteins. Sodium dodecyl sulfate (SDS) also participates in the denaturing process and in addition gives negative charge to the proteins in proportion of their mass. In this way, the detergent provides all proteins with a uniform charge-to-mass ratio, depending the migration into the gel only on the length [178].

To analyze the NP biofunctionalization process with this type of analysis, what is usually loaded is the supernatants of the reaction after separating the NPs from the reaction media. Once the electrophoresis took place, the gel can be stained with Coomassie blue or with silver nitrate depending on the sensitivity needed to visualize the proteins bands. The staining with Coomassie blue allows the detection of protein bands up to 100–1000 ng while the silver staining is more sensitive allowing to detect up to 0.1 ng [175]. In order to analyze the functionalization of NPs with proteins or peptides by this method, the pure protein and the supernatant obtained after the incubation of the AuNPs with the protein must be loaded on the gel. In this way, once the migration is finished, the difference in intensity of the protein bands corresponding to the molecular weight shows the attached biomolecule and allows to quantify the amount of protein linked. If immobilization to the NPs took place, it is expected that less or no protein will be present in the supernatant of the reaction. The percentage of union can be estimated using image analysis software such as Image J [179]. This technique gives the same info than the blot assay but the main difference is that the dot assay has higher sensitivity (less amount of target protein could be detected) but does not give quantification. So, SDS–PAGE is a good option if primary antibodies are not available to carry out a dot-blot assay.

As an example, an image corresponding to an SDS–PAGE electrophoresis gel is shown below (Fig. 9.5). It can be noticed that a successful

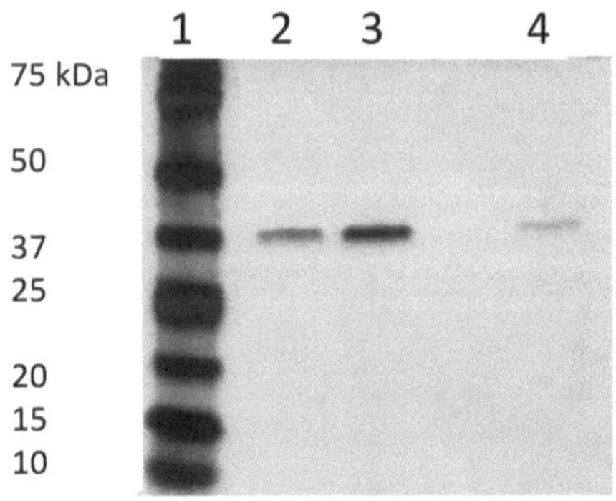

FIGURE 9.5 SDS-gel electrophoresis. 1, 45 μg/mL of E-cadherin; 2, 100 μg/mL of E-cadherin; 3, supernatant obtained after incubation with 45 μg/mL of E-cadherin; and 4, supernatant obtained after incubation with 100 μg/mL of E-cadherin.

functionalization has been achieved with both concentration of E-cadherin used: 45 μg/mL (3) and 100 μg/mL (4). The highest concentration of protein used (100 μg/mL) caused the saturation of the AuNPs' surface as there is still protein on the supernatant of the immobilization reaction, which means that not all the offered protein was bounded.

This technique allows the calculation, with great precision, of the binding percentage and thus the number of biomolecules attached per NP. The quantification of the protein using this methodology is quite fast (15 min) and quite simple to be carried out. The quantification of the amount of protein bound is much more precise with this assay than by image analysis using SDS–PAGE. The sensitivity reached by this reagent is very high, being able to detect as low as 1 μg present on the NPs' supernatant [180].

9.5.9 Other assays

9.5.9.1 Dot-blot immunoassay

In order to monitor the biofunctionalization of photoactive NPs, immunoassays are a valid characterization method. In this case, the recognition of the biomolecule attached to the NPs surface is detected by the binding of a specific antibody labeled by an enzyme or a fluorescent tag. A simple way to take advantage of this technique for analyzing the biofunctionalization of NPs is using a dot-blot indirect immunoassay. It implies first the direct deposition of the NPs (biofunctionalized or not) onto nitrocellulose membranes. Then a specific antibody that recognizes the biomolecule used in the functionalization is added (detection antibody). After several washing steps, a secondary antibody usually labeled with horseradish peroxidase is used for detection of the immunocomplex (biomolecule-antibody) [181]. The secondary antibody has to have specificity for both the particular species and the isotype of the detection antibody. Both antibodies provide specificity to the assay while the HRP enzyme bounded to the secondary antibody produces a detectable signal in the presence of a substrate (e.g., light production when

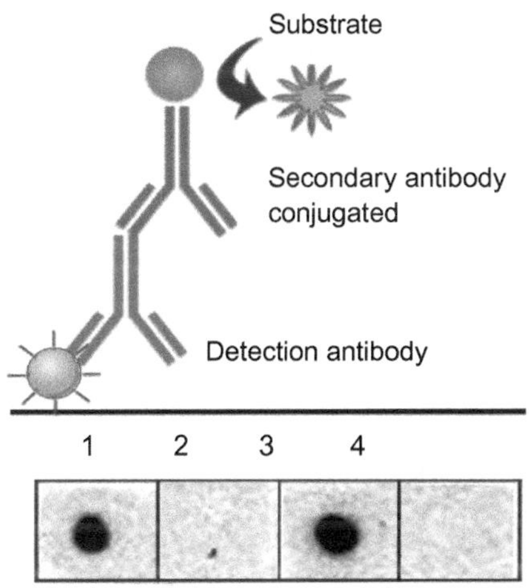

FIGURE 9.6 Indirect immunoassay with 40 nm gold nanoparticles. 1, AuNPs + 45 μg/mL E-cadherin; 2, AuNPs without E-cadherin; 3, positive control (only primary and secondary antibodies on the membrane); and 4, negative control; only secondary antibody in solution.

catalyzes the oxidation of a chemiluminescent substrate such as Luminol) [182,183]. Unlike the analysis by SDS–PAGE, this is a direct way not only to detect whether the biomolecule was attached to the NPs but also to check its binding through a certain orientation if the detection antibody used is a monoclonal antibody recognizing a specific area of the biomolecule surface. Besides, it is complementary to agarose electrophoresis as it allows confirming that the change on the mobility of the functionalized NPs is due to the biomolecule binding and not due to aggregation during the functionalization process.

Below (Fig. 9.6) are the results of an indirect immunoassay performed to check the biofunctionalization of the abovementioned 40 nm AuNPs with E-cadherin recombinant fragments. This assay demonstrates the effectiveness of functionalization since the signal showing the presence of the detection antibody is only developed on the spot where the biofunctionalized NPs (1) were deposited. However, a negative result was obtained in the spot where nonfunctionalized NPs (control NPs, 2) were deposited. This confirms the functionalization with E-cadherin fragments, as it shows that the primary antibody has attached in a specific way to the biofunctionalized NPs due to the presence of E-cadherin on their surface.

Therefore this is a simple technique that could be used to confirm NPs biofunctionalization, the only requirement is to have a detection antibody able to detect the biomolecule selected for the functionalization. Besides, as it was mentioned, it could also give information about the biomolecule orientation once it is attached to the NP surface. For this, it is necessary to use a monoclonal detection antibody instead of a polyclonal one (Scheme 9.6). Polyclonal antibodies (pAbs) are antibodies that are secreted by different B-cell lineages. Thus, they are a collection of immunoglobulin molecules that

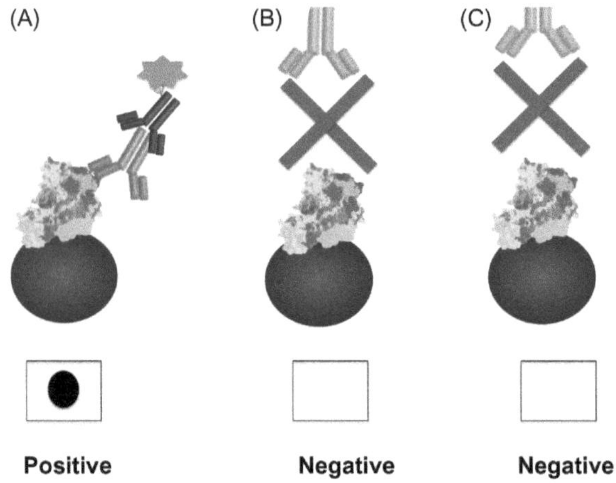

SCHEME 9.6 Scheme showing how this assay could be used to confirm a specific orientation of the attached biomolecules. (A) The detection antibody (light grey (purple in online version)) recognizing its specific epitope could attach to the NP as this epitope is available and free towards the reaction media. (B) and (C) Both detection antibodies (light grey and dark grey (orange and green in online version)) are not able to bind their specific epitopes due to steric hindrance that is a consequence of the biomolecule orientation once attached to the NP.

react against the same specific antigen but each identifying a different epitope. If a detection pAb is used for this analysis, it will be possible to confirm the attachment of the biomolecule to the NPs but it will be not possible to confirm the attachment through a preferential orientation. However, as monoclonal antibodies (mAbs) come from a single-cell lineage, they recognize a specific epitope that is placed at a specific site within the target biomolecule. Therefore the use of mAbs in this assay is useful to analyze the accessibility of the specific epitope, which gives information about the orientation of the attached biomolecule.

9.5.9.2 Bradford assay

The Bradford protein assay is used to measure spectroscopically the concentration of total protein in a sample [184] and it is commonly used to quantify the amount of protein attached onto the biofunctionalized NPs. The quantification occurs in an indirect way measuring protein present on the supernatants obtained after incubating the particles with the desired protein, like in the SDS–PAGE assay. The assay is based on an absorbance shift of the dye Coomassie Brilliant Blue G-250 under acidic conditions due to the binding of the dye to basic amino acid residues, arginine, lysine, and histidine. The binding of the dye results in a color change from brown to blue. An advantage over other colorimetric assays used to quantify protein is that its reducing agents (i.e., DL-Ditiotreitolo and beta-mercaptoethanol) and metal

(μg/ml) Offered	(μg/ml) Supernatant	(μg/ml) Linked	Union (%)
3	0	3	100
6	2	4	66
13	7	6	46
15	8	7	46

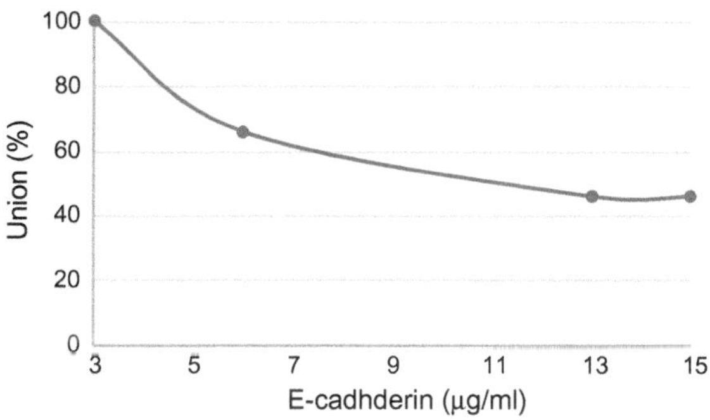

FIGURE 9.7 Percentage of union of E-cadherin onto 40 nm gold nanoparticles' surface depending on the amount of protein offered.

chelators (i.e., Ethylenediaminetetraacetic acid, Ethylene glycol-bis(β-aminoethyl ether)-N,N,N′,N′-tetraacetic acid) at low concentration do not cause interference. Fig. 9.7 shows the data obtained after analyzing with Bradford the supernatants obtained after incubation of 40 nm AuNPs with increasing amounts of E-cadherin. The more protein is offered, the higher percentage of protein remains unbound due to saturation of the NPs' surface.

9.6 Conclusion and perspectives

The use of the photoactive biomolecule-modified NPs has rapidly grown in the last years for their application in nanomedicine. In particular, they have already shown their potential to have a great impact on the development of new diagnosis and therapeutic strategies. However, advances in nanoscience are still needed, and in particularly in nanochemistry. A better understanding and control of the NP–biomolecule interface is required to be able to ensure a general and reproducible success in diagnosis and therapy. In particular, the biofunctionalization strategy selected could affect several key aspects including targeting efficiency, biodistribution, pharmacokinetics and even degradation of biofunctionalized NPs intended to be used as novel nanomedicines.

Acknowledgment

The authors acknowledge the support of Aragon Government-FSE (Research Group E93) and PIE201660I012 (CSIC).

References

[1] B. Pelaz, et al., Tailoring the synthesis and heating ability of gold nanoprisms for bioapplications, Langmuir 28 (2012) 8965–8970.

[2] E.C. Dreaden, A.M. Alkilany, X. Huang, C.J. Murphy, M.A. El-Sayed, The golden age: gold nanoparticles for biomedicine, Chem. Soc. Rev. 41 (2012) 2740–2779.

[3] M.-C. Daniel, D. Astruc, Gold nanoparticles: assembly, supramolecular chemistry, quantum-size-related properties, and applications toward biology, catalysis, and nanotechnology, Chem. Rev. 104 (2004) 293–346.

[4] R. Weissleder, A clearer vision for in vivo imaging, Nat. Biotechnol. 19 (2001) 316–317.

[5] R. Bardhan, S. Lal, A. Joshi, N.J. Halas, Theranostic nanoshells: from probe design to imaging and treatment of cancer, Acc. Chem. Res. 44 (2011) 936–946.

[6] S. Oldenburg, R. Averitt, S. Westcott, N. Halas, Nanoengineering of optical resonances, Chem. Phys. Lett. 288 (1998) 243–247.

[7] N.R. Jana, L. Gearheart, C.J. Murphy, Seed-mediated growth approach for shape-controlled synthesis of spheroidal and rod-like gold nanoparticles using a surfactant template, Adv. Mater. 13 (2001) 1389–1393.

[8] S.R. Nicewarner-Peña, et al., Submicrometer metallic barcodes, Science 294 (2001) 137–141.

[9] Y. Sun, Y. Xia, Shape-controlled synthesis of gold and silver nanoparticles, Science 298 (2002) 2176–2179.

[10] Y. Xia, et al., Gold nanocages: from synthesis to theranostic applications, Acc. Chem. Res. 44 (2011) 914–924.

[11] Z. Liang, A. Susha, F. Caruso, Gold nanoparticle-based core-shell and hollow spheres and ordered assemblies thereof, Chem. Mater. 15 (2003) 3176–3183.

[12] F. Kim, S. Connor, H. Song, T. Kuykendall, P. Yang, Platonic gold nanocrystals, Angew. Chemie Int. Ed 43 (2004) 3673–3677.

[13] M.S. Yavuz, et al., Gold nanocages covered by smart polymers for controlled release with near-infrared light, Nat. Mater. 8 (2009) 935–939.

[14] J. Aaron, et al., Polarization microscopy with stellated gold nanoparticles for robust, in-situ monitoring of biomolecules, Opt. Express 16 (2008) 1999–2004.

[15] A.M. Gobin, et al., Near-infrared resonant nanoshells for combined optical imaging and photothermal cancer therapy, Nano. Lett. 7 (2007) 1929–1934.

[16] Jaesook, P. *et al.* Two-photon-induced photoluminescence imaging of tumors using near-infrared excited gold nanoshells, in: Quantum Electronics and Laser Science Conference on Lasers and Electro-Optics, CLEO/QELS, 2008, San Jose, CA.

[17] G. Von Maltzahn, et al., SERS-coded cold nanorods as a multifunctional platform for densely multiplexed near-infrared imaging and photothermal heating, Adv. Mater. 21 (2009) 3175–3180.

[18] X. Wang, et al., Noninvasive laser-induced photoacoustic tomography for structural and functional in vivo imaging of the brain, Nat. Biotechnol. 21 (2003) 803–806.

[19] H.F. Zhang, K. Maslov, G. Stoica, L.V. Wang, Functional photoacoustic microscopy for high-resolution and noninvasive in vivo imaging, Nat. Biotechnol. 24 (2006) 848–851.

[20] N.L. Rosi, C.A. Mirkin, Nanostructures in biodiagnostics, Chem. Rev. 105 (2005) 1547–1562.

[21] T.A. Taton, T.A. Taton, C.A. Mirkin, R.L. Letsinger, Scanometric DNA array detection with nanoparticle probes, Science 289 (2000) 1757–1760.

[22] L. Andrew Lyon, M.D. Musick, M.J. Natan, Colloidal Au-enhanced surface plasmon resonance immunosensing, Anal. Chem. 70 (1998) 5177–5183.

[23] X. Liu, et al., A one-step homogeneous immunoassay for cancer biomarker detection using gold nanoparticle probes coupled with dynamic light scattering, J. Am. Chem. Soc. 130 (2008) 2780–2782.

[24] R.L. Phillips, O.R. Miranda, C.-C. You, V.M. Rotello, U.H.F. Bunz, Rapid and efficient identification of bacteria using gold-nanoparticle–poly(para-phenyleneethynylene) constructs, Angew. Chemie Int. Ed 47 (2008) 2590–2594.

[25] P. Baptista, et al., Gold nanoparticles for the development of clinical diagnosis methods, Anal. Bioanal. Chem. 391 (2008) 943–950.

[26] H. He, X. Xu, H. Wu, Y. Zhai, Y. Jin, In situ nanoplasmonic probing of enzymatic activity of monolayer-confined glucose oxidase on colloidal nanoparticles, Anal. Chem. 85 (2013) 4546–4553.

[27] R.J. Watts, S. Kong, M.P. Orr, G.C. Miller, B.E. Henry, Photocatalytic inactivation of coliform bacteria and viruses in secondary wastewater effluent, Water Res. 29 (1995) 95–100.

[28] L. Zan, W. Fa, T. Peng, Z.K. Gong, Photocatalysis effect of nanometer TiO_2 and TiO_2-coated ceramic plate on hepatitis B virus, J. Photochem. Photobiol. B Biol 86 (2007) 165–169.

[29] P. Hajkova, P. Spatenka, J. Horsky, I. Horska, A. Kolouch, Photocatalytic effect of TiO_2 films on viruses and bacteria, Plasma. Process. Polym. 4 (2007) S397–S401.

[30] S. Mura, et al., Multifunctionalization of wool fabrics through nanoparticles: a chemical route towards smart textiles, J. Colloid. Interface. Sci. 456 (2015) 85–92.

[31] M. Zhou, et al., A chelator-free multifunctional [64Cu]CuS nanoparticle platform for simultaneous micro-PET/CT imaging and photothermal ablation therapy, J. Am. Chem. Soc. 132 (2010) 15351–15358.

[32] Z. Li, J.C. Barnes, A. Bosoy, J.F. Stoddart, J.I. Zink, Mesoporous silica nanoparticles in biomedical applications, Chem. Soc. Rev. 41 (2012) 2590.

[33] M.W. Ambrogio, C.R. Thomas, Y.L. Zhao, J.I. Zink, J.F. Stoddart, Mechanized silica nanoparticles: a new frontier in theranostic nanomedicine, Acc. Chem. Res. 44 (2011) 903–913.

[34] S. Angelos, E. Choi, F. Vögtle, L. De Cola, J.I. Zink, Photo-driven expulsion of molecules from mesostructured silica nanoparticles, J. Phys. Chem. C 111 (2007) 6589–6592.

[35] Q. Liu, et al., Sub-10 nm hexagonal lanthanide-doped $NaLuF_4$ upconversion nanocrystals for sensitive bioimaging in vivo, J. Am. Chem. Soc. 133 (2011) 17122–17125.

[36] F. Auzel, Upconversion and anti-Stokes processes with f and d ions in solids, Chem. Rev. 104 (2004) 139–173.

[37] F. Wang, X. Liu, Recent advances in the chemistry of lanthanide-doped upconversion nanocrystals, Chem. Soc. Rev. 38 (2009) 976.

[38] L. Xiong, et al., High contrast upconversion luminescence targeted imaging in vivo using peptide-labeled nanophosphors, Anal. Chem. 81 (2009) 8687–8694.

[39] C. Salthouse, S. Hildebrand, R. Weissleder, U. Mahmood, Design and demonstration of a small-animal up-conversion imager, Opt. Express 16 (2008) 21731–21737.

[40] C.T. Xu, et al., Autofluorescence insensitive imaging using upconverting nanocrystals in scattering media, Appl. Phys. Lett. 93 (2008) 171103.

[41] M. Nyk, R. Kumar, T.Y. Ohulchanskyy, E.J. Bergey, P.N. Prasad, High contrast in vitro and in vivo photoluminescence bioimaging using near infrared to near infrared up-conversion in Tm 3+ and Yb 3+ doped fluoride nanophosphors, Nano. Lett. 8 (2008) 3834–3838.

[42] L. Xiong, T. Yang, Y. Yang, C. Xu, F. Li, Long-term in vivo biodistribution imaging and toxicity of polyacrylic acid-coated upconversion nanophosphors, Biomaterials. 31 (2010) 7078–7085.

[43] J. Zhou, et al., Dual-modality in vivo imaging using rare-earth nanocrystals with near-infrared to near-infrared (NIR-to-NIR) upconversion luminescence and magnetic resonance properties, Biomaterials. 31 (2010) 3287–3295.

[44] L. Cheng, K. Yang, M. Shao, X. Lu, Z. Liu, Pharmacokinetics, long-term biodistribution and toxicology study of functionalized upconversion nanoparticles in mice, Nanomedicine. 6 (2011) 1327–1340.

[45] H.S. Qian, H.C. Guo, P.C.L. Ho, R. Mahendran, Y. Zhang, Mesoporous-silica-coated up-conversion fluorescent nanoparticles for photodynamic therapy, Small. 5 (2009) 2285–2290.

[46] J. Shan, et al., Pegylated composite nanoparticles containing upconverting phosphors and meso-tetraphenyl porphine (TPP) for photodynamic therapy, Adv. Funct. Mater. 21 (2011) 2488–2495.

[47] C. Wang, H. Tao, L. Cheng, Z. Liu, Near-infrared light induced in vivo photodynamic therapy of cancer based on upconversion nanoparticles, Biomaterials. 32 (2011) 6145–6154.

[48] Y. Jin, C. Jia, S.W. Huang, M. O'Donnell, X. Gao, Multifunctional nanoparticles as coupled contrast agents, Nat. Commun. 1 (2010) 1–8.

[49] J. Sun, H. Wu, Y. Jin, Synthesis of thiolated Ag/Au bimetallic nanoclusters exhibiting an anti-galvanic reduction mechanism and composition-dependent fluorescence, Nanoscale 6 (2014) 5449.

[50] C. Mateo, J.M. Palomo, G. Fernandez-Lorente, J.M. Guisan, R. Fernandez-Lafuente, Improvement of enzyme activity, stability and selectivity via immobilization techniques, Enzyme Microb. Technol. 40 (2007) 1451–1463.

[51] G. Wu, et al., Remotely triggered liposome release by near-infrared light absorption via hollow gold nanoshells, J. Am. Chem. Soc. 130 (2008) 8175–8177.

[52] E. Mahon, A. Salvati, F. Baldelli Bombelli, I. Lynch, K.A. Dawson, Designing the nanoparticle-biomolecule interface for 'targeting and therapeutic delivery', J. Control Release 161 (2012) 164–174.

[53] R.A. Sperling, W.J. Parak, Surface modification, functionalization and bioconjugation of colloidal inorganic nanoparticles, Philos. Trans. R. Soc. A Math. Phys. Eng. Sci. 368 (2010) 1333–1383.

[54] J. Kimling, et al., Turkevich method for gold nanoparticle synthesis revisited, J. Phys. Chem. B 110 (2006) 15700–15707.

[55] N.G. Bastús, J. Comenge, V. Puntes, Kinetically controlled seeded growth synthesis of citrate-stabilized gold nanoparticles of up to 200 nm: size focusing versus ostwald ripening, Langmuir 27 (2011) 11098–11105.

[56] C. Ziegler, A. Eychmüller, Seeded growth synthesis of uniform gold nanoparticles with diameters of 15-300 nm, J. Phys. Chem. C 115 (2011) 4502–4506.

[57] F. Schulz, et al., Little adjustments significantly improve the Turkevich synthesis of gold nanoparticles, Langmuir 30 (2014) 10779–10784.

[58] K. Aslan, C.C. Luhrs, V.H. Pérez-Luna, Controlled and reversible aggregation of biotinylated gold nanoparticles with streptavidin, J. Phys. Chem. B 108 (2004) 15631–15639.

[59] Y. Zheng, et al., Preparation of solvent-free gold nanofluids with facile self-assembly technique, Chemphyschem 11 (2010) 61–64.

[60] S. Rana, A. Bajaj, R. Mout, V.M. Rotello, Monolayer coated gold nanoparticles for delivery applications, Adv. Drug Deliv. Rev. 64 (2012) 200–216.

[61] Y.-C. Yeh, B. Creran, V.M. Rotello, Gold nanoparticles: preparation, properties, and applications in bionanotechnology, Nanoscale 4 (2012) 1871–1880.

[62] S. Fraga, et al., Influence of the surface coating on the cytotoxicity, genotoxicity and uptake of gold nanoparticles in human HepG2cells, J. Appl. Toxicol. 33 (2013) 1111–1119.

[63] Z. Wang, et al., Design of polymeric stabilizers for size-controlled synthesis of monodisperse gold nanoparticles in water, Langmuir 23 (2007) 885–895.

[64] D. Zopes, B. Stein, S. Mathur, C. Graf, Improved stability of 'naked' gold nanoparticles enabled by in situ coating with mono and multivalent thiol PEG ligands, Langmuir 29 (2013) 11217–11226.

[65] J.S. Kang, T.A. Taton, Oligothiol graft-copolymer coatings stabilize gold nanoparticles against harsh experimental conditions, Langmuir 28 (2012) 16751–16760.

[66] L.-N. Ma, D.-J. Liu, Z.-X. Wang, Synthesis and applications of gold nanoparticle probes, Chinese J. Anal. Chem. 38 (2010) 1–7.

[67] C.J. Bueno-Alejo, et al., Ultraclean derivatized monodisperse gold nanoparticles through laser drop ablation customization of polymorph gold nanostructures, Langmuir 28 (2012) 8183–8189.

[68] A. Lanterna, E. Pino, A. Doménech-Carbó, M. González-Béjar, J. Pérez-Prieto, Enhanced catalytic electrochemical reduction of dissolved oxygen with ultraclean cucurbituril[7]-capped gold nanoparticles, Nanoscale 6 (2014) 9550–9553.

[69] N. Jornet-Martínez, M. González-Béjar, Y. Moliner-Martínez, P. Campíns-Falcó, J. Pérez-Prieto, Sensitive and selective plasmonic assay for spermine as biomarker in human urine, Anal. Chem. 86 (2014) 1347–1351.

[70] M. Tréguer-Delapierre, J. Majimel, S. Mornet, E. Duguet, S. Ravaine, Synthesis of non-spherical gold nanoparticles, Gold Bull. 41 (2008) 195–207.

[71] M.A. Malik, P. O'Brien, N. Revaprasadu, A simple route to the synthesis of core/shell nanoparticles of chalcogenides, Chem. Mater. 14 (2002) 2004–2010.

[72] M. Bruchez, M. Moronne, P. Gin, S. Weiss, A.P. Alivisatos, Semiconductor nanocrystals as fluorescent biological labels, Science (80) 281 (1998) 2013–2016.

[73] J.L. Lyon, D.A. Fleming, M.B. Stone, P. Schiffer, M.E. Williams, Synthesis of Fe oxide Core/Au shell nanoparticles by iterative hydroxylamine seeding, Nano. Lett. 4 (2004) 719–723.

[74] Z. Xu, Y. Hou, S. Sun, Magnetic core/shell Fe_3O_4/Au and Fe_3O_4/Au/Ag nanoparticles with tunable plasmonic properties, J. Am. Chem. Soc. 129 (2007) 8698–8699.

[75] J.J. Zhu, X.H. Liao, X.N. Zhao, H.Y. Chen, Preparation of silver nanorods by electrochemical methods, Mater. Lett. 49 (2001) 91–95.

[76] J.P. Abid, A.W. Wark, P.F. Brevet, H.H. Girault, Preparation of silver nanoparticles in solution from a silver salt by laser irradiation, Chem. Commun. (2002) 792–793.

[77] J. Talebi, R. Halladj, S. Askari, Sonochemical synthesis of silver nanoparticles in Y-zeolite substrate, J. Mater. Sci. 45 (2010) 3318–3324.

[78] S.M. Hosseinpour-Mashkani, M. Ramezani, Silver and silver oxide nanoparticles: synthesis and characterization by thermal decomposition, Mater. Lett. 130 (2014) 259–262.

[79] Q. Zhang, N. Li, J. Goebl, Z. Lu, Y. Yin, A systematic study of the synthesis of silver nanoplates: Is citrate a 'magic' reagent? J. Am. Chem. Soc. 133 (2011) 18931–18939.

[80] S.U. Ganaie, T. Abbasi, S.A. Abbasi, Green synthesis of silver nanoparticles using an otherwise worthless weed mimosa (*Mimosa pudica*): feasibility and process development toward shape/size control, Part. Sci. Technol. 33 (2015) 638–644.

[81] X.-F. Zhang, Z.-G. Liu, W. Shen, S. Gurunathan, Silver nanoparticles: synthesis, characterization, properties, applications, and therapeutic approaches, Int. J. Mol. Sci. 17 (2016) 1534.

[82] V. Deepak, et al., Biointerfaces synthesis of gold and silver nanoparticles using purified URAK, Colloid. Surfaces B Biointerfaces 86 (2011) 353–358.

[83] K.G. Stamplecoskie, J.C. Scaiano, Light emitting diode irradiation can control the morphology and optical properties of silver nanoparticles, J. Am. Chem. Soc. 132 (2010) 1825–1827.

[84] E.I. Alarcon, et al., The biocompatibility and antibacterial properties of collagen-stabilized, photochemically prepared silver nanoparticles, Biomaterials. 33 (2012) 4947–4956.

[85] Y. Jin, X. Gao, Spectrally tunable leakage-free gold nanocontainers, J. Am. Chem. Soc. 131 (2009) 17774–17776.

[86] Sergeev, B.M., Kasaikin, V.A., Litmanovich, E.A., Sergeev, B., Prusov, A.N. Cryochemical synthesis and properties of silver nanoparticle dispersions stabilised by poly (2-dimethylaminoethyl methacrylate). Mendeleev Commun. **9**, 130–131 (1999).

[87] S. Gurunathan, J. Woong Han, J. Kim, Green chemistry approach for the synthesis of biocompatible graphene, Int. J. Nanomed. 8 (2013) 2719–2732.

[88] M. Brust, J. Fink, D. Bethell, D.J. Schiffrin, C. Kiely, Synthesis and reactions of functionalised gold nanoparticles, J. Chem. Soc. Chem. Commun. ()) (1995) 1655. Available from: https://doi.org/10.1039/c39950001655.

[89] S. Gurunathan, et al., Reduced graphene oxide-silver nanoparticle nanocomposite: a potential anticancer nanotherapy, Int. J. Nanomed. 10 (2015) 6257–6276.

[90] K. Kalimuthu, R. Suresh Babu, D. Venkataraman, M. Bilal, S. Gurunathan, Biosynthesis of silver nanocrystals by *Bacillus licheniformis*, Colloid. Surfaces B Biointerfaces 65 (2008) 150–153.

[91] K. Kalishwaralal, V. Deepak, S. Ramkumarpandian, H. Nellaiah, G. Sangiliyandi, Extracellular biosynthesis of silver nanoparticles by the culture supernatant of *Bacillus licheniformis*, Mater. Lett. 62 (2008) 4411–4413.

[92] M.D. Mullen, et al., Bacterial sorption of heavy metals, Appl. Environ. Microbiol. 55 (1989) 3143–3149.

[93] T. Klaus, R. Joerger, E. Olsson, C.-G. Granqvist, Silver-based crystalline nanoparticles, microbially fabricated, Proc. Natl. Acad. Sci. 96 (1999) 13611–13614.

[94] K. Kalishwaralal, et al., Biosynthesis of silver and gold nanoparticles using *Brevibacterium casei*, Colloid. Surfaces B Biointerfaces 77 (2010) 257–262.

[95] B. Nair, T. Pradeep, Coalescence of nanoclusters and formation of submicron crystallites assisted by *Lactobacillus* strains, Cryst. Growth Des. 2 (2002) 293–298.

[96] M. Nadeem, B.H. Abbasi, M. Younas, W. Ahmad, T. Khan, A review of the green syntheses and anti-microbial applications of gold nanoparticles, Green Chem. Lett. Rev. 10 (2017) 216–227.

[97] C. Matea, et al., Quantum dots in imaging, drug delivery and sensor applications, Int. J. Nanomedicine. 12 (2017) 5421–5431.

[98] S. Krizkova, et al., Metallothionein—a promising tool for cancer diagnostics, Bratislava Med. J. 110 (2009) 93–97.

[99] K. Tmejova, et al., Structural effects and nanoparticle size are essential for quantum dots-metallothionein complex formation, Colloid. Surfaces B Biointerfaces 134 (2015) 262–272.

[100] S.-H. Wu, C.-Y. Mou, H.-P. Lin, Synthesis of mesoporous silica nanoparticles, Chem. Soc. Rev. 42 (2013) 3862.

[101] J. Song, et al., Plasmonic vesicles of amphiphilic gold nanocrystals: self-assembly and external-stimuli-triggered destruction, J. Am. Chem. Soc. 133 (2011) 10760–10763.

[102] S. Duraiswamy, S.A. Khan, Plasmonic nanoshell synthesis in microfluidic composite foams, Nano. Lett. 10 (2010) 3757–3763.

[103] W.F. Oliveira, et al., Functionalization of titanium dioxide nanotubes with biomolecules for biomedical applications, Mater. Sci. Eng. C 81 (2017) 597–606.

[104] J. Goossens, et al., Functionalization of gold nanoparticles with nanobodies through physical adsorption, Anal. Method. 9 (2017) 3430–3440.

[105] S. Puertas, et al., Taking advantage of unspecific interactions to produce highly active magnetic nanoparticle antibody conjugates, ACS Nano 5 (2011) 4521–4528.

[106] L.R. Khoury, J. Kost, G. Enden, Effects of surface coating on nanoparticle-protein adsorption selectivity, Regen. Eng. Transl. Med. 4 (2018) 1–13.

[107] R. Torres, et al., Supports coated with PEI as a new tool in chromatography, Enzyme Microb. Technol. 39 (2006) 711–716.

[108] V.S. Meka, et al., A comprehensive review on polyelectrolyte complexes, Drug Discov. Today 22 (2017) 1697–1706.

[109] S. Meerod, N. Deepuppha, B. Rutnakornpituk, M. Rutnakornpituk, Reusable magnetic nanocluster coated with poly(acrylic acid) and its adsorption with an antibody and an antigen, J. Appl. Polym. Sci. 135 (2018) 46160.

[110] A.F. Palomec-Garfias, K.V. Jardim, M.H. Sousa, C. Márquez-Beltrán, Influence of polyelectrolyte chains on surface charge and magnetization of iron oxide nanostructures, Colloid. Surfaces A Physicochem. Eng. Asp. 549 (2018) 13–24.

[111] M.H. Jazayeri, H. Amani, A.A. Pourfatollah, H. Pazoki-Toroudi, B. Sedighimoghaddam, Various methods of gold nanoparticles (GNPs) conjugation to antibodies, Sens. Bio-Sensing Res. 9 (2016) 17–22.

[112] J. Conde, et al., Design of multifunctional gold nanoparticles for in vitro and in vivo gene silencing, ACS Nano 6 (2012) 8316–8324.

[113] K.H. Lim, H. Huang, A. Pralle, S. Park, Engineered streptavidin monomer and dimer with improved stability and function, Biochemistry 50 (2011) 8682–8691.

[114] J.D. Swartz, C.P. Gulka, F.R. Haselton, D.W. Wright, Development of a histidine-targeted spectrophotometric sensor using Ni(II)NTA-functionalized Au and Ag nanoparticles, Langmuir 27 (2011) 15330–15339.

[115] F. Aldeek, M. Safi, N. Zhan, G. Palui, H. Mattoussi, Understanding the self-assembly of proteins onto gold nanoparticles and quantum dots driven by metal-histidine coordination, ACS Nano 7 (2013) 10197–10210.

[116] J.M. Abad, M. Gass, A. Bleloch, D.J. Schiffrin, Direct electron transfer to a metalloenzyme redox center coordinated to a monolayer-protected cluster, J. Am. Chem. Soc. 131 (2009) 10229–10236.

[117] K.E. Sapsford, et al., Kinetics of metal-affinity driven self-assembly between proteins or peptides and CdSe-ZnS quantum dots, J. Phys. Chem. C 111 (2007) 11528–11538.

[118] X.-Y. Li, et al., High gold nanoparticles stability towards DNA modification and efficient hybridization via surfactant-free peptide route, Chem. Commun. 53 (2017) 11909–11912.

[119] J. Zong, S.L. Cobb, N.R. Cameron, Peptide-functionalized gold nanoparticles: versatile biomaterials for diagnostic and therapeutic applications, Biomater. Sci. 5 (2017) 872–886.

[120] P. García Calavia, et al., Targeted photodynamic therapy of breast cancer cells using lactose-phthalocyanine functionalized gold nanoparticles, J. Colloid. Interface. Sci. 512 (2018) 249–259.

[121] S. Boden, K. Wagner, M. Karg, L. Hartmann, Presenting precision glycomacromolecules on gold nanoparticles for increased lectin binding, Polymers (Basel) 9 (2017) 716.

[122] X. Wang, Z. Mei, Y. Wang, L. Tang, Gold nanorod biochip functionalization by antibody thiolation, Talanta 136 (2015) 1–8.

[123] M.E. Aubin-Tam, Conjugation of nanoparticles to proteins, Methods Mol. Biol. 1025 (2013) 19–27.

[124] K. Hamad-Schifferli, P. Bergese, Nanomaterial Interfaces in Biology: Methods and Protocols, Humana Press, 2013.

[125] M. McCully, et al., Significance of the balance between intracellular glutathione and polyethylene glycol for successful release of small interfering RNA from gold nanoparticles, Nano Res. 8 (2015) 3281–3292.

126] L. Maus, O. Dick, H. Bading, J.P. Spatz, R. Fiammengo, Conjugation of peptides to the passivation shell of gold nanoparticles for targeting of cell-surface receptors, ACS Nano 4 (2010) 6617–6628.

[127] M. Borzenkov, et al., Thermal and chemical stability of thiol bonding on gold nanostars, Langmuir 31 (2015) 8081–8091.

[128] A. Friedman, S. Claypool, R. Liu, The smart targeting of nanoparticles, Curr. Pharm. Des. 19 (2013) 6315–6329.

[129] R.M. Fratila, S.G. Mitchell, P. Del Pino, V. Grazu, J.M. De La Fuente, Strategies for the biofunctionalization of gold and iron oxide nanoparticles, Langmuir 30 (2014) 15057–15071.

[130] R.C. Rodrigues, C. Ortiz, Á. Berenguer-Murcia, R. Torres, R. Fernández-Lafuente, Modifying enzyme activity and selectivity by immobilization, Chem. Soc. Rev. 42 (2013) 6290–6307.

[131] E. Polo, et al., Tips for the functionalization of nanoparticles with antibodies, Methods Mol. Biol. 1051 (2013) 149–163.

[132] C. Parolo, et al., Design, preparation, and evaluation of a fixed-orientation antibody/gold-nanoparticle conjugate as an immunosensing label, ACS Appl. Mater. Interfaces 5 (2013) 10753–10759.

[133] E. Polo, P. del Pino, B. Pelaz, V. Grazu, J.M. de la Fuente, Plasmonic-driven thermal sensing: ultralow detection of cancer markers, Chem. Commun. (Camb). 49 (2013) 3676–3678.

[134] S. Puertas, et al., Designing novel nano-immunoassays: antibody orientation versus sensitivity, J. Phys. D. Appl. Phys. 43 (2010) 474012.

[135] J.M. Montenegro, et al., Controlled antibody/(bio-) conjugation of inorganic nanoparticles for targeted delivery, Adv. Drug Deliv. Rev. 65 (2013) 677–688.

[136] N. Tajima, M. Takai, K. Ishihara, Significance of antibody orientation unraveled: well-oriented antibodies recorded high binding affinity, Anal. Chem. 83 (2011) 1969–1976.

[137] N.R. Mohamad, N.H.C. Marzuki, N.A. Buang, F. Huyop, R.A. Wahab, An overview of technologies for immobilization of enzymes and surface analysis techniques for immobilized enzymes, Biotechnol. Biotechnol. Equip. 29 (2015) 205–220.

[138] F. Secundo, Conformational changes of enzymes upon immobilisation, Chem. Soc. Rev. 42 (2013) 6250.

[139] A. Sanchez, et al., Inactivation of immobilized trypsin under dissimilar conditions produces trypsin molecules with different structures, RSC Adv. 6 (2016) 27329–27334.

[140] V. Grazú, et al., Promotion of multipoint covalent immobilization through different regions of genetically modified penicillin G acylase from *E. coli*, Process Biochem. 45 (2010) 390–398.
[141] C. Mateo, et al., Immobilization of enzymes on heterofunctional epoxy supports, Nat. Protoc. 2 (2007) 1022–1033.
[142] C.A. Godoy, et al., Glyoxyl-disulfide agarose: a tailor-made support for site-directed rigidification of proteins, Biomacromolecules 12 (2011) 1800–1809.
[143] C.A. Godoy, et al., Changes on enantioselectivity of a genetically modified thermophilic lipase by site-directed oriented immobilization, J. Mol. Catal. B Enzym. 87 (2013) 121–127.
[144] P. Ciaurriz, F. Fernández, E. Tellechea, J.F. Moran, A.C. Asensio, Comparison of four functionalization methods of gold nanoparticles for enhancing the enzyme-linked immunosorbent assay (ELISA), Beilstein J. Nanotechnol. 8 (2017) 244–253.
[145] J.C. Breger, et al., Quantum dot display enhances activity of a phosphotriesterase trimer, Chem. Commun. 51 (2015) 6403–6406.
[146] W. Kang, et al., Cascade biocatalysis by multienzyme–nanoparticle assemblies, Bioconjug. Chem. 25 (2014) 1387–1394.
[147] I. Ardao, J. Comenge, M.D. Benaiges, G. Álvaro, V.F. Puntes, Rational nanoconjugation improves biocatalytic performance of enzymes: aldol addition catalyzed by immobilized rhamnulose-1-phosphate aldolase, Langmuir 28 (2012) 6461–6467.
[148] L.P. Kreuzer, M.J. Männel, J. Schubert, R.P.M. Höller, M. Chanana, Enzymatic catalysis at nanoscale: enzyme-coated nanoparticles as colloidal biocatalysts for polymerization reactions, ACS Omega 2 (2017) 7305–7312.
[149] W. Liu, S. Zhang, P. Wang, Nanoparticle-supported multi-enzyme biocatalysis with in situ cofactor regeneration, J. Biotechnol. 139 (2009) 102–107.
[150] M.D. Blankschien, et al., Light-triggered biocatalysis using thermophilic enzyme–gold nanoparticle complexes, ACS Nano 7 (2013) 654–663.
[151] K. Min, Y.J. Yoo, Recent progress in nanobiocatalysis for enzyme immobilization and its application, Biotechnol. Bioprocess Eng. 19 (2014) 553–567.
[152] S. Ding, A.A. Cargill, I.L. Medintz, J.C. Claussen, Increasing the activity of immobilized enzymes with nanoparticle conjugation, Curr. Opin. Biotechnol. 34 (2015) 242–250.
[153] A.K. Patri, M.A. Dobrovolskaia, S.T. Stern, S.E. McNeil, Preclinical characterization of engindered nanoparticles intended for cancer therapeutics, Nanotechnology for Cancer Therapy., *CRC Press*, 2006, pp. 105–138.
[154] Z.L. Wang, Characterization of nanophase materials, Characterization of Nanophase Materials. Transmission Electron Microscopy and Spectroscopy of Nanoparticles, Wiley-VCH, 2001, pp. 37–80.
[155] B. Dubertret, et al., In vivo imaging of quantum dots encapsulated in phospholipid micelles, Science (80). 298 (2002) 1759–1762.
[156] P.C. Lin, S. Lin, P.C. Wang, R. Sridhar, Techniques for physicochemical characterization of nanomaterials, Biotechnol. Adv. 32 (2014) 711–726.
[157] A. Parracino, et al., Photonic immobilization of bsa for nanobiomedical applications: creation of high density microarrays and superparamagnetic bioconjugates, Biotechnol. Bioeng. 108 (2011) 999–1010.
[158] I.V. Gmoshinski, et al., Nanomaterials and nanotechnologies: methods of analysis and control, Russ. Chem. Rev. 82 (2013) 48–76.
[159] S. Inagaki, R. Ghirlando, R. Grisshammer, Biophysical characterization of membrane proteins in nanodiscs, Methods 59 (2013) 287–300.

[160] K.E. Sapsford, K.M. Tyner, B.J. Dair, J.R. Deschamps, I.L. Medintz, Analyzing nanomaterial bioconjugates: a review of current and emerging purification and characterization techniques, Anal. Chem. 83 (2011) 4453–4488.

[161] S.K. Brar, M. Verma, Measurement of nanoparticles by light-scattering techniques, TrAC - Trends Anal. Chem. 30 (2011) 4–17.

[162] G. Rischitor, et al., Quantification of the cellular dose and characterization of nanoparticle transport during in vitro testing, Part. Fibre. Toxicol. 13 (2016) 47.

[163] L. Shang, G.U. Nienhaus, Small fluorescent nanoparticles at the nano–bio interface, Mater. Today 16 (2013) 58–66.

[164] J. Gross, S. Sayle, A.R. Karow, U. Bakowsky, P. Garidel, Nanoparticle tracking analysis of particle size and concentration detection in suspensions of polymer and protein samples: influence of experimental and data evaluation parameters, Eur. J. Pharm. Biopharm. 104 (2016) 30–41.

[165] M. Wright, Nanoparticle tracking analysis for the multiparameter characterization and counting of nanoparticle suspensions, Nanoparticles Biol. Med. 906 (2012) 511–524.

[166] V. Filipe, A. Hawe, W. Jiskoot, Critical evaluation of nanoparticle tracking analysis (NTA) by NanoSight for the measurement of nanoparticles and protein aggregates, Pharm. Res. 27 (2010) 796–810.

[167] L. Zu, S.Y. Han, K. Gu, X.G. Cui, Preparation of ultrafine polyethylene-silica composite particle with core-shell structure, Adv. Mater. Res 557–559 (2012) 554–557.

[168] R. Khaydarov, O. Gapurova, Y. Estrin, A novel method of continuous fabrication of aqueous dispersions of silver nanoparticles, Artic. Int. J. Nanoparticles 3 (2010) 77–91.

[169] F. Vogel, F. Träger, F. Hubenthal, A new route for mass production of uniform metal nanoparticles in water by means of laser light induced processes, J. Nanosci. Nanotechnol. 11 (2011) 2368–2375.

[170] V. Spampinato, M.A. Parracino, R. La Spina, F. Rossi, G. Ceccone, Surface analysis of gold nanoparticles functionalized with thiol-modified glucose SAMs for biosensor applications, Front. Chem. 4 (2016) 8.

[171] D. Philip, Synthesis and spectroscopic characterization of gold nanoparticles, Spectrochim. Acta A Mol. Biomol. Spectrosc. 71 (2008) 80–85.

[172] W.C.W. Chan (Ed.), Advances in Experimental Medicine and Biology. Springer.

[173] B. Hötzer, I.L. Medintz, N. Hildebrandt, Fluorescence in nanobiotechnology: sophisticated fluorophores for novel applications, Small. 8 (2012) 2297–2326.

[174] V. Poderys, M. Matulionyte, A. Selskis, R. Rotomskis, Interaction of water-soluble CdTe quantum dots with bovine serum albumin, Nanoscale Res. Lett. 6 (2010) 9.

[175] D.L. Nelson, M. Cox, Lehninger: Principles of Biochemistry, Omega, 2017.

[176] D. Zanchet, C.M. Micheel, W.J. Parak, D. Gerion, A.P. Alivisatos, Electrophoretic isolation of discrete Au nanocrystal/DNA conjugates, Nano. Lett. 1 (2001) 32–35.

[177] T. Pellegrino, R.A. Sperling, A.P. Alivisatos, W.J. Parak, Gel electrophoresis of gold-DNA nanoconjugates, J. Biomed. Biotechnol. 2007 (2007) 1–9.

[178] P.G. Righetti, M. Hamdan, A. Bossi, Electrophoresis of proteins and peptides, J. Chromatogr. Libr. Vol. 69B (2004) 633–668.

[179] ImageJ software. Dot Blot Analysis. Available from: <https://imagej.nih.gov/ij/docs/examples/dot-blot/>.

[180] M.M. Bradford, A rapid and sensitive method for the quantitation microgram Quantities of protein utilizing the principle of protein-dye binding, Anal. Biochem. 72 (1976) 248–254.

[181] R. Hnasko, ELISA : Methods and Protocols, Humana Press, 2015.

[182] P.V. Hornbeck, Enzyme-linked immunosorbent assays, Curr. Protoc. Immunol. 2.1.1–2.1.23 (Suppl.) (2015) 1–23.

[183] Abyntek Biopharma, ELISA tipo sandwich: cómo poner a punto tu inmunoensayo, Anticuerpos Aplicaciones Investigación (2017). Available from: http://www.abyntek.com/elisa-tipo-sandwich/.

[184] C.V. Sapan, et al., Colorimetric protein assay techniques, Biotechnol. Appl. Biochem. 108 (1999) 99–108.

Index

Note: Page numbers followed by "*f*" and "*t*" refer to figures and tables, respectively.

C

M

N

O

P

V

W

X

Y

Z

FSC
www.fsc.org
MIX
Papier | Fördert
gute Waldnutzung
FSC® C083411

Kolibri 360 GmbH
Ferdinand-Jühlke-Straße 7
99095 Erfurt, Deutschland
produktsicherheit@kolibri360.de